International Trade and Health Protection

ELGAR INTERNATIONAL ECONOMIC LAW

This new monograph series is intended to provide a point of convergence for high quality, original work on various aspects of international economic and WTO law, ranging from established subject matter, such as international agricultural trade or the application of core trade disciplines such as MFN, to cross-cutting issues involving the interaction of international standards in the fields of investment, tax, competition, food safety and consumer protection with international trade law or the relationship of horizontal exceptions such as the general exception to domestic regulatory barriers. Theoretically rigorous, these books will take an analytical and discursive approach to the field, wherever possible drawing on insights from disciplines other than law, such as economics and politics, in an attempt to arrive at a genuinely inter-disciplinary perspective. Proposals are encouraged that primarily engage with new and previously under-developed themes in the field, or alternatively offer an innovative analysis of areas of uncertainty in the existing law.

Bringing together work from both established authors – academics and practitioners alike – and from a new generation of scholars, the *Elgar International Economic Law Series* aims to play an important role in the development of thinking in the field.

International Economic Law and the Digital Divide
A New Silk Road?
Rohan Kariyawasam

The Law and Economics of Contingent Protection in the WTO
Petros C. Mavroidis, Patrick A. Messerlin and Jasper M. Wauters

International Trade and Health Protection
A Critical Assessment of the WTO's SPS Agreement
Tracey Epps

International Trade and Health Protection

A Critical Assessment of the WTO's SPS Agreement

Tracey Epps

Lecturer, Faculty of Law, University of Otago, New Zealand

ELGAR INTERNATIONAL ECONOMIC LAW

Edward Elgar
Cheltenham, UK • Northampton, MA, USA

Published by
Edward Elgar Publishing Limited
The Lypiatts
15 Lansdown Road
Cheltenham
Glos GL50 2JA
UK

Edward Elgar Publishing, Inc.
William Pratt House
9 Dewey Court
Northampton
Massachusetts 01060
USA

A catalogue record for this book
is available from the British Library

Library of Congress Cataloguing in Publication Data
Epps, Tracey, 1973–
International trade and health protection : a critical assessment of the WTO's SPS agreement /c by Tracey Epps.
p. cm.—(Elgar international economic law series)
Includes bibliographical references and index.
1. Food adulteration and inspection—Law and legislation. 2. Plant inspection—Law and legislation. 3. Animals—Inspection—Law and legislation. 4. Agreement on the Application of Sanitary and Phytosanitary Measures (1995) I. Title.
K3631.E67 2008
344.04′2—dc22

2008023877

ISBN 978 1 84720 657 2

Typeset by Cambrian Typesetters, Camberley, Surrey
Printed and bound in Great Britain by MPG Books Ltd, Bodmin, Cornwall

Contents

Acknowledgements

I owe my thanks and gratitude to a number of individuals and organizations. To Professor Michael Trebilcock for his superb supervision during my doctoral studies where this book began life; it was an absolute pleasure and an honour to work under his direction. To Professors Andrew Green and Don Dewees for their valuable contributions and guidance provided so willingly as members of my advisory committee. To the CIHR Training Program in Health Law and Policy and the Faculty of Law at the University of Toronto for providing financial assistance and institutional support. To my colleagues at the Faculty of Law, University of Otago, for providing such a supportive environment in which to continue work on this project. A special mention and thank you must also be made here to my parents without whose encouragement and sacrifices throughout the years I never would have come this far. Finally, to Greig for his ongoing support, proof-reading, and acting as a sounding board on many occasions.

Abbreviations

AQIS	Australian Quarantine and Inspection Service
BSE	Bovine Spongiform Encephalopathy
CDC	Centers for Disease Control and Prevention
CODEX	Codex Alimentarius Commission
DSB	Dispute Settlement Body
DSU	Dispute Settlement Understanding
EPA	Environmental Protection Agency
FAO	Food and Agriculture Organization of the United Nations
FDA	Food and Drug Administration
GATT	General Agreement on Tariffs and Trade
GM	genetically modified
ICCPR	International Convention on Civil and Political Rights
ICESCR	International Convention on Economic, Social, and Cultural Rights
IPPC	International Plant Protection Convention
MRL	Maximum Residue Level
NGOs	Non-government Organizations
NRC	National Research Commission (US)
OECD	Organization for Economic Cooperation and Development
OIE	International Office of Epizootics/International Animal Health Organization
SPS	Sanitary and phytosanitary measures
SPS Agreement	Agreement on the Application of Sanitary and Phytosanitary Measures
vCJD	variant Creutzfeldt-Jakob Disease
WHO	World Health Organization
WTO	World Trade Organization

PART I

Background and overview

1. Introduction

1.1 INTRODUCTION

At the conclusion of the Uruguay Round in 1994, Members of the World Trade Organization (WTO) adopted the Agreement on the Application of Sanitary and Phytosanitary Measures (SPS Agreement). The SPS Agreement stipulates rules that Members must follow when adopting and enforcing sanitary and phytosanitary (SPS) measures. These are measures adopted to protect humans, animals, and plants against risks arising from, *inter alia*, pests, additives, contaminants, toxins, or diseases. The SPS Agreement's rules are intended to further its key objectives of ensuring that no Member should be prevented from adopting or enforcing measures necessary to protect human, animal, or plant life or health; and that the negative impacts of domestic SPS measures on trade are minimized. By these objectives, the SPS Agreement recognizes the important role played by both domestic health protection and trade liberalization in advancing global and domestic welfare.

Achieving both objectives is complicated by the fact that, even where there is no protectionist intent on the part of lawmakers when they adopt SPS measures, differences between countries in regulatory or standard-setting regimes can impede trade.[1] The SPS Agreement seeks to overcome this apparent conflict by recognizing the importance of health protection as a domestic regulatory objective and permitting trade-restrictive SPS measures where they are 'necessary' to protect health. This proviso is subject to a number of substantive and procedural rules regarding the SPS measures that countries may enact.

The SPS Agreement encourages Member countries to base their SPS measures on international standards, guidelines, or recommendations, where they exist.[2] Where countries wish to introduce or maintain SPS measures that result in a higher level of protection than would be achieved by measures based on the relevant international standards, guidelines, or recommendations, they must ensure that the SPS measure chosen 'is applied only to the extent

[1] See Michael J. Trebilcock and Robert Howse, *The Regulation of International Trade*, 3rd ed. (London: Routledge, 2005).

[2] Article 3.1.

necessary to protect human, animal or plant life or health, is based on scientific principles and is not maintained without sufficient scientific evidence'.[3] Further, SPS measures must be based on an assessment of the risks to human, animal, or plant life or health.[4] The implication of these provisions is that unless scientific justification can be provided for a trade-restrictive SPS measure, it will not be justifiable under the SPS Agreement, even if it is enacted with the genuine intention of protecting human, animal, or plant health. The rationale for this approach is that requiring scientific justification for standards that deviate from international norms makes it more difficult for countries to shelter domestic industries behind restrictive health regulations or to disguise protectionist strategies under the cloak of health regulations.[5]

The SPS Agreement's negotiators apparently believed that these scientific evidence requirements would make it a straightforward task to deal with conflict between health and trade by enabling WTO panels and the Appellate Body to determine *when* a measure is necessary. Once a determination of necessity is made, the Agreement allows health to be privileged over trade. However, the task is not always straightforward. Rather, there are a range of cases where it is difficult, if not impossible, to determine by any objective scientific standard whether or not a measure is necessary. In these cases, conflict between trade and health persists. It is in such cases that WTO panels and the Appellate Body face difficulties – both conceptual and factual – in balancing the competing objectives of health protection and trade liberalization. The difficulties are intensified where countries diverge in their scientific opinions, and where public opinion and scientific opinion are at odds.

The SPS Agreement's reliance on scientific evidence as a benchmark to determine when measures are necessary to protect health has drawn much criticism from commentators and scholars. These critics tend to follow one of two strands of thought. The first strand (in the minority) is concerned with the integrity of the Agreement's trade liberalization objectives. These critics contend that the use of a scientific benchmark allows Member countries too much discretion in their regulatory decision-making. They argue that 'science' should not be relied upon as a 'neutral arbiter'[6] because it is neither objective nor politically untainted as is often presumed, but is in fact socially constructed.[7] This view of science might conceivably lead to the conclusion

---

[3] Article 2.2.

[4] Article 5.1.

[5] Tim Josling, Donna Roberts and David Orden, *Food Regulation and Trade* (Washington DC: Institute for International Economics, 2004).

[6] See generally Vern Walker, 'The Myth of Science as a "Neutral Arbiter" for Triggering Precautions' (2003) 26:2 B.C. Int'l & Comp. L. Rev. 197.

[7] See for example, Robert Hudec, 'Science and "Post-Discriminatory" WTO

that the provisions requiring scientific justification fail to achieve the drafters' goal of preventing protectionism.

In contrast, the second strand of thought finds that the provisions impose too much of a straitjacket on governments. Sykes, for example, suggests that the scientific benchmark represents undue hurdles for regulators who sincerely pursue objectives other than protectionism.[8] Others are concerned with the SPS Agreement's apparent exclusion of non-scientific justifications for measures, arguing that reliance on science is misplaced because it precludes any consideration of broad social, cultural, and ethical concerns.[9] They consider that social factors such as public opinion should be regarded as a legitimate reason to restrict trade through SPS measures. Such arguments gain support from social science accounts of risk perception as well as trends towards greater public participation in public policy decision-making regarding risk.

These various criticisms raise the question of what factors influence domestic regulatory decision-making concerning SPS measures, and why there are regulatory differences between countries. It is axiomatic to note that regulatory measures differ between countries, even where the objective is the same (that is, the protection of the life and health of humans, animals, and plants).[10] What is perhaps not so obvious is *why* such regulatory divergence exists. This book identifies a number of reasons for regulatory divergence, including the role played by 'public sentiment', a term adopted as shorthand for the public expression of cultural, social, individual, ethical, and/or political values and beliefs. Public sentiment often influences (directly or indirectly) regulatory decision-making in the face of risk and differs between countries for a wide variety of reasons, including underlying political, cultural, ethical, religious, and social circumstances.

1.2 AIMS AND LIMITATIONS

This book asks whether the SPS Agreement allows an appropriate balance to

Law' (2003) 26:2 B.C. Int'l. & Comp. L. Rev. 185 at 189. Jeffery Atik, 'Science and International Regulatory Convergence' (1996–7) 17 Nw. J. Int'l L. & Bus. 736 at 758.

[8] Alan O. Sykes, 'Domestic Regulation, Sovereignty, and Scientific Evidence Requirements: A Pessimistic View' (2002) 3:2 Chicago J. Int'l L. 353 at 354.

[9] See for example, Dayna Nadine Scott, *Nature/Culture Clash: The Transnational Trade Debate over GMOs* (New York: Hauser Global Law School Program, Global Law Working Paper 06/05, 2005) at 42. See also David Winickoff, *et al.*, 'Adjudicating the GM Food Wars: Science, Risk, and Democracy in World Trade Law' (2005) 30 Yale J. Int'l L. 81.

[10] M. Gregg Bloche, 'WTO Deference to National Health Policy: Toward an Interpretive Principle' (2002) 5:4 J. Int'l Econ. L. 825.

be struck between conflicting domestic health protection and trade liberalization objectives. It takes stock of the voluminous literature on the subject and advances the discussion by focusing on difficult situations not yet fully contemplated either in the literature or by the WTO's Dispute Settlement Body (DSB). These are situations where there is scientific uncertainty as to the existence of a risk and the necessity of the chosen regulatory measure, and those where public sentiment is an influencing factor demanding regulation at the national level.

The book's central argument is that while aspects of the rationale underlying the SPS Agreement's science-based framework are questionable, such an approach is the most appropriate means available of dealing with conflict between health and trade. It is argued that this conclusion is consistent with a normative position that recognizes the importance of both health protection and trade liberalization, but accepts that in some cases the welfare gains from trade should properly be foregone in order to protect health (including in some cases where there is lack of clarity as to the necessity of the SPS measures taken). The conclusion is in large part driven by a positive analysis of domestic regulatory decision-making which notes the potential for regulatory capture by domestic protectionist interests and the consequent importance of ensuring that decisions are made on a sound and principled basis. It is supported by an examination of decision-making regarding risk which finds that the science-based framework provides countries with more flexibility to respond to scientific uncertainties and public sentiment than many critics contend.

Having addressed the critics' concerns surrounding the SPS Agreement's science-based framework, it is argued further that if the WTO is to command legitimacy among the public, panels and the Appellate Body must take a principled approach to disputes that reflects the vital importance of health and carefully considers how it can be balanced against trade liberalization interests. It will be argued that such an approach has not been sufficiently evidenced to date, but will be particularly critical in cases where it is difficult to determine the necessity of a health protection measure, for example, where there is only a low probability of a risk eventuating, but the potential consequences are serious, or where there is genuine scientific uncertainty combined with strong public sentiment demanding regulation. Given the complexity and indeterminacy of science, the value of public sentiment, and importance of sovereignty issues, it is argued that panels should take a relatively deferential approach to the decisions of domestic regulatory agencies. However, deference cannot be total and still allow detection of protectionism and thus it is argued that panels should focus on reviewing the procedural integrity of a country's decision-making rather than second-guessing their final scientific conclusions.

Some limitations must be noted at the outset. Hudec described the SPS

Agreement as 'post-discriminatory' because its scientific evidence requirements do not require proof of discrimination, but essentially call for an international tribunal to judge the rationality of a regulatory judgment at the national level.[11] It is unlike the General Agreement on Tariffs and Trade (GATT) where discrimination (or some other violation of the Agreement) must be found before the regulating country is required to justify its actions under the exceptions found in Article XX. This, Hudec argued, is an extension of WTO oversight into regulatory areas not impacting trade, and the amount of information required to justify regulatory actions is therefore much higher than under the GATT regime where only a certain subset of measures is subject to such scrutiny.[12]

The 'post-discriminatory' nature of the SPS Agreement is open to criticism on several fronts, including that it goes beyond the WTO's mandate of promoting international trade and that the rules give foreign traders a higher set of legal rights than is given to the domestic producers with whom they compete.[13] However, the book will not address this issue. Proceeding on the basis that political economy factors augur against reform of the post-discriminatory orientation of the SPS Agreement, it seeks to work within the current framework to suggest methods of interpretation that will provide countries with sufficient policy space to regulate to protect health while minimizing opportunities for protectionism.

Second, it is recognized that countries at unequal levels of development have different policy priorities and different levels of capacity to adopt or comply with SPS measures. Issues arising out of this reality will be acknowledged, but it is beyond the scope of the book to explore the problem in the full detail that it deserves.

1.3 ORGANIZATION

The book is organized as follows. The chapters in Part I provide an historical and factual background to the issues. Chapter 2 explains the linkage between health and international trade. Chapter 3 explores the extent to which health regulations have conflicted with international trade liberalization – both in the past and currently. It also describes efforts taken to allay this conflict through multilateral negotiations and discusses key aspects of the SPS Agreement. Finally, Chapter 4 identifies factors that point to likely future conflicts, thereby highlighting the importance of the questions raised in the book.

---

11 Hudec, *supra* note 7 at 2.

12 *Ibid.*

13 *Ibid.* at 3.

Part II establishes the normative basis which frames the book's critique of the SPS Agreement's science-based framework and its interpretation by panels and the Appellate Body. It does so by outlining the theoretical foundations underlying both international trade theory and domestic regulatory theory, and identifying situations where conflict between health and trade objectives persists despite the SPS Agreement's tolerance of trade-restrictive measures where its requirements are met. The final chapter in Part II summarizes the normative basis that will be adopted throughout the remainder of the book. While acknowledging the value of both international trade liberalization and human, animal, and plant health, it suggests that in certain cases, panels and the Appellate Body ought to adopt a bias towards health protection where there is conflict or tension between the two.

As Perez observes, the power to strike out national health regulations has transformed the WTO into a key player in the global debate about risks.[14] Part III explores aspects of this debate, beginning with a discussion of subsidiarity in which it questions at what level health regulations are most appropriately made, and asks whether one way in which conflict or tension between health and trade objectives might be avoided is by negotiating an end to regulatory differences through 'regulatory harmonization'.[15] It then investigates the rationale behind the Agreement's science-based framework by examining regulatory decision-making in the context of risks to health, including the role of both scientific and non-scientific factors (including public sentiment).

Part IV begins by exploring criticisms of the SPS Agreement's science-based framework and then seeks to counter these criticisms in a section entitled 'validating the science-based approach'. This section sets out an approach which it is argued that panels and the Appellate Body should follow when reviewing domestic agency decisions. It then goes on to critique interpretation of the SPS Agreement by panels and the Appellate Body by reference to the suggested approach. It also considers decisions under the GATT in cases where the panel and/or Appellate Body's analysis therein is relevant to interpreting the SPS Agreement.

Finally, in Part V, a conclusion sums up the arguments made in the book.

---

[14] Oren Perez, *Ecological Sensitivity and Global Legal Pluralism* (Oxford: Hart Publishing, 2004) at 116.

[15] Alan O. Sykes, 'The (Limited) Role of Regulatory Harmonization in International Goods and Services Markets' (1999) 2 J. Int'l Econ. L. 49 at 50.

2. What's health got to do with it? The linkage between health and international trade

2.1 INTRODUCTION

SPS measures are a critical component of a country's regulatory response to risks that might negatively impact human, animal, or plant health or life. Measures often vary considerably between countries, even when dealing with the same subject-matter. As Trebilcock and Soloway note, 'one nation's bunch of grapes is another nation's repository of carcinogenic pesticide residue'.[1] A sound basis often exists for SPS measures, but in some cases, suspicions are raised that measures are based on dubious grounds, existing to protect domestic producers or industry, or responding to unjustified consumer pressures.[2] Trade tensions may arise in such cases when exporters feel that they are being adversely affected by measures that become non-tariff barriers to trade by either restricting or unduly raising the cost of market access. These cases raise the issue of when divergent domestic regulations should be justifiable in international trade law and it is this issue that lies at the heart of the SPS Agreement and of this book.

This chapter begins with an examination of the meaning of human, animal, and plant health as it is crucial that these concepts be clearly understood before continuing with the analysis. It then examines the critical phrase 'non-tariff barriers to trade' and describes the nature of SPS measures, exploring how they may come to constitute non-tariff barriers to trade.

---

[1] Michael Trebilcock and Julie Soloway, 'International Policy and Domestic Food Safety Regulation: The Case for Substantial Deference by the WTO Dispute Settlement Body under the SPS Agreement' in Daniel Kennedy and James Southwick, eds, *The Political Economy of International Trade* (Cambridge, UK: Cambridge University Press, 2002) at 1.

[2] Tim Josling, Donna Roberts and David Orden, *Food Regulation and Trade* (Washington DC: Institute for International Economics, 2004) at 13.

2.2 WHAT IS 'HEALTH'? KEEPING HUMANS, ANIMALS, AND PLANTS ALIVE AND HEALTHY

The Constitution of the World Health Organization (WHO) defines 'health' as 'a state of complete physical, mental and social well-being and not merely the absence of disease or infirmity'.[3] Regulations aimed to guard against risk are typically directed at protection of physical health. It is, however, arguable that regulations are sometimes also required to protect mental well-being, both for its own sake and for the impact mental well-being can have on physical well-being. A report by the European Commission, for example, adopts a definition which finds good health to be 'the personal independent ability [of people] to control their life conditions, to adapt to accidental modifications of their environment and to refuse intolerable environments'.[4] In this context, the Commission argues that new risks of today's technological and anxiogenic civilization damage health because people cannot face the hazards in an autonomous manner.[5] To date, no WTO panel has been called upon to interpret the meaning of health. It seems likely, however, that the Agreement's negotiators did not anticipate a conception of health that extends beyond the physical. The acceptance of such a broad conception by a panel would arguably give countries significantly more scope to enact trade-restrictive SPS measures in the name of health protection than the negotiators intended.

SPS measures designed to protect human health are part of the particular branch of health known as 'public health'. Public health is distinguishable from 'medicine' in that it focuses on the health of populations, whereas 'medicine' focuses on the health of individuals.[6] A classic definition of public health states: 'as one of the objects of the police power of the state, the "public health" means the prevailingly healthful or sanitary condition of the general body of people or the community in mass, and the absence of any general or widespread disease or cause of mortality'.[7]

More specifically, public health is seen as 'the array of interventions directed to health promotion campaigns, the things done to prevent the spread

---

[3] Constitution of the WHO (1946), available online at www.who.org (date accessed: 25 April 2005). Currently 192 nations are members of the WHO.

[4] European Commission – Health and Consumer Protection Directorate-General, *Final Report on Setting the Scientific Frame for the Inclusion of New Quality of Life Concerns in the Risk Assessment Process* (Brussels, 2003) at 3.

[5] *Ibid.* at 4.

[6] Lawrence O. Gostin, 'Public Health Law in a New Century: Law as a Tool to Advance the Community's Health' (2000) 283:21 Journal of the American Medical Association 2837 at 2839.

[7] B.A. Garner, ed., *Black's Law Dictionary* 7th ed. (St Paul, MN: West Group, 1999). Cited in Gostin, *ibid.*

of communicable disease, the food and sanitation requirements and the pollution controls written into our environment laws'. Less obviously, it is also a series of initiatives in areas of product safety, the regulation of drugs and therapeutics, and a range of initiatives that aim to promote and protect health in the sense envisaged by the WHO's definition.[8]

SPS measures are critical to public health. Many countries, particularly those in the industrialized West, maintain a vast number of such measures, many of which have the potential to restrict the flow of international trade. SPS measures to protect human health may be either direct (protection from disease-carrying organisms and poisonings) or indirect (protection of crops and domestic animals from diseases – zoonoses – that would impact on human health).

The terms 'animal health' and 'plant health', unlike human health, are generally used in a more narrow sense to refer to the absence of physical disease. With respect to animals, however, there is room for debate as to the scope of the meaning of the term 'health'. The European Commission argues, for example, that any alteration of an animal's normal functions is potentially harmful.[9] In this respect, it argues that 'the welfare of an animal is what is achieved when it is in harmony with its environment and with itself, both physically and psychologically'.[10] This remains a controversial contention, however, and to date the general consensus internationally has been that animal health refers to physical diseases.[11]

Measures to protect animal health may apply to four categories of animal subjects: livestock;[12] aquatic animals; wild animals (fauna); and domestic animals (pets). Animal health measures have various goals and objectives.[13] First, effective measures are critical to human health by (i) ensuring food safety, and (ii) minimizing the transmission of zoonotic[14] infections from animals to humans. Second, livestock/fisheries industries have a strong

---

[8] Christopher Reynolds and Genevieve Howse, *Public Health: Law and Regulation* (Sydney: The Federation Press, 2004) at 3.

[9] European Commission, *supra* note 4 at 17.

[10] *Ibid.*

[11] See also generally Cass R. Sunstein and Martha R. Nussbaum, *Animal Rights: Current Debates and New Directions* (New York: Oxford University Press, 2004).

[12] Livestock are animals reared in captivity for the commercial production of meat, milk, eggs, and by-products. Livestock include poultry, swine, cattle, sheep, and goats. Wild animals may also be considered livestock where they are reared in captivity, for example, deer, elk, bison, llamas, alpacas, ostriches, and emu. Robert F. Kahrs, *Global Livestock Health Policy: Challenges, Opportunities, and Strategies for Effective Action* (Ames, Iowa: Iowa State Press, 2004) at 3.

[13] Kahrs, *ibid.* at 250.

[14] Zoonotic infections are those that are transmissible from animals to humans.

economic interest in maintaining animal health; accordingly, measures are sought to meet the goal of maximizing profits for these industries. Third, food security is maintained by keeping a healthy livestock/fisheries population. Fourth, measures may aim to protect the health of wild animals (fauna).

To date, animal health measures that have attracted the most controversy are those that deal with livestock and/or fisheries by regulating diseases, vaccines, feed additives, and the conditions under which animals are reared and processed.

Regarding plant health, regulations apply to one of two key plant categories, commercial crops, or natural vegetation (wild flora). Much of the science and work of plant health regulation is focused on commercial crops, aiming to prevent or minimize the spread and establishment of plant pests in new areas, and eradicate or control them if they exist. Such regulation will usually take the form of SPS measures. As well, measures may aim to protect food security and also the health of wild fauna.

2.3 WHEN HEALTH PROTECTION MEASURES BECOME NON-TARIFF BARRIERS TO TRADE

Non-tariff barriers are government measures other than tariffs that restrict trade flows.[15] Actions to protect health that take the form of SPS measures may or may not restrict trade flows; where a measure does so, it will constitute a *non-tariff trade barrier*. Trade-restricting SPS measures thus form a subset of non-tariff barriers to trade and may or may not violate WTO rules. As noted, SPS measures are the subject of the WTO's SPS Agreement and are measures designed to protect human, animal, or plant life or health. They most often take the form of government regulations, for example, food and product safety standards and quarantine restrictions. The goods most widely affected by SPS measures are food products for human or animal consumption, agricultural commodities, live animals, and horticultural products.

There are at least four broad categories of situations where SPS measures may constitute non-tariff trade barriers and thus give rise to trade disputes.[16] These are: (i) prohibitions on the sale of an imported product on health grounds; (ii) positive requirements for imported products (for example, certi-

---

[15] Walter Goode, ed., *Dictionary of Trade Policy Terms* (Cambridge, UK: World Trade Organization/Cambridge University Press, 2003) at 253.

[16] With respect to the first three situations listed, see Alan O. Sykes, *Product Standards for Internationally Integrated Goods Markets* (Washington, DC: Brookings Institution, 1995) at 17.

fication requirements) that discriminate against and/or result in a burden being placed on foreign suppliers; (iii) exclusion of imported products from compulsory approval for marketing; and (iv) market-driven requirements for compliance with voluntary standards.

SPS measures commonly take the form of *import prohibitions* (or *bans*), with plant health (phytosanitary) measures often tending to be more trade restrictive than food safety (sanitary) measures.[17] In order to reduce the risk of introducing plant pests or diseases, countries often prohibit imports altogether. On the other hand, food safety measures tend to take the form of product or process regulations that may increase foreign producers' costs, but do not restrict imports outright.[18]

Import prohibitions may be either total or partial.[19] Total prohibitions might be adopted when alternative measures to reduce risks are technically infeasible, for example, bans on diseased livestock such as those infected with foot and mouth or mad cow disease.[20] Partial bans include seasonal or regional restrictions and are often used where it is possible to target the measure to reduce risks to appropriate levels. For example, until recently, the US sought to protect domestic orchards from certain pests by only allowing imports of Mexican avocados to come from a certain region (Michoacán state) where the pests are not common and only from November to February when they are less prevalent.[21]

Where a country imposes prohibitions on the sale of imported products, a common source of dissension is the regulator's interpretation of either the available evidence on risk or over the tolerable risk level.[22] As we will see, this

---

17 Donna Roberts and Barry Krissoff, *Regulatory Barriers in International Horticultural Markets* (Washington DC: United States Department of Agriculture, 2004) at 2.

18 *Ibid.*

19 Donna Roberts, 'Analyzing Technical Trade Barriers in Agricultural Markets: Challenges and Priorities' (1999) 15:3 Agribusiness 335 at 339.

20 *Ibid.*

21 Julie Soloway, 'Institutional Capacity to Constrain Suboptimal Welfare Outcomes from Trade – Restricting Environmental, Health and Safety Regulation under NAFTA' (University of Toronto, 2000, unpublished) at 170. In December 2005, the US Department of Agriculture made a final ruling that from 1 February 2007, its Animal and Plant Health Inspection Service would allow importation of commercial shipments of fresh Haas avocados from approved orchards and municipalities in Michoacán to all US states on a year-round basis. Restrictions remain in Puerto Rico and other US territories. USDA, Animal and Plant Health Inspection Service, Industry Alert, February 2007, available online: USDA, http://www.aphis.usda.gov/publications/plant_health/content/printable_version/ia_hass_avocados.pdf (accessed 13 October 2007).

22 Sykes, *supra* note 16 at 16.

has been the case in a number of disputes under the SPS Agreement, including the *EC – Beef Hormones*[23] and *Japan – Apples*[24] cases.

There is an infinite array of *positive requirements* that may cause trade disputes by imposing a burden on foreign producers. While in theory any exporter can spend the necessary resources and meet other countries' specified requirements, in practice the cost of meeting such requirements may be prohibitive if the requirement is particularly stringent, or varies significantly from a domestic or international standard. Requirements may also be written in such a way as to favour domestic producers, for example, by requiring use of an input more commonly available in the home country than in exporting countries.[25] Examples of positive requirements that have caused concern in the context of SPS measures since 1995 are Canada's requirement that cheese be pasteurized or produced from pasteurized milk in order to be marketed and sold in Canada;[26] the Czech Republic's requirement that warehouses and silos for animal feeds be under state control for the purposes of quality assurance;[27] and Korea's requirement for imported grains, fruits, and vegetables to be subject to an annual maximum residue level test for the presence of various agricultural chemicals.[28]

The third situation, *exclusion of imported products from compulsory approval*, also occurs fairly frequently and may be the subject of disputes. For example, in one instance, Venezuela claimed that Colombia had failed to grant sanitary certificates for Venezuelan potatoes, fresh mushrooms, fresh tomatoes, fertile eggs, day-old chicks, and meat products.[29] The sanitary certifi-

---

[23] *EC – Measures Concerning Meat and Meat Products (Hormones) Complaint by the United States* (1997), WTO Doc. WT/DS26/R/USA (Panel Report). *EC – Measures Concerning Meat and Meat Products (Hormones) Complaint by Canada* (1997), WTO Doc. WT/DS48/R/CAN (Panel Report). *European Communities – Measures Affecting Meat and Meat Products* (1998), WTO Doc. WT/DS26/AB/R, WT/DS48/AB/R (Appellate Body Report).

[24] *Japan – Measures Affecting the Importation of Apples* (2003), WTO Doc. WT/DS245/R (Panel Report). *Japan – Measures Affecting the Importation of Apples* (2003), WTO Doc. WT/DS245/AB/R (Appellate Body Report).

[25] Josling, Roberts and Orden, *supra* note 2 at 23.

[26] A concern was raised by the EC regarding this measure in May 1996. WTO Committee on Sanitary and Phytosanitary Measures, Summary of the Meeting held on 29–30 May 1996, WTO Doc. G/SPS/R/5.

[27] A concern was raised by the EC regarding this measure in October 1997. WTO Committee on Sanitary and Phytosanitary Measures, Summary of the Meeting held on 15-16 October 1997, WTO Doc. G/SPS/R/9/Rev.1.

[28] A concern was raised by the US regarding this measure in October 2003. WTO Committee on Sanitary and Phytosanitary Measures, Summary of the Meeting Held on 29–30 October 2003, WTO Doc. G/SPS/R/31.

[29] WTO Committee on Sanitary and Phytosanitary Measures, Summary of the Meeting Held on 19–21 March 2002, WTO Doc. G/SPS/R/26.

cates were granted on a discretionary basis and Venezuela claimed that the restrictions seemed designed to protect Colombian producers. In another case, the EC presented a list of substances, including food additives, aromas, food ingredients, and extract solvents which had been excluded from a list of products authorized by Japanese law, thus presenting a barrier to the importation of food products into Japan.[30]

Voluntary standards may become non-tariff barriers to trade in certain circumstances such as where compliance with quality standards is important for the marketing of products to meet consumer preferences. The issue of voluntary, private sector standards has been raised in the SPS Committee. St Vincent and the Grenadines first raised the issue in June 2005 (supported by Argentina, Ecuador, Jamaica, and Peru), with respect to requirements for exporting bananas and other products to European supermarkets imposed by the Euro-Retailer Produce Working Group.[31] St Vincent and the Grenadines raised the issue again in October 2006, when it was joined by Argentina, Belize, Cuba, Dominica, Egypt, Indonesia, Kenya, and South Africa.

In cases such as this, consumer pressure is causing constraints on trade rather than government-led initiatives, but this may cause tension if it is felt that the pressure is unjustified and/or provoked by groups with an interest in keeping out imports. Trade restrictions may also arise from voluntary standards where government procurement is undertaken with reference to voluntary standards.[32] The question of whether voluntary standards are subject to the SPS Agreement is a contested one and will not be dealt with in this book.[33] However, it is likely to be an issue of continuing importance as supermarkets and other retailers increase their market presence and power globally and thus the ability to demand high standards of suppliers.

[30] WTO Committee on Sanitary and Phytosanitary Measures, Summary of the Meeting Held on 7–8 November 2002, WTO Doc. G/SPS/R/28.

[31] WTO Committee on Sanitary and Phytosanitary Measures, Summary of the Meeting Held on 29–30 June, 2005, WTO Doc. G/SPS/R/37 at paras 16 to 20.

[32] Sykes, *supra* note 16 at 14.

[33] See working paper, Tracey Epps, 'Demanding Perfection: Private Food Standards and WTO Rules' (University of Otago, 2008).

3. Through the (historical) looking-glass: health and trade in context

3.1 INTRODUCTION

This chapter provides an historical overview of conflicts between health and trade which, it turns out, have been with us for longer than one might expect. This overview will serve to highlight the importance and continuing relevance of the issues discussed here. The chapter then looks at the development of trade rules concerning health regulations in the twentieth century, culminating with the implementation of the SPS Agreement in 1994. Finally, it considers the trade impact of SPS measures and looks at the nature of disputes that have arisen to date under the SPS Agreement.

3.2 DISPUTES OVER TRADE AND HEALTH IN THE 1800s

International trade was revolutionized in the nineteenth century with developments in technology and transport. Until the mid-nineteenth century, the consumption of fresh foods was essentially limited to locally produced items.[1] However, new preservation techniques such as canning allowed fresh foods to be preserved and shipped long distances,[2] while advances in land and ocean transportation meant quicker transit times and cheaper freight costs.[3] The advent of refrigeration opened up even greater possibilities to transport fresh produce and meat. In 1876, the first ship equipped with a refrigeration system – 'la Frigorique' – was built in France and used to transport meat from Argentina to France. This was followed by the first crossings from Australia and the US to France and England. By 1890, refrigeration equipment became

1 Albert Sonnenfeld, ed., *Food: A Culinary History from Antiquity to the Present* (New York: Columbia University Press, 1999) at 463.

2 Preservation by way of canning was developed in the early nineteenth century but did not develop significantly until later. Sonnenfeld, *ibid.* at 487.

3 *Ibid.* at 464.

available for overland transportation, and freight terminals were equipped with storage and ripening facilities.

Improved transportation was accompanied by policy changes which impacted trading patterns. Repeal of the English Corn Laws in 1846 signalled Britain's unilateral adoption of free trade.[4] This was followed by the Anglo-French commercial treaty of 1860 and a subsequent series of bilateral tariff agreements between European countries incorporating the most-favoured nation clause. These agreements led to a substantial reduction in tariffs and gave rise to the term 'low-tariff era' to describe the period from 1846 to the 1880s. The 1860s, in particular, were years of considerable expansion of world trade. By the late 1870s, however, there were signs of a return to protectionism as European countries began introducing new tariffs in response to a flood of North American grains onto the European market.[5]

As trade was expanding in the 1850s, countries were becoming industrialized and doctrines of public health were evolving. Public health developed out of nineteenth-century sanitary reforms in England, France, Germany, the US, and other countries feeling the effects of the Industrial Revolution. Public health was an inherent problem of industrialization; where people lived in towns, and worked in factories, health conditions deteriorated. In the 1850s, particular attention began to be paid in Britain to the transference of disease from animals to humans. Interest in the conditions surrounding the sale and processing of meat was part of the increasing anxiety about urban public health as people became aware of the filthy conditions in which meat was processed and the consequent sale of meat from diseased cattle and other livestock.[6]

The impacts of trade on plant health also came to be felt during the 1800s. One of the first international measures against a plant pest was enacted in the 1870s following the establishment and spread of a American vine louse in French vineyards. Commonly known as *Phylloxera vastatrix*, the pest spread quickly throughout the vine-growing districts of first France and then the rest of Europe. Losses in France alone were assessed as the equivalent of £50

---

4 The Corn Laws placed restrictions on the import of corn in order to protect British farmers. Following their repeal, over 1000 Customs Acts were repealed as Britain moved towards free trade.

5 Leonard Gomes, *The Economics and Ideology of Free Trade* (Cheltenham, UK and Northampton, MA: Edward Elgar, 2003) at 296.

6 See generally, Richard Perren, *The Meat Trade in Britain 1840–1914* (London: Routledge & Kegan Paul, 1978) at 53. See also Upton Sinclair, *The Jungle* (Middlesex, UK: Penguin Modern Classics, 1906). The latter aroused the ire of the American public, who demanded improved sanitary conditions in slaughterhouses.

million sterling in 1875. This was an enormous loss in an industry of great commercial importance.[7]

Thus by the 1870s, the stage was set for trade disputes similar to those that we see today. Public health was a legitimate concern of government; veterinary and medical measures meant more official intervention in the affairs of the meat trade; trade in all manner of foodstuffs and plants was possible and happening on a global basis;[8] there was uncertainty about the unsettling effects of globalization; and there was tension between those who wanted to see the protection of local producers and those who favoured free trade.

The late 1800s saw two notable disputes that have a number of commonalities with those that have occurred since the signing of the Uruguay Round Agreements. In the 1840s, Great Britain had suffered from the introduction of several livestock diseases. One of these, pleuro-pneumonia, was noted by the Veterinarian Inspector to the Privy Council to be 'one of the most insidious diseases with which we are acquainted . . . the cause of greater losses to British stock-owners and dairymen than any other single disease'.[9] In 1878, the British government adopted a Foreign Animals Order regulating cattle imports following the appearance of pleuro-pneumonia in a shipload of live cattle exported from the US to Great Britain and the publication of a Canadian inspector's report on the presence of pleuro-pneumonia in cattle in Virginia and New York.[10] The Order prohibited imports from several European countries. However, it provided exceptions from other countries, including the US, on condition first, that animals were slaughtered within ten days of importation if they showed signs of diseases, and, second, that healthy animals were

---

[7] See discussion in Robert F. Kahrs, *Global Livestock Health Policy: Challenges, Opportunities, and Strategies for Effective Action* (Ames, Iowa: Iowa State Press, 2004).

[8] Imports of meat into the UK increased significantly following the Free Trade budget in 1842, when both meat and livestock imports were freed from a considerable portion of the duties they had previously carried. Perren, *supra* note 6 at 216.

[9] Thomas Walley, *The Four Bovine Scourges: Pleuro-Pneumonia, Foot and Mouth Disease, Cattle Plague, Tubercle (Scrofula) with an Appendix on the Inspection of Live Animals and Meat* (Edinburgh: MacLachlan and Stewart, 1879). Cited in Justin Kastner and Doug Powell, *Food Safety, International Trade, and History Repeated* (Guelph, Ontario: The Food Safety Network, 2001).

[10] This Order was declared under the *Contagious Diseases (Animals) Act 1869*. D. McEachran, Montreal Veterinary College. Extract Report of the Minister of Agriculture for the Department's Special Investigation into Existence of Cattle Disease in the United States. Quoted in Appendix of Thomas Walley, *The Four Bovine Scourges: Pleuro-Pneumonia, Foot and Mouth Disease, Cattle Plague, Tubercle (Scrofula) with an Appendix on the Inspection of Live Animals and Meat* (Edinburgh: MacLachlan and Stewart, 1879) and cited in Kastner and Powell, *ibid.*

only released into British commerce after the Inspector of the Privy Council had inspected them and certified them as healthy.

This situation caused a considerable amount of dissension in Great Britain, with some arguing that for the sake of public health, American cattle and meat should be banned altogether. Others, however, argued that it was hardly justifiable 'to stop so great a trade and to prevent the supply of food from reaching Great Britain on account of a few isolated cases of pleuro-pneumonia'.[11] The result was that the British government never went so far as to invoke a permanent ban on US cattle or meat imports. Yet historian John Perren notes that while there was no permanent ban, there was a constant tension in Britain and other European states during the nineteenth century between providing regulatory measures to guard against disease and doing so simply to protect domestic producers.[12]

A longer-lasting dispute arose with respect to European imports of US pork. Following the reported discovery of the parasite *trichinella spiralis* in US pork imports, a number of European governments adopted or considered adopting bans on US pork. In 1878, a letter from a professor of anatomy at the University of Vienna, Richard Heschl, appeared in print in the European press. Professor Heschl claimed to have found *trichinella spiralis*, the parasitic nematode worm that causes trichinosis,[13] in 20 per cent of the imported hams he randomly inspected. Despite a subsequent examination of American pork in which Dr Heschl pronounced it sound, the letter was telegraphed to the London *Times* and rumours spread of diseased American pork.[14] In 1879, Italy was the first country to ban the importation of American pork, 'in consequence of trichinosis having manifested itself there'.[15]

Italy's exclusion was followed over the next decade by a number of other European nations. Despite protests and threats of retaliatory trade restrictions by five successive US presidents, by 1891, American pork products had been excluded, either in part or fully, from Italy, Portugal, Greece, Spain, Germany, France, Austria-Hungary, Turkey, Romania, and Denmark. Each government stated officially that the ban was a precaution against the introduction of disease, particularly trichinosis.

---

[11] 'The Importation of Cattle from America' *The Times* (8 April 1879). Cited in Kastner and Powell, *ibid.*

[12] See Perren, *supra* note 6.

[13] Trichinosis is an illness caused by eating raw or undercooked meat of animals infected with the larvae of a species of worm called *Trichinella.*

[14] John L. Gignilliat, 'Pigs, Politics, and Protection: The European Boycott of American Pork, 1879–1891' (1961) 35 Agricultural History 3 at 4.

[15] Benjamin Moran to James G. Blaine, 25 July 1881, *Foreign Relations* 1881. Cited in Gignilliat, *ibid.*

In Great Britain, much alarm was raised following the release of a report by the British consul in Philadelphia, Mr Crump, which described in some detail an alleged case of trichinosis in a man in Kansas:[16]

> Trichinae were found; worms were in his flesh by the million, being scraped and squeezed from the pores of the skin. They are felt creeping through his flesh and are literally eating up his substance. The disease is thought to have been contracted by eating sausages.

The public release of the report in the London *Times* in 1881 predictably caused panic among British consumers and served to complicate the trade dispute, even though the case of trichinosis in Kansas was never confirmed. As Secretary of State, James G. Blaine, stated, 'had it been Mr Crump's specific purpose to cause a panic among British consumers, he could hardly have framed his report more appropriately'.[17]

Despite the report which was supported by the Foreign Office, the British government refused to exclude imports of what was, at the time, an important source of cheap food. The situation was politically sensitive enough, however, that the Americans suggested inspection of exports by the American government, lest 'the theory of protection may be disguised as a legitimate carefulness of the public health'.[18]

Gignilliat argues that while genuine fear for public health in European countries played a role in the dispute, the possibility of protectionism should not be overlooked. Indeed, this was the view of many at the time. The American Consul-General in Frankfurt, for example, considered sanitary precaution to be a mere excuse for the German exclusion, noting that: 'the large and growing imports of American pork and canned meats are viewed by the German dealers in German meats with extreme jealousy and no pretext is unemployed for prejudicing consumers against American articles'.[19] Gignilliat argues that 'the general fear of trichinosis was a godsend for European protectionists. Exclusion based on sanitary grounds had certain political advantages over exclusion based on open protectionism. A government keeping out cheap food would be resented by the poor; a government posing as the protector of its people's health could not be blamed.' As he notes, there was enough scientific uncertainty surrounding the matter that both those supporting the prohi-

---

16 This was reported in *The Times*, 19 February 1881. Cited in Gignilliat, *ibid.*

17 Blaine to J.R. Lowell, 21 March 1881. Cited in Gignilliat, *ibid.* at 6.

18 Lowell to Blaine, 9 April 1881 *Foreign Relations* 1881. Cited in Gignilliat, *ibid.*

19 John M. Wilson to Seward, 3 June 1879, *Ex. Doc*, 200. Cited in Gignilliat, *ibid.* at 5.

bition, and those opposing it, could find an 'expert' opinion to support their case.[20]

The French too, appear to have been motivated by more than public health. Like its European neighbours, the French banned imports of American pork in 1882, on the grounds of the risks to public health from *trichinella spiralis*. The US argued that the exclusion was protectionism under another name; the American consul at Le Havre claiming that inspectors sent to the US from Paris had instructions to find *trichinella spiralis* in at least 25 per cent of all American meat they examined.[21] The French foreign minister suggested that there was more than sanitation at stake when he stated that high tariffs in the US presented 'very numerous difficulties' to the settlement of the pork dispute.[22] Ferrières argues that American meats had never threatened health on the Continent – 'at the end of their transatlantic voyage, any possible trichina were dead or nearly so and hardly capable of causing the slightest infection'.[23]

The dispute dragged on until 1891 when the US government passed a bill that authorized microscopic inspection of pork intended for export. However, while the passing of the Meat Inspection Act removed the pretext for exclusion of pork by European countries, it did not settle the issue of protection as several nations moved to place high tariffs on American pork.

Similar forces can be seen at play in a recent domestic Canadian case with roots in the nineteenth century. In 1866, Napoleon III offered a prize for the invention of a healthy, economical, easy-to-preserve substitute for butter after the Franco-Prussian War had resulted in a shortage of butter for the French troops. The prize was won by Hippolyte Mège-Mouriés for his invention 'oleomargarine', a product made from rendered beef emulsified with a mixture of water and casein prior to churning. The product entered the market after he sold his formula to two Dutch butter manufacturers.

In Canada, the government imposed high import duties on the new product, declaring it to be a 'glaring fraud on the public'.[24] This was followed by pressure from Ontario's 'butter belt' to promote legislation to ban the new product. The butter producers were supported by the *Toronto Globe* newspaper which argued in an editorial in 1886 that 'a ban on this bare counterfeit of butter' is

[20] Gignilliat, *ibid.* at 4.

[21] J.A. Bridgland to Hitt, 14 June 1881, *House Doc.*, 209. Cited in Gignilliat, *ibid.* at 7.

[22] Barthelemy St Hilaire to Morton, 3 November 1881, *Foreign Relations* 1881, 435. Cited in Gignilliat, *ibid.*

[23] Madeleine Ferrières, *Sacred Cow Mad Cow* (New York: Columbia University Press, 2006).

[24] Ronald L. Doering, *Margarine Mayhem* (Guelph, Ontario: Food Safety Network, 2004).

justified to protect the dairy industry from competition, to protect the consumer from fraud, and to protect public health from a 'compound of the most villainous character, which is often poisonous'.[25] Parliament subsequently passed An Act to Prohibit the Manufacture and Sale of Certain Substitutes for Butter, which prohibited margarine in Canada for the next 60 years. In a 1950 decision, the Privy Council declared the margarine law as *ultra vires* the federal government.[26] As a result, various provinces either banned the sale of margarine, or imposed prohibitive restrictions that forbade margarine to be yellow (and thus potentially confused with butter). Most recently, a manufacturer of margarine, Unilever Canada Inc., brought an (unsuccessful) legal challenge against a Quebec law that prohibits the sale or offering for sale in Quebec of yellow coloured margarine.[27] While not an international trade dispute, this case is a useful illustration of how public health can be used as a pretext for protectionism, and has been so used since the nineteenth century.

These nineteenth-century disputes have a number of modern parallels, including difficulties in sifting out genuine intentions to protect health from protectionist intentions, the existence of scientific uncertainty, and the importance of economic considerations and public sentiment. In terms of risk perception, a particularly interesting parallel is the use of techniques to stir up public fear that may support protectionist concerns. In the pork dispute discussed above, the publication of a story about an alleged case of trichinosis in Kansas stirred up fear among the British public. Similar tactics are common today; for example, they have been used by groups opposed to genetically modified (GM) food, contributing to public concerns which played a role in the *EC – Biotech* dispute.[28]

---

[25] *Toronto Globe*, 13 May 1886. Cited in Doering, *ibid.*

[26] *Reference re Validity of Section 5(a) of the Dairy Industry Act; Canadian Federation of Agriculture v. Attorney-General of Quebec et al; Margarine Case* [1950] 4 D.L.R. 689. Affirming decision of the Canadian Supreme Court: *In the Matter of a Reference as to the Validity of Section 5(a) of the Dairy Industry Act, R.S.C. 1927, Chapter 45* [1949] S.C.R. 1.

[27] Unilever's claim that the regulation in question is null, unconstitutional, invalid, inoperable, unreasonable, and contrary to Canadian federalism was dismissed in the Superior Court, District of Montreal (26 May 1999) and its appeal declined by the Quebec Court of Appeal: *UL Canada Inc. v. Procureur Général de Québec* [2003] J.Q. no. 13505 (C.A.). A subsequent appeal to the Supreme Court of Canada was also dismissed: *UL Canada Inc v. Attorney General of Quebec, et al.* [2005] S.C.C. case no. 30065.

[28] *European Communities – Measures Affecting the Approval and Marketing of Biotech Products* (2006), WTO Doc. WT/DS291/R, WT/DS292/R, WT/DS293/R (Panel Report).

In many ways, broader current issues also bear a resemblance to those faced in the nineteenth century: rapid technological developments; an awareness of and concern for human, animal, and plant health; uncertainty about the implications of globalization; and ever-present protectionist interests. The old adage 'history repeats itself' is pertinent. History tells us that these kinds of disputes are not isolated incidents; rather, it shows that when blatant protectionism is not politically or legally viable, governments will resort to more subtle measures to protect local interests. It also points to the importance of trade rules that are capable of dealing with the problem in all its complexities. This is especially the case today, as disputes are likely to be even more complex than those in the past, involving complicated and uncertain science, and competing interests and perspectives, such as a more sophisticated view of health, risk regulation, interests of 'civil society', and political and economic interests.

3.3 DEVELOPMENT OF RULES ON NON-TRADE BARRIERS

The primary focus of the GATT 1947 was to reduce tariffs; consequently, it did not deal directly with the issue of non-tariff barriers such as health regulations. Rather, it simply disciplined such regulations through various non-discrimination requirements, prohibiting discrimination between trading partners (the most-favoured nation obligation in Article I) and between foreign suppliers and domestic suppliers (the national treatment obligation in Article III); as well as prohibitions on quantitative restrictions (Article XI.I). The GATT did, however, recognize that countries may sometimes have valid reasons that justify discrimination against goods from certain sources. To this end, Article XX provides that discriminatory measures may be justified as, *inter alia*, necessary to protect human, animal, or plant life or health, so long as they are not 'applied in a manner which would constitute arbitrary or unjustifiable discrimination between countries where the same conditions prevail, or a disguised restriction on international trade'.

As the international trading community secured successes in reducing tariffs, its attention turned to non-tariff barriers. In 1967, GATT Members commenced a study aimed at establishing a comprehensive inventory of existing non-tariff barriers to international trade. They noted that: '[g]overnments and economic circles are paying increasing attention to non-tariff barriers. These barriers have tended to grow in significance as customs tariffs have been reduced over the years. If GATT can consider that it has partly fulfilled its mission so far as customs tariffs are concerned . . . its duty now is also to turn to this other category of measures that distort the free conduct of trade:

namely, non-tariff barriers'.[29] The study reviewed approximately 800 notifications received from Member nations and categorized non-tariff barriers under five headings, including that of 'standards applicable to imported and domestic products (product standards or testing rules for health, security or other reasons, packaging, labelling and marking rules)'.[30]

The impetus to tackle non-tariff barriers at a multilateral level gained strength in the late 1960s as awareness of environmental issues increased, giving birth to a raft of environmental and health and safety regulations across Western countries. Very often, these regulations diverged from each other, with the potential to impede trade. Recognizing the trade implications of this trend, in 1972 the Organization for Economic Cooperation and Development (OECD) produced its Guiding Principles Concerning the International Economic Aspects of Environmental Policies. These principles represented an attempt to balance the imposition of standards, which related to the valid environmental protection measures of a country, against the consequent impacts on trade.[31] The principles recognized that valid reasons exist for divergent standards, such as different social objectives or levels of industrialization. The principles also recognized that harmonization, while desirable, would be difficult to achieve in practice.[32] To this end, key principles included:

> Where valid reasons for differences do not exist, governments should seek harmonization of environmental policies, for instance with respect to timing and the general scope of regulation for particular industries to avoid the unjustified disruption of international trade patterns and of the international allocation of resources which may arise from the diversity of national environmental standards;
>
> Measures taken to protect the environment should be framed as far as possible in such a manner to avoid the creation of non-tariff barriers to trade;
>
> Where products are traded internationally and where there could be significant obstacles to trade, governments should seek common standards for polluting products and agree on the timing and the general scope of regulations for particular products.

---

[29] GATT, *The Activities of GATT 1967/68*, at 10 (1969).

[30] The other headings were: government participation in trade; customs and administrative entry procedures; specific limitations on imports and exports; and limitations on imports and exports through price mechanisms. Terence P. Stewart, *GATT Uruguay Round: A Negotiating History (1986–1992), Volume 1: Commentary* (Boston, MA: Kluwer Law International, 1993) at 706.

[31] Michael J. Trebilcock and Robert Howse, *The Regulation of International Trade*, 3rd ed. (London: Routledge, 2005) at 204.

[32] OECD Guiding Principles Concerning the International Economic Aspects of Environment Policies. Cited in Trebilcock & Howse, *ibid.*

Lasting from 1973 to 1979, the Tokyo Round of negotiations was the first serious attempt to tackle non-tariff trade barriers on a multilateral basis. The negotiations recognized that implicit discrimination against imports often existed in technical regulations specifying characteristics to which products must conform. Further, it was expected that product standards would grow in importance as governments adopted more regulations to protect the health and safety of their citizens, safeguard the environment, and protect consumers.[33]

The 1967 GATT study played an important role in the Tokyo Round negotiations[34] and, in 1979, the Agreement on Technical Barriers to Trade (otherwise known as the 'Standards Code') was adopted. The Standards Code applied to 'all products, including industrial and agricultural products'[35] and introduced a number of new disciplines aimed at minimizing trade distortions arising from divergent national regulations.

The Standards Code encouraged countries to work towards and adopt internationally harmonized standards. It obligated them to accept such standards unless inappropriate for reasons including, *inter alia*, the protection of human, animal, and plant health.[36] Article 2.1 required Parties to ensure that no technical regulations or standards, or their application, would have the effect of creating unnecessary obstacles to international trade. However, it did not provide any criteria that would assist in drawing the line between necessary and unnecessary obstacles to trade. This weakened the Code's effectiveness and placed a difficult onus on the complaining Party who had to prove 'deliberate protectionist intent, or to demonstrate that the measure went beyond what was necessary'.[37]

Another deficiency in the Standards Code was that only 39 countries had signed up to it, which significantly weakened its effectiveness and precluded a number of standards-related disputes from being brought before a GATT dispute settlement panel.[38] Even where two countries had signed the Standards Code, the pre-Uruguay Round consensus-based dispute settlement system meant that either country could block a request to convene a panel or block adoption of a report. It is therefore not surprising that no disputes were decided under the Standards Code.

---

33 Stewart, *supra* note 30 at 1067.
34 *Ibid.* at 707.
35 Article 1.3.
36 Article 2.2.
37 Michael J. Trebilcock and Robert Howse, *The Regulation of International Trade*, 2nd ed. (London: Routledge, 1999) at 141.
38 D. Roberts, 'Preliminary Assessment of the Effects of the WTO Agreement on Sanitary and Phytosanitary Trade Regulation' (1998) J. Int'l Econ. L. 377 at 380.

In the decade following the Tokyo Round, consensus emerged that the GATT and the Standards Code had failed to curtail disruptions of trade caused by proliferating non-tariff barriers.[39] Disruptions were particularly visible in respect of agricultural products. Most notably, the US and EC had been at loggerheads over the EC's ban on imports of hormone-treated beef throughout the 1980s and had failed to reach a resolution under the GATT legal infrastructure. Avery *et al.* suggest that this failure led the US to increase efforts to harmonize national health and safety standards, leading to negotiation of the SPS Agreement.[40]

The Punta del Este Ministerial Declaration which launched the Uruguay Round in 1986 included non-tariff measures among its subjects for negotiation, stating that 'negotiations shall aim to reduce or eliminate non-tariff measures'.[41] Agriculture was a key component of the negotiations and this focus created an additional incentive to deal with non-tariff barriers as some negotiators feared that if an agriculture agreement was reached, governments would resort to regulatory compensation, in the form of trade-restrictive SPS measures, to appease domestic producers in this politically sensitive sector.[42]

SPS measures thus formed a crucial part of the negotiations on agriculture during the Uruguay Round and, like the rest of the negotiations, attracted a considerable amount of dissension among the negotiating parties. In November 1990 the Working Group on SPS Regulations produced a draft text of the SPS Agreement. The main points of agreement in the draft included:

- Measures should be non-discriminatory as regards particular nations and should not constitute disguised trade barriers;
- Measures should be harmonized, in accordance with generally accepted scientific principles;
- Special consideration should be taken of developing nations and their difficulties in meeting standards; and
- Parties should develop procedures for transparency in setting regulations, and for resolving disputes concerning their validity.

---

[39] *Ibid.*

[40] Natalie Avery, Martine Drake and Tim Lang, 'Codex Alimentarius: Who is Allowed in? Who Is Left Out?' (1993) 23:3 The Ecologist 110 at 110.

[41] Terence P. Stewart, *The GATT Uruguay Round: A Negotiating History (1986–1992), Volume III: Documents* (Boston, MA: Kluwer Law International, 1993) at 1.

[42] Timothy E. Josling, Stefan Tangermann and T.K. Warley, *Agriculture in the GATT* (London: Macmillan Press, 1996).

Several important areas were bracketed in the November 1990 draft as they remained subject to dispute. Key areas were, first, a failure to agree as to whether and under what conditions countries could impose more stringent SPS standards than those established by international organizations. This topic was particularly controversial in the US where environmental and consumer groups claimed that agreeing to standardize health and safety measures would severely reduce their effectiveness.[43] The US and the EC had argued for the right to impose more stringent standards, but the Cairns Group of agricultural exporting nations[44] had disputed the validity of such national flexibility in the context of an effort to harmonize standards.

Second, there was some dispute as to whether 'other economic considerations and genuine consumer concerns' could be taken into account as legitimate factors in the risk assessment undertaken in reviewing imposed or proposed SPS standards.[45] Japan, for example, had proposed that a country's geographical conditions and 'dietary customs' should be considered during any review of whether a regulation violated trade rules.[46] On the other hand, the Cairns Group had suggested that 'food grading, consumer preferences, consumer information, animal welfare and religious and moral issues . . . are not SPS matters and should not be dealt with in the context [of an SPS Agreement]'.[47]

The first of these issues was eventually resolved by allowing countries to enact more stringent national regulations than provided for by international standards. The second was dealt with more ambiguously, meaning that it has remained a source of contention. The Agreement did not explicitly make provision for countries to take 'other economic considerations and genuine consumer concerns' into account in the risk assessment undertaken in reviewing imposed or proposed SPS standards.[48] Sandford notes, however, that it is not clear that the Cairns Group approach of separating issues such as consumer concerns or animal welfare from the imposition of SPS measures

---

[43] Stewart, *supra* note 30 at 201.

[44] The Cairns Group at the time consisted of the following agricultural exporting countries: Argentina; Australia; Brazil; Canada; Chile; Colombia; Hungary; Indonesia; Malaysia; New Zealand; the Philippines; Thailand; and Uruguay.

[45] Stewart, *supra* note 30 at 201.

[46] *Negotiating Group on Agriculture: Submission by Japan,* GATT Doc. No. MTN.GNG/NG5/W/131 (6 December 1989), at 2. Cited in Stewart, *ibid.* at 188.

[47] *Negotiating Group on Agriculture – Working Group of Sanitary and Phytosanitary Issues – Supplementary Communication of the Cairns Group*, GATT Doc. No. MTN.GNG/NG5/W/164. Cited in Ian Sandford, 'Hormonal Imbalance? Balancing Free Trade and SPS Measures after the Decision in Hormones' (1999) 29 V.U.W.L.R. 389 at 396.

[48] *Ibid.* at 213.

had prevailed. There was, in fact, no language in the final draft to say that such factors could not be taken into account.[49]

3.4 THE FINAL PRODUCT: THE SPS AGREEMENT

The SPS Agreement's objectives are clear from the Preamble which states that 'no Member should be prevented from adopting or enforcing measures necessary to protect human, animal or plant life or health' and also notes the Members' desire to minimize such measures' negative effects on trade.

The SPS Agreement defines 'sanitary or phytosanitary measures' in Annex A as:

> Any measure applied:
>
> (a) to protect animal or plant life or health within the territory of the Member from risks arising from the entry, establishment or spread of pests, diseases, disease-carrying organisms or disease-causing organisms;
> (b) to protect human or animal or plant life or health within the territory of the Member from risks arising from additives, contaminants, toxins or disease-causing organisms in foods, beverages or feedstuffs;
> (c) to protect human life or health within the territory of the Member from risks arising from diseases carried by animals, plants or products thereof, or from the entry, establishment or spread of pests; or
> (d) to prevent or limit other damage within the territory of the Member from the entry, establishment or spread of pests.

The SPS Agreement recognizes that discrimination may actually be necessary where goods from a particular source pose a greater risk than those from other sources. To this end, discrimination *per se* is not prohibited except where it is arbitrary or unjustifiable as between Members where 'identical or similar conditions prevail'.[50]

The SPS Agreement was negotiated on the basis that domestic SPS measures based on international norms could reduce trade conflicts and lower transactions costs.[51] As such, it aims to further the use of harmonized SPS measures between Members, on the basis of international standards, guidelines, and recommendations developed by relevant international organizations, including the Codex Alimentarius Commission (Codex), the International

---

[49] Sandford, *supra* note 47 at 396. This issue was later to take on significant importance in the *EC – Hormones* case as discussed in Chapter 12.

[50] Article 5.5.

[51] Tim Josling, Donna Roberts and David Orden, *Food Regulation and Trade* (Washington DC: Institute for International Economics, 2004) at 40.

Office of Epizootics (OIE), and organizations operating within the framework of the International Plant Protection Convention (IPPC). Measures which conform to relevant international standards are presumed to be in compliance with the requirements of the SPS Agreement, and Members may only introduce or maintain measures 'which result in a higher level of . . . protection than would be achieved by measures based on the relevant international standards . . . if there is a scientific justification'.[52] The SPS Agreement contains detailed requirements relating to scientific justification, including in Article 5.1 the requirement that Members must ensure that their SPS measures are 'based on an assessment, as appropriate to the circumstances, of the risks to human, animal or plant life or health'. Article 5.2 states that the risk assessment must take into account, *inter alia*, 'available scientific evidence'.

As well as harmonization, the SPS Agreement seeks to encourage the concept of equivalency. Article 4 requires Members to accept the SPS measures of other Members as equivalent, even if these measures differ from their own or from those used by other Members trading in the same product, if the exporting Member objectively demonstrates to the importing member that its measures achieve the importing Member's appropriate level of SPS protection. For this purpose, Members are required to give, upon request, reasonable access to the importing Member for inspection, testing, and other relevant procedures. Article 4 also encourages Members to enter into bilateral and multilateral agreements on the recognition of the equivalence of specified SPS measures.

While it intrudes on a country's decision-making by requiring scientific evidence of a risk before an SPS measure may be enacted, the SPS Agreement aims to respect regulatory sovereignty with respect to risk tolerance. It thus recognizes the right of Members to set their appropriate level of protection once scientific evidence has established the presence of a risk. However, several constraints are placed on this right. Article 2.3 requires Members to ensure that their SPS measures do not arbitrarily or unjustifiably discriminate between Members where identical or similar conditions prevail. Relatedly, Article 5.5 requires Members to avoid arbitrary or unjustifiable distinctions in the levels of SPS protection it considers appropriate in different situations, if such distinctions result in discrimination or a disguised restriction on international trade.

Article 5.4 of the SPS Agreement provides that Members should, when determining the appropriate level of SPS protection, take into account the objective of minimizing negative trade effects. Article 5.6 requires Members to ensure that SPS measures are 'not more trade-restrictive than required to

---

[52] Article 3.3.

achieve their appropriate level of sanitary or phytosanitary protection, taking into account technical and economic feasibility'.

The SPS Agreement exists alongside the GATT 1994 which incorporates the original GATT agreement of 1947 and thus maintains the Article XX(b) exception for measures that are necessary, *inter alia*, to protect human, animal, or plant life or health. It is possible that the same measure could give rise to a claim that it violates both the SPS Agreement and provisions of the GATT, especially Article I (Most-Favoured Nation), Article III (National Treatment), and/or Article XI (Elimination of Qualitative Restrictions). In such a case, a panel is required to examine the measure first against the more specific rules of the SPS Agreement and then consider the more general provisions of the GATT.[53]

3.5 THE TRADE IMPACT OF SPS MEASURES

3.5.1 Quantifiable Impact

Economists generally concur that as traditional trade barriers such as tariffs and quotas have been progressively lowered, non-tariff barriers have increased.[54] A number of studies have sought to establish what impact these non-tariff barriers have on trade. Of course, not all SPS measures will be barriers to trade. Iacovone identifies some of the positive effects of standards which can, he argues, act as facilitators of trade.[55] These include: conveying information to consumers in a consistent and understandable way; reduction of transaction costs; reduction of consumer uncertainty; ease of comparison; increased demand for complementary goods; and the elasticity of substitution between similar goods. However, there are also a number of studies which suggest that – for developing countries in particular – SPS measures may have a largely negative effect.[56] In a recent study, Disdier, Fontagné, and Mimouni examined the impact of both SPS and technical barriers on agricultural trade.

---

[53] Trebilcock and Howse, *supra* note 31 at 207.

[54] Keith E. Maskus and John S. Wilson, eds, *Quantifying the Impact of Technical Barriers to Trade: Can it be Done?* (Ann Arbor, MI: The University of Michigan Press, 2001). R. Baldwin, 'Frictional Trade Barriers, Developing Nations and a Two-Tiered World Trading System' (Graduate Institute of International Studies, Geneva, 2000). Cited in Leonardo Iacovone, 'The Analysis and Impact of Sanitary and Phytosanitary Measures' (2005) 22 Integration and Trade 97 at 104.

[55] Iacovone, *ibid.* However, Iacovone also notes, at p. 117, that economists lack precise comparable quantifications of these benefits.

[56] For a comprehensive discussion of economic studies and literature dealing with the impacts of SPS measures, see Iacovone, *ibid.*

They found that, on the whole, they have a negative impact on trade in agricultural products. In particular, they found that exports from developing countries and least developed countries are negatively and significantly affected.[57]

A number of studies have looked at the impact of isolated standards on developing countries. In a 2000 study, for example, Otsuki, Wilson, and Sewadeh sought to quantify the impact of the EU's recently harmonized aflatoxin standard on food exports from Africa. They concluded that the implementation of the new aflatoxin standards would have a negative impact on African exports of cereals, dried fruits, and nuts to Europe. The standards, which would reduce health risk by approximately 1.4 deaths per billion a year, would decrease these African exports by 64 per cent or US$670 million.[58]

In another study, Wilson and Otsuki examined the impact of standards on pesticide residues on food exports by analysing the relationship between pesticide residue standards in 11 OECD countries and banana exports from 21 developing countries.[59] Focusing on one of the most commonly used pesticides in worldwide banana production (chlorpyrifos), they concluded that a 1 per cent increase in regulatory stringency – tighter restrictions on chlorpyrifos – leads to a decrease of banana imports by 1.63 per cent.[60] They estimated that US$5.5 billion would be lost in exports per year if an international standard were set at the EU levels of regulatory stringency in contrast to a world standard set by Codex at the internationally recommended level.[61]

In a study published in 2005, Kimball, Wong, and Taneda find further evidence of harm to developing countries as a result of high standards in export markets. In 1997, there was a cholera outbreak in Mozambique, Kenya, Tanzania, and Uganda. In response to the outbreak, the EC placed restrictions on fish imports from those states. The authors looked at the impact of these restrictions and found that they were significant. Each of the affected states suffered a loss in trade which as a percentage of GDP rose from 0.26 per cent in 1997 to 0.96 per cent in 2002.[62] As the authors point out, these losses were

---

57 Anne-Célia Disdier, Lionel Fontagné and Mondher Mimouni, *The Impact of Regulations on Agricultural Trade: Evidence From SPS and TBT Agreements* (Paris: Centre d'Etude Prospectives et d'Informations Internationales – Working Paper No 2007–04, 2007).

58 Tsunehiro Otsuki, John S. Wilson and Mirvat Sewadeh, 'Saving Two in a Billion: Quantifying the Trade Effect of European Food Safety Standards on African Exports' (2001) 26 Food Policy 495 at 495.

59 J.S. Wilson and Tsunehiro Otsuki, 'To Spray or Not to Spray: Pesticides, Banana Exports, and Food Safety' (2004) 29 Food Policy 131.

60 *Ibid.* at 144.

61 *Ibid.*

62 A.M. Kimball, K.-Y. Wong and K. Taneda, 'An Evidence Base for International Health Regulations: Quantitative Measurement of the Impacts of

significant for the exporting countries because they are poor and highly dependent on fisheries exports.

Thus, the economic evidence is suggestive of negative impacts for developing countries in particular and highlights the difficult question of how to approach the trade-off between appropriate levels of risk to health, and the costs to international trade of differing levels of sanitary and phytosanitary protection.[63] It certainly augurs for strict disciplines on the use of SPS measures to prevent them being used as a guise for protectionism.

3.5.2 Disputes under the SPS Agreement

Article 12 of the SPS Agreement establishes the WTO Committee on SPS Measures and mandates it with providing a regular forum for consultations. In 1996, the Committee agreed that Members should be encouraged to discuss trade problems at its meetings whenever possible before initiating formal dispute settlement procedures, without prejudice to their rights under that process.[64] Since 1995, a total of approximately 245 specific trade concerns have been brought to the Committee's attention.[65] Of these, approximately 40 per cent relate to animal health and zoonoses, 29 per cent relate to plant health, 27 per cent relate to food safety, and 4 per cent relate to other issues such as certification requirements or translation. Within the category of animal health and zoonoses, over one-third of concerns raised (39 per cent) related to transmissible spongiform encephalopathy (TSE).

Predictably, the countries to have raised the highest number of concerns to date are the world's major food and commodity exporters including the US (64), the EC (54), Argentina (31), and Canada (20). Developing country Members have raised approximately 120 trade concerns (in many cases, the same concern is raised by more than one Member); and have supported other countries in their concerns on over 170 occasions. Those who have raised the most concerns are China (15), Thailand (nine), Brazil (nine), and Chile (seven). Three least-developed countries have raised concerns (Côte d'Ivoire, Gambia, and Senegal).

Those countries whose measures have most often been the subject of complaint are the industrialized countries of the EC (on approximately 56

Epidemic Disease on International Trade' (2005) 24:3 OIE Scientific and Technical Review 825 at 829.

[63] Iacovone, *supra* note 54 at 117.

[64] WTO Committee on Sanitary and Phytosanitary Measures, Summary of the Meeting held on 29–30 May 1996, WTO Doc. G/SPS/R/5 at 6.

[65] This is the number cited in a summation paper by the Secretariat: WTO Secretariat, Specific Trade Concerns – Note by the Secretariat (2007), WTO Doc. G/SPS/GEN/204/Rev.7.

occasions), Japan (on 21 occasions), the US (on 23 occasions), and Australia (on 15 occasions). Developing countries have been the subject of complaint on just over 100 occasions, including since 2004, Brazil (on 10 occasions); China and Korea (nine each); India, Argentina, Panama, and Indonesia (seven each); and Mexico and Venezuela (five each). No trade concerns have been raised regarding measures maintained by least-developed countries. A number of the countries most often the subject of complaint rank among the world's top ten net importers of agricultural products, raising suspicions that they are more likely to act to block imports in order to protect their domestic producers: in 2004, Japan ranked first on this list, China third, Korea fourth, the EC sixth,[66] and Mexico seventh). The US is also a major agricultural importer.

The majority of the specific trade concerns raised have been resolved,[67] with 28 requests made for formal consultations under the Disputes Settlement Understanding. Of these, five disputes have been heard by a Disputes Panel.[68]

The most common kinds of measures to have been the subject of concern are those that: impose outright import prohibitions (for example, Australia's ban on imports of North American salmon); allow imports but impose restrictions (for example, Canadian and US import restrictions on Hungarian meat products due to incidences of foot and mouth disease in Europe); require certification or notification with respect to a particular product (for example, Panama's requirement that imports of rice be certified free from a particular fungus); specify certain production methods (for example, a French imposition of a specific production method for all gelatin exported to France); and set maximum residue levels (MRLs) (for example, the EC's proposal to set new MRLs for aflatoxins in foodstuffs).

---

[66] The EC is in the top ten of both net importers and major exporters. See statistics available at http://comtrade.un.org.

[67] The Secretariat reported in February 2007 that out of the 245 trade concerns raised in the SPS Committee, 66 had been reported resolved, with 15 partially resolved. No solutions had been reported for 164 trade concerns. WTO Secretariat, Specific Trade Concerns – Note by the Secretariat (2007), WTO Doc. G/SPS/GEN/204/Rev.7 at 5.

[68] *EC – Measures Concerning Meat and Meat Products (Hormones) Complaint by the United States* (1997), WTO Doc. WT/DS26/R/USA (Panel Report). *EC – Measures Concerning Meat and Meat Products (Hormones) Complaint by Canada* (1997), WTO Doc. WT/DS48/R/CAN (Panel Report). *Australia – Measures Affecting Importation of Salmon* (1998), WTO Doc. WT/DS18/R (Panel Report). *Japan – Measures Affecting the Importation of Apples* (2003), WTO Doc. WT/DS245/R (Panel Report). *Japan – Measures Affecting Agricultural Products* (1998), WTO Doc. WT/DS76/R (Panel Report). *European Communities – Measures Affecting the Approval and Marketing of Biotech Products* (2006), WTO Doc. WT/DS291/R, WT/DS292/R, WT/DS293/R (Panel Report).

3.6 CONCLUSION

This chapter has shown that disputes over trade-restrictive domestic health regulations are not a new development. In the 1870s, trade disputes arose over health and safety regulations for many of the same reasons that we see today, including a concern for public health; a degree of uncertainty about the unsettling effects of globalization; and tension between those who wanted to see the protection of local producers and those who favoured free trade. History tells us that health-related trade disputes are not isolated incidents; rather, it shows that when blatant protectionism is not politically viable, governments will resort to more subtle measures. This was recognized by WTO Members when they negotiated the SPS Agreement during the Uruguay Round, and is evidenced by the number of trade concerns that have arisen to date. The next chapter looks to the future to assess the likelihood of continuing friction in this area.

4. Looking to the future: forces of change

4.1 INTRODUCTION

This chapter examines emerging forces that are contributing or are likely to contribute to an increase in the number and/or complexity of disputes over trade-restrictive SPS measures. The forces identified are: (i) the increasing use of non-tariff barriers in lieu of direct tariffs (referred to as the 'substitution effect'); (ii) new and increasing food safety risks; (iii) an increasing emphasis in Western democracies on public participation in regulatory decision-making; (iv) differing approaches between countries to regulatory decision-making; and (v) North–South conflicts.

4.2 TARIFF SUBSTITUTION

Given the GATT's success over the years in reducing tariffs, some commentators expect that countries will seek to substitute repealed tariffs with non-tariff barriers such as health regulations that are capable of achieving the same result of protecting domestic interests.[1] This is especially a concern in the case of agricultural goods where the 1994 Agreement on Agriculture required Members to convert all non-tariff barriers into tariff equivalents and to begin to reduce tariff levels.[2] It has also been argued that because domestic regula-

---

[1] In the context of the North American Free Trade Agreement (NAFTA), for example, Julie Soloway, 'Institutional Capacity to Constrain Suboptimal Welfare Outcomes from Trade – Restricting Environmental, Health and Safety Regulation under NAFTA' (University of Toronto, 2000, unpublished) finds that of 25 environmental, health, and safety-related trade irritants or disputes, there was evidence in 24 cases that the regulation was disguised protection. There was only one case in which it was definitely established that there was an actual, serious environmental, health or safety issue at stake. Soloway, *supra* note 21 at 439. See also: Silvia Weyerbrock and Tian Xia, 'Technical Trade Barriers in US/Europe Agricultural Trade' (2000) 16:2 Agribusiness 235 at 235. World Health Organization (WHO) and WTO, *WTO Agreements and Public Health: A Joint Study by the WHO and the WTO Secretariat* (Geneva: WHO and WTO, 2002) at 63.

[2] At the time of writing, the Doha Round negotiations are still under way and it is not clear to what extent agricultural trade will be further liberalized. However, it is

tions are less transparent than tariffs or quotas, it is possible to tweak them to make them stronger than necessary for achieving optimal levels of protection.[3]

Difficulties will inevitably arise in recognizing where such substitution has taken place, as illustrated in two recent cases of trade friction. In the first, New Zealand and Australia stopped imports of Swiss hard cheeses made from unpasteurized milk on the grounds that pasteurization was necessary to ensure the inactivation of pathogenic micro-organisms, particularly E. coli. In the second, Australia enacted regulations that effectively banned imports of Roquefort cheese on the grounds that its risk assessment had identified potential problems with pathogenic micro-organisms, including E. coli. While on their face, these regulations were enacted for health reasons, Josling, Roberts, and Orden argue that it is tempting to see in these cases some reflection of the trade relations between the parties involved, given that the EU was widely blamed in Australia and New Zealand for depressing world dairy prices due to its export subsidies. As they note, 'there must be little enthusiasm in Canberra or Wellington for easing the way for French cheeses to enter the domestic market'.[4]

4.3 FOOD SAFETY

Food safety has been recognized as important for millennia but emerged as a particularly salient issue in the late twentieth century with the increased movement of food products, animals, and people across borders. For much of history, food safety standards were local in character and as such were vastly different from today. Prior to the twentieth century, standards focused mainly on 'adulteration'.[5] Today, however, standards generally focus on safety in a broader sense. Food safety issues can usefully be divided into two categories: those that revolve around the regulation of *content attributes*, and those that revolve around the regulation of *process attributes*. The following highlights

---

likely that where there is further liberalization, domestic pressure in liberalizing countries for tariff substitution may increase.

[3] Prema-chandra Athukorala and Sisira Jayasuriya, 'Food Safety Issues, Trade and WTO Rules: A Developing Country Perspective' (2004) 26:9 The World Economy 1395 at 7.

[4] Tim Josling, Donna Roberts and David Orden, *Food Regulation and Trade* (Washington DC: Institute for International Economics, 2004) at 111.

[5] 'Adulteration' has been defined as the action of being 'adulterated, corrupted or debased by spurious admixture'. Lawrence Busch, 'Virgil, Vigilance, and Voice: Agrifood Ethics in an Age of Globalization' (2002) 16 Journal of Agricultural and Environmental Ethics 459 at 467. *Oxford English Dictionary Online*, *s.v.* 'adulteration'.

the most important emerging issues that have had and will continue to have trade implications.

4.3.1 Content Attributes

The use of health regulations that deal with the content attributes of food safety has exploded in recent years. Regulations focus on two kinds of health hazards: pathogenic agents in the environment ('*natural*' risks) and those that arise from the use of productivity-enhancing inputs and technologies in the food production process ('*man-made*' risks).[6] *Man-made* risks include a range of residues found in foods, including chemicals (for example, elevated levels of the carcinogen acrylamide, introduced through the cooking process into some foods such as potato chips), agro-chemicals (for example, nitrates and pesticides), veterinary drugs, growth promoters, and packaging components.[7]

Natural food safety risks include natural residues (for example, mycotoxins) and environmental contaminants (for example, mercury in fish, and dioxins). Particularly problematic from a public health standpoint are the risks posed by pathogenic micro-organisms which have been found to be far greater than some of the factors that consumers identify as being of greatest concern, namely, use of additives, chemical pesticides, and food irradiation.[8] Recent years have seen outbreaks of diseases such as salmonellosis, cholera, and enterohaemorrhagic Escherichia coli infections in both developed and developing countries. In the US alone, the Centers for Disease Control and Prevention has estimated that foodborne diseases cause approximately 76 million illnesses, 325 000 hospitalizations, and 5000 deaths each year.[9] Causes of the increase in foodborne illnesses are wide-ranging and include changes in the food supply system such as mass production and distribution, environmental conditions, social situations, behaviour, and lifestyles, and health systems and infrastructure.[10] It has been argued that the increase in international trade in foodstuffs is not an impor-

---

6 Josling, Roberts and Orden, *supra* note 4 at 101.

7 Anne Wilcock, *et al.*, 'Consumer Attitudes, Knowledge and Behaviour: A Review of Food Safety Issues' (2004) 15 Trends in Food Science & Technology 56 at 57.

8 Kenneth F. Kiple and Kriemhild Conee Ornelas, eds, *The Cambridge World History of Food* (Cambridge, UK: Cambridge University Press, 2000) at 1663. Food irradiation involves irradiating food to reduce or eliminate pathogens from food and extend its shelf life.

9 Paul S. Mead, *et al.*, 'Food-Related Illness and Death in the United States' (1999) 5:5 Emerging Infectious Diseases 607.

10 Yasmine Motarjemi and Fritz Kaferstein, 'Food Safety, Hazard Analysis and Critical Control Point and the Increase in Foodborne Diseases: A Paradox?' (1999) 10 Food Control 325 at 326. See also generally F. Kaferstein and M. Abdussalam, 'Food Safety in the 21st Century' (1999) 77:4 Bulletin of the World Health Organization 347.

tant factor, as statistics suggest that safety of imported products is on average similar to that of domestic products. However, highly publicized cases linked to food imports have spread the idea that globalization magnifies risks related to food safety, thus increasing consumer pressure on governments to regulate.[11] Recent scares about food and other products originating from China provide a vivid example of this. *The Economist* newspaper has reported, for example, that scandals involving tainted pet food, toxic toothpaste, toys, and seafood, have contributed to concerns among consumers in importing countries.[12]

Food may also be infected with viruses, such as norovirus and hepatitis A. Food-borne viral outbreaks are usually traced to food that has been manually handled by an infected food handler, rather than to industrially processed foods. The viral contamination of food can occur at any stage of the chain from 'farm to fork' and even where it occurs prior to processing, residual viral infectivity may be present after some processing.[13]

The WTO Committee on SPS Measures has seen numerous concerns raised in relation to these various food safety threats. Concerns have been raised in regard to all manner of trade-restricting food safety measures, including the 'natural' threats of mycotoxins (for example, EC restrictions on Brazil nuts from Bolivia due to aflatoxin residues[14]), pathogenic micro-organisms (for example, China's zero tolerance regulations for E. coli contamination of raw meats and poultry products[15]), environmental contaminants (for example, Malaysia's dioxin-related restrictions on imports of European goods[16]), and viruses (Australian regulations concerning the import of chicken meat from Thailand[17]).

---

[11] J.C. Bureau and W. Jones, 'Issues in Demand for Quality and Trade' (Paper presented to the International Agricultural Trade Research Consortium: Global Food Trade and Consumer Demand for Quality, Montreal, 2000).

[12] 'The Diddle Kingdom' *The Economist* (17 September 2007).

[13] Marion Koopmans and Erwin Duizer, 'Foodborne Viruses: An Emerging Problem' (2004) 90 International Journal of Food Microbiology 23 at 23.

[14] Bolivia argued that the measure would severely restrict trade while not resulting in a significant reduction in health risk to consumers: WTO Committee on Sanitary and Phytosanitary Measures, Summary of the Meeting Held on 25–6 June 2002, WTO Doc. G/SPS/R/27.

[15] The US claimed that the zero-tolerance rule standard was not achievable: WTO Committee on Sanitary and Phytosanitary Measures, Summary of the Meeting Held on 7–8 November 2002, WTO Doc. G/SPS/R/28.

[16] The EC raised concerns over Malaysia's restrictions on European foodstuffs: WTO Committee on Sanitary and Phytosanitary Measures, Summary of the Meeting Held on 7–8 July 1999, WTO Doc. G/SPS/R/15.

[17] Thailand argued that the regulations were in excess of what was needed for the purpose of protecting human or animal life: WTO Committee on Sanitary and Phytosanitary Measures, Summary of the Meeting Held on 15–16 September 1998, WTO Doc. G/SPS/R/12.

In terms of *'man-made'* threats, concerns have been raised regarding chemicals (for example, EC regulations concerning maximum levels of 3-MCPD in soy sauce[18]); agro-chemicals (for example, the EC's maximum residue levels for pesticides in fruits and vegetables[19]); veterinary drugs and growth promoters (for example, Swiss regulations restricting the import of meat from animals treated with hormones, antibiotics, and similar products[20]). Many chemicals have not yet been fully tested for human health effects and where there have been, there are often uncertainties about the risks posed to health.[21]

4.3.2 Process Attributes

Process attributes present unique problems for the international trading system which largely assumes that goods have characteristics that are identifiable in the final, traded product, and therefore susceptible to inspection. Given that process attributes are not readily apparent and available for inspection, and that internationally recognized standards generally do not exist for their regulation, there is more possibility for trade distortion through regulatory confusion and protectionist capture.[22] Emerging issues around process attributes include carbon footprinting, the growing market for organic foods and the problems this raises for the compatibility of national regulations; concerns over the welfare of farm animals; and regulation of the trade and labelling of GM crops and food products.[23]

These increased food safety risks have shifted the way we view food from simply an agricultural/trade commodity to an important public health issue.[24]

---

[18] Thailand argued that the level set was too low to be practicable and an unnecessary barrier to trade: WTO Committee on Sanitary and Phytosanitary Measures, Summary of the Meeting Held on 14–15 March 2001, WTO Doc. G/SPS/R/21.

[19] Côte d'Ivoire argued that the new EC maximum residue levels (MRLs) for pesticides would affect its exports of pineapples, mangoes, papayas, cashew nuts, passion fruits, and green beans and that they were not consistent or based on a pertinent risk assessment: WTO Committee on Sanitary and Phytosanitary Measures, Summary of the Meeting Held on 10–11 July 2001, WTO Doc. G/SPS/R/22.

[20] The US and Australia questioned the public health validity of Switzerland's regulations: WTO Committee on Sanitary and Phytosanitary Measures, Summary of the Meeting Held on 15–16 September 1998, WTO Doc. G/SPS/R/12.

[21] Aaron Cosbey, *A Forced Evolution? The Codex Alimentarius Commission, Scientific Uncertainty and the Precautionary Principle* (Winnipeg: International Institute for Sustainable Development, 2002) at 7.

[22] Josling, Roberts and Orden, *supra* note 4 at 153.

[23] *Ibid.* at 152.

[24] See F.K. Kaferstein, 'Actions to Reverse the Upward Curve of Foodborne Illness' (2003) 14 Food Control 101 at 102. See also Kaferstein and Abdussalam, *supra* note 10 at 351.

In 2000, the WHO adopted a resolution on food safety urging Member States to integrate food safety as one of their essential public health functions.[25] The resolution also requested the Director-General to support the inclusion of health considerations in international trade in food. Food safety is thus a key public health issue for health policy-makers, and, with the enormous volume of food traded internationally, one that will likely continue to be the cause of trade friction.

4.4 PUBLIC PARTICIPATION IN REGULATORY DECISION-MAKING

What role the public can and should play in regulatory policy decision-making is a source of debate in many countries. The overall trend in many Western democracies is for greater public participation, although such participation may take various forms along a continuum, ranging from simple communication by government about how decisions were arrived at, to the active participation of public representatives in the decision-making process itself. The traditional view has been that decisions involving technical or scientific content should be left in the hands of experts and scientists, with public participation limited to the lesser end of the continuum.[26] However, this view is increasingly being challenged, and governmental, scientific, and industrial bodies in many countries are beginning to pay greater heed to the public, seeking to involve it in regulatory decision-making involving science.[27]

This departure from the traditional view is prompted by a number of factors, including a practical recognition that implementing unpopular policies may result in protest and reduced support for government. It also has a more normative basis, with public participation arguably being an ingredient in enhancing the legitimacy of regulations. The concept of legitimacy is a central one in the field of administrative law which is largely concerned with ensuring that the exercise of authority is justified.[28] Lindseth sees it as 'a broad,

[25] World Health Organization, *The Fifty-third World Health Assembly – Resolution WHA53.15 (Food Safety)* (2000).

[26] Busch, *supra* note 5 at 472.

[27] See Gene Rowe and Lynne J. Frewer, 'Public Participation Methods: A Framework for Evaluation' (2000) 25:1 Science, Technology and Human Values 3. See also Brian Martin and Evelline Richards, 'Scientific Knowledge, Controversy, and Public Decision-Making' in Sheila Jasanoff, *et al.*, eds, *Handbook of Science and Technology Studies* (Thousand Oaks, CA: Sage, 1995) 506.

[28] Allison Marston Danner, 'Enhancing the Legitimacy and Accountability of Prosecutorial Discretion at the International Criminal Court' (2003) 97 American Journal of International Law 510 at 535.

empirically determined societal acceptance of the system'.[29] In a democracy, legislative branches of government obtain their legitimacy from accountability to the electorate. Administrative agencies, however, have no such direct accountability, and the question of legitimacy therefore becomes more contested. Thus, issues of procedure loom large. In his well-known work, *Why People Obey the Law*, Tyler reported his research findings that while people are interested in getting what they want, they are more concerned with being able to exercise control over the process in ways they deem fair.[30] A key means of exercising such control is through public participation in the relevant processes. In the area of regulatory decision-making, governments throughout the world have sought to increase opportunities for public participation, particularly areas of social importance such as environment and health policy.[31] Ensuring public participation in regulatory decision-making is seen as a means to enhance government legitimacy.[32]

Arguments for greater public participation in regulatory decision-making around health measures are in many cases predicated on the notion that while scientists may be able to tell the public what the risks are, they have no particular expertise in decisions as to whether risks are worth taking; rather, in those cases, public participation is necessary to preserve the legitimacy of science.[33] It is also argued by some that the public have a legitimate role in determining whether or not a risk exists.[34]

---

[29] Peter L. Lindseth, 'Democratic Legitimacy and the Administrative Character of Supranationalism: The Example of the European Union' (1999) 99 Columbia Law Review 628 at 645.

[30] Tom Tyler, *Why People Obey the Law* (New Haven: Yale University Press, 1990).

[31] David L. Markell, 'Slack in the Administrative State and its Implications for Governance: The Issue of Accountability' (2005) 84 Oregon Law Review 1.

[32] David L. Markell and Tom Tyler, 'Using Empirical Research to Design Government Citizen Participation Processes: A Case Study of Citizens' Roles in Environmental Compliance and Enforcement' (FSU College of Law, Public Law Research Paper No. 270, 2nd Annual Conference on Empirical Legal Studies Paper, FSU College of Law, Law and Economics Paper No. 07-014). Markell, *ibid.* Note, however, the contrary view that rather than legitimating administrative decisions, participation initiatives may do no more than empower the already empowered and negatively impact on the efficacy of decision-making. Markell, *ibid.* Citing Jerry L. Mashaw, *Due Process in the Administrative State* (New Haven, Yale University Press: 1985).

[33] Busch, *supra* note 5 at 473. See also Martin and Richards, *supra* note 27, and David Kriebel and Joel Tickner, 'Reenergizing Public Health Through Precaution' (2001) 91:9 American Journal of Public Health 1351.

[34] See, for example, Cindy G. Jardine, *et al.*, 'Risk Management Frameworks for Human Health and Environmental Issues' (2003) Part B:6 Journal of Toxicology and Environmental Health 569. See also discussion in Chapter 10.

Sunstein distinguishes between a *technocratic* and a *populist* approach to regulatory decision-making.[35] He describes a technocratic approach as one where emphasis is placed on the notion that ordinary people are often ill-informed and regulators are thus urged to 'follow science and evidence, not public opinion'.[36] As such, the technocratic approach is underscored by a commitment to the role of science and risk assessment. This is the approach that, on its face, the SPS Agreement requires countries to adopt. The populist approach on the other hand holds that in a democracy, government should follow the will of the people rather than that of a technocratic elite. That is, 'law and policy should reflect what people actually fear, not what scientists, with their inevitably fallible judgments, urge society to do'.[37] Sunstein argues that both of these positions are too simple.[38] For example, he rejects the notion that a government should respond to excessive fear by enacting legislation that cannot be justified by any kind of rational accounting. Rather, he maintains that it is important to ensure a large role for specialists in the regulatory process.[39] It is, however, useful to view regulatory decision-making along a spectrum where there is lesser or greater emphasis on specialist, scientific reasoning, and/or on public opinion. The more the decision-making process leans towards being populist in its approach, the greater the potential influence of public sentiment on regulatory decision-making. As will be discussed in later chapters, much of the opposition to the SPS Agreement can be interpreted as a rejection of a more technocratic-leaning approach and a call for recognition in WTO law of the freedom of countries to choose a more populist-leaning approach, taking into account public sentiment concerning risks to health.

Public participation raises a concern that resulting regulation may be less than optimal, from both a domestic and global welfare point of view. Green argues, for example, that in the context of environmental policy, the US system of public participation has demonstrated the potential to shift the incentives of policy-makers in inefficient and irrational directions, leading to the 'pollutant of the month' syndrome where stringent regulations are put in place due to public demand whether or not such controls are necessary or desirable.[40] Chang argues along similar lines that where interest groups create public fears, in the absence of supporting evidence, in order to obtain regula-

---

[35] Cass R. Sunstein, 'Book Review – The Laws of Fear' (2001) 115 Harv. L. Rev. 1119 at 1120.

[36] *Ibid.*

[37] *Ibid.*

[38] Cass R. Sunstein, *Laws of Fear – Beyond the Precautionary Principle* (Cambridge: Cambridge University Press, 2005) at 126.

[39] *Ibid.*

[40] Andrew J. Green, 'Public Participation and Environmental Policy Outcomes' (1997) 23:4 Can. Pub. Pol'y 435 at 436.

tions responding to those fears, it is likely that the regulatory response will be less than efficient. This, he argues, is problematic from a trade perspective if the producers burdened by the regulation are disproportionately foreign.[41]

The challenges posed to the international trading system by a move towards greater public participation and input into decision-making on health issues will be discussed in later chapters. These challenges are particularly complex, given the fact that consumers from different countries are often differentiated by their level of risk perception and the risks that they are willing to take. The difficulty in the context of divergent SPS measures is determining when a trade-restrictive measure ought to be justified under international trade rules, particularly where the exporting country contends that the regulation is excessive due to lack of scientific evidence and/or unfounded public fears.

4.5 REGULATORY DIVERGENCE

Countries vary both in their approaches to regulatory decision-making in the face of risk and in the regulatory outcomes that result. Regulatory divergence (or 'regulatory regionalism' as it has been called by some authors[42]) has been an important factor in a number of SPS-related trade disputes. Whether divergence is accepted by trading partners or causes a dispute depends on the circumstances. Bureau and Marette note for example that it seems straightforward to acknowledge the cultural right of Islamic nations to erect trade barriers to pork imports, or Israel to reject non-kosher products, but the US will not accept the fact that a large percentage of European consumers may have a 'cultural aversion' to eating beef produced with growth-enhancing hormones or antibiotic drugs.[43]

A useful illustration of regulatory divergence is the case of cheese. Cheese made from unpasteurized milk is more likely to contain pathogenic bacteria such as salmonella or listeria, than cheese made from pasteurized milk. Raw milk cheeses are, however, popular in countries such as France and Italy where

---

41 Howard F. Chang, 'Risk Regulation, Endogenous Public Concerns, and the Hormones Dispute: Nothing to Fear But Fear Itself?' (2004) 77 S. Cal. L. Rev. 743 at 32.

42 Grant E. Isaac and William A. Kerr, 'Genetically Modified Organisms at the World Trade Organization: A Harvest of Trouble' (2003) 37:6 J. World Trade 1083 at 1085.

43 Jean-Christophe Bureau and Stephan Marette, 'Accounting for Consumers' Preferences in International Trade Rules' in National Research Council, ed., *Incorporating Science, Economics, and Sociology in Developing Sanitary and Phytosanitary Standards in International Trade: Proceedings of a Conference* (Washington DC: National Academy Press, 2000).

people value the tradition and taste of raw milk cheese. The risk associated with raw milk cheeses in those countries is considered comparable to that presented by pasteurized cheese when improperly stored. The preferred approach to safety is therefore to control the risk by regulating dairy production and warning vulnerable populations (such as pregnant women) about the potential risks. In part, this approach reflects the fact that the cheeses have been produced and eaten for hundreds of years with few perceived public health consequences.[44] On the other hand, the US considers raw milk cheeses to be risky and prefers to eliminate the risk entirely by requiring mandatory pasteurization of milk before cheese production.[45] Yet, while Europe supports traditional processes in the case of cheese (as well as in other instances such as in respect of traditionally cured meats) it is less receptive towards new or 'novel' products and technologies and this is seen in the cases of irradiation and genetic engineering, two modern technologies that have been more readily accepted in the US.[46]

A number of explanations have been put forward to account for regulatory divergence among countries. They include economic interests which play a role in regulatory divergence when the burdens and benefits of regulation fall differently in different national contexts;[47] historical explanations;[48] local realities which dictate the use of one method of risk prevention over another;[49]

---

[44] Marsha A. Echols, 'Food Safety Regulation in the European Union and the United States: Different Cultures, Different Laws' (1998) 4 Colum. J. Eur. L. 525 at 532.

[45] This has also emerged as an issue within the EU where recent regulations have tightened the requirements for cheese production. The president of the Association Fromage du Terroir argues in this regard that 'pasteurizing makes for bland and mediocre cheese, but because of unfounded hysteria over bacteria such as listeria, which is not dangerous except to particular vulnerable groups, we are killing a fabulous product'. Kim Willsher, 'EU hygiene regulations threaten traditional French cheeses' *The Telegraph* (10 July 1995).

[46] Jonathon B. Wiener, 'Whose Precaution After All? A Comment on the Comparison and Evolution of Risk Regulatory Systems' (2003) 13:207 Duke J. Comp. & Int'l L. 207 at 242.

[47] Sheila Jasanoff, 'Technological Risk and Cultures of Rationality' in National Research Council, ed., *Incorporating Science, Economics, and Sociology in Developing Sanitary and Phytosanitary Standards in International Trade: Proceedings of a Conference* (Washington DC: National Academy Press, 2000) 65 at 66. Jasanoff cites the example of environmental regulation where countries such as the UK and the US whose pollution has caused acid rain in countries such as Norway and Canada have been predictably less aggressive in seeking to control the emission of sulphur oxides.

[48] Jasanoff, *ibid*. She cites the example of Germany's hostility to biotechnology in the 1980s as being a likely reaction to experiences from the Nazi era.

[49] Bureau and Marette, *supra* note 43. Bureau and Marette cite the example of

and differences in legal systems for protecting consumers from health risks.[50] Divergence may also be attributable to the way in which power is formally divided in society and between government agencies.[51] For example, the UK's Ministry of Agriculture, Fisheries and Food is thought to have underestimated the transmissibility of mad cow disease because its primary goals were to help the beef industry and prevent public panic. In another example cited by Jasanoff, US environmentalists in the early 1970s successfully lobbied to have pesticide regulation removed from the Department of Agriculture, where agribusiness interests dominated, to the newly formed and politically less committed Environmental Protection Agency. She notes that, in a more general sense, the way that power is organized has an impact on the ways in which non-governmental actors can seek to influence policy decisions. In parliamentary democracies, for example, political parties play a key role in influencing decisions. In the US, by contrast, citizen interest groups tend to play a more important role.[52]

Differences in *political culture* among countries also account for regulatory divergence.[53] One such feature is the regulatory style adopted by governments, for example, how and whether they solicit input from interested parties, the opportunities afforded for public participation, the relative transparency of regulatory processes, and the strategies employed for resolving or containing conflict.[54]

---

the trade impacts of Californian regulations maintaining low tolerance levels for pesticide residues which restrict the use of procymidone, a fungicide used in the wine industry. There was low tolerance because wine manufacturers had never questioned the issue due to the fact that the Californian climate renders the use of procymidone unnecessary. The opposite is true in Europe, however, where procymidone is widely used. This caused a problem for European wine producers seeking to export their products to California. Trade was stopped temporarily while US procedures were amended.

[50] For example, in the US, product liability laws and accompanying tort claims play an important role in encouraging firms to adhere to high safety standards, while in some other countries such as France there are few product liability law suits. See Bureau & Marette, *ibid.*

[51] Jasanoff, *supra* note 47 at 76. Jasanoff notes for example that the US system where agency decisions are subject to review by the courts to ensure that they are not arbitrary or capricious has led to a preference for more explicit and formal analytic techniques (for example, quantitative risk assessment of chemical carcinogens or cost-benefit analysis of proposed projects) than those used by policy-makers in other countries.

[52] *Ibid.*

[53] Jasanoff, *ibid.* at 67. The term 'political culture' describes those features of politics that seem, in the aggregate, to give governmental actions a distinctively national flavour, even in countries that share similar social, political, and economic philosophies.

[54] *Ibid.* at 70. See also generally Grace Skogstad, 'Contested Political Authority,

Jasanoff argues that we can expect to see more and greater regulatory divergences between countries as risk debates are globalized, engaging more disparate societies. She also argues that cultural differences are likely to arise when a risk domain touches upon issues that are basic to a society's conceptions of itself, such as constitutional relations between science and the state or religious and philosophical ideas about what is 'natural'.[55] Likewise, it is in such cases that public sentiment is likely to differ, demanding different regulations in different countries. Regulatory divergences present difficult challenges for international trade law in trying to balance a nation's right to enact health regulation with the objectives of trade liberalization.

4.6 NORTH–SOUTH CONFLICT

The development gap between countries is brought into clear view in the area of health and environmental standards. As Wiener notes (in the context of environmental regulations): 'Debates between the United States and Europe over who is "more precautionary than thou" may look baffling and hairsplitting to the billions of people who live in countries with less stringent environmental standards as compared to either the United States or Europe, less institutional capacity to enforce those standards, less scientific capacity to detect and warn of remote future risks, and much more pressing immediate crises in hunger, health, and environmental quality'.[56]

Many developing countries do not have detailed SPS legislation, and lack the regulatory infrastructure to enforce controls they do have. This has the potential to lead to trade disputes as consumers in Western, industrialized nations demand higher standards from their own governments with which suppliers from developing countries must then comply. Developing countries have complained, for example, that developed countries use SPS measures as a form of disguised protectionism.[57] Issues of protectionism aside, developing countries can (as noted in Chapter 3) be disproportionately impacted by stringent measures enacted by developed countries, even where the regulations are enacted to advance legitimate public health objectives. Developing countries have exhibited frustration with the strict standards

Risk Society, and the Transatlantic Divide in Genetic Engineering Regulation' (2005) (unpublished, archived at University of Toronto). See also discussion of public participation above.

[55] Jasanoff, *supra* note 47.

[56] Wiener, *supra* note 46 at 253.

[57] See Graham Mayeda, 'Developing Disharmony? The SPS and TBT Agreements and the Impact of Harmonisation on Developing Countries' (2003) 7:4 J. Int'l Econ. L. 737. See also Athukorala and Jayasuriya, *supra* note 3.

faced by their exports.[58] Their concern is that such standards will impede their participation in world trade, regardless of any other successes achieved in improving market access. Aside from the quantifiable impact noted in Chapter 3, developing countries may be impacted in more indirect ways. For example, Zambia and other African countries have rejected US aid in the form of GM corn, partly on the basis that such corn might cross-pollinate their own corn and make future exports to the EU difficult as they would violate restrictions on imports of GM goods.[59]

As Zarrilli notes, developing countries lack complete information on the number of measures that affect their exports; they are not sure whether these measures are consistent or inconsistent with the SPS Agreement; they do not have reliable estimates on the impact such measures have on their exports; they experience serious problems with respect to scientific research, testing, conformity assessment, and equivalency. Further, they have problems in effectively participating in the international standard-setting process, and therefore, face difficulties in meeting requirements in foreign markets based on international standards.[60]

One way to minimize trade tensions concerning health regulations may be to encourage the greater harmonization of standards in accordance with the SPS Agreement. Yet this solution raises concerns for developing countries in terms of their ability to meet the harmonized standards and lack of involvement in the standard-setting process. The concept of special and differential treatment has been incorporated in the SPS Agreement, yet many issues remain about how to promote the integration of developing countries into the international trading system while recognizing their developmental situations and objectives. The group of developing and least developed countries known as the G-90 has called upon WTO Members to exercise restraint in applying SPS measures to their products, and to provide increased technical assistance.[61]

---

[58] Laurian Unnevehr and Donna Roberts, 'Food Safety and Quality: Regulations, Trade, and the WTO' (Paper presented to the International Conference on Agricultural Policy Reform and the WTO: Where Are We Heading? Capri, Italy, 2003). They note that this frustration has been expressed in a number of developing country proposals submitted leading up to and following the Doha Ministerial Conference in November 2001.

[59] Brenda Zulu, 'As Drought takes Hold, Zambia's Door Stays Shut to GM' 22 April 2005, SciDev.Net (on file with author).

[60] See Simonetta Zarrilli, *WTO Sanitary and Phytosanitary Agreement: Issues for Developing Countries, part of the South Centre Trade-Related Development and Equity (TRADE) Working Papers Series, No. 3* (1999) at 1.

[61] Report from the G-90 Trade Ministers Meeting, Grand Baie, Mauritius, 10 July 2004, 'Elements of a G-90 Platform on the Doha Work Programme', online at Government of Mauritius: http://www.gov.mu/portal/sites/ncb/acp/english/doc4.htm (date accessed: 25 October 2007).

4.7 CONCLUSION

Each of the forces of change discussed above points to a very strong likelihood of continued friction over trade-restrictive SPS measures. The remainder of this book will focus on the difficulties faced by WTO panels and the Appellate Body in resolving this friction in cases where one country disputes the basis for another's decision to adopt an SPS measure.

PART II

Health and trade: conflicting objectives?

5. Foundations of tension between health protection and trade liberalization

5.1 INTRODUCTION

Tension between health protection and trade liberalization objectives arises in large part due to the different values and interests underlying, on one hand, domestic regulatory agendas, and on the other, trade liberalization. In Section 5.2, this chapter highlights key strands of the economic case for trade and notes objections to liberalization. It then examines health regulation, looking first at normative theories of regulation (including, specifically, health protection regulation), followed by positive theories explaining regulatory outcomes. This analysis forms the foundation for identification of tension between health and trade in Chapter 6, and in Chapter 7 it will be called upon to establish a normative framework to guide WTO adjudicating bodies in balancing countries' sovereign rights to protect health with the interests of the multilateral trading community in pursing trade liberalization.

5.2 THE CASE FOR TRADE

WTO Agreements reflect a widespread acceptance among the international community of nations of the benefits to be gained from liberalizing trade. The Preamble to the Agreement Establishing the WTO[1] reflects the Parties' belief that the Agreement's objectives will be furthered by reducing tariffs and other barriers to trade and eliminating discriminatory treatment in international trade relations'. Stated objectives include raising standards of living, ensuring full employment and a growing volume of real income and effective demand, as well as sustainable development. The GATT refers to the contribution of liberalized trade to 'raising standards of living, ensuring full employment and a large and steadily growing volume of real income and effective demand,

[1] *Marrakesh Agreement Establishing the World Trade Organization*, 15 April 1994, in *The Legal Texts: The Results of the Uruguay Round of Multilateral Trade Negotiations* (Cambridge, UK: Cambridge University Press, 1999).

developing the full use of the resources of the world and expanding the production and exchange of goods'.[2] The SPS Agreement implicitly assumes the benefits of trade through its goal of minimizing the negative effects on trade of SPS measures. The discussion here begins with an examination of the economic case for international trade. Put more simply, why do nations trade?

5.2.1 The Economic Case for Trade

Recognition of the benefits of trade liberalization flows from acceptance of the underlying economic case for international trade. This is a case founded on the theory of *comparative advantage* as developed by David Ricardo in *The Principles of Political Economy*, published in 1817. Ricardo showed that all countries can benefit from international trade even if they have no absolute advantage in anything.[3] Every country derives a benefit from specializing in producing the goods in which it has a comparative advantage – which arises where the opportunity cost of producing a good (in terms of other goods) is lower in the home country than it is in other countries.[4] Ricardo's theory has been refined and added to over the years[5] but the underlying principle remains sound. Economists widely accept not only that countries are better off specializing, but that international trade liberalization produces benefits even where implemented unilaterally.

There are a number of qualifications to the economic case for trade, some with a rich history in the scholarly literature. They include optimal tariff; infant industries; national security considerations; strategic trade theory; and externalities. [6] These qualifications will not be discussed here as the essential aspects of the case for trade are well accepted by economists and form the foundation for trade liberalization efforts under the umbrella of the WTO Agreements. However, note will be taken of critiques of the gains from trade as well as vari-

---

[2] *General Agreement on Tariffs and Trade*, 30 October 1947, 58 U.N.T.S. 187, Can. T.S. 1947 No. 27 (entered into force 1 January 1948) [GATT].

[3] Adam Smith had earlier argued that every country would have an advantage in something, its 'absolute advantage': Adam Smith, *The Wealth of Nations* (New York: The Modern Library Classics, 1776).

[4] Gerber uses the term 'Ricardian model' in describing the basic model of production and trade based on the theory of comparative advantage. James Gerber, *International Economics*, 2nd ed. (Boston, MA: Addison-Wesley, 2002) at 41.

[5] Including by Heckscher and Ohlin – the 'Heckscher-Ohlin' model or 'Factor Proportions Hypothesis'; Vernon – the 'technology-gap' idea or 'Product Cycle Theory'; and Krugman – models of increasing returns as a cause of international trade. See Gerber, *ibid.*

[6] For a concise discussion of these and other qualifications to trade, see Michael J. Trebilcock and Robert Howse, *The Regulation of International Trade*, 3rd ed. (London: Routledge, 2005).

ous objections to trade as it is these which arguably have potential for greater influence on the way in which trade agreements are, or should be, interpreted, particularly where matters of public interest such as health are impacted.

5.2.2 The Gains from Trade

The economic case for trade liberalization holds that government intervention into international trade flows creates undesirable forms of economic 'inefficiency' which should be avoided. Modern welfare economics uses 'cost-benefit' analysis to compare the favourable and unfavourable effects of policy changes and determine whether the net impact of a change is efficient or inefficient. Policy changes that produce a surplus of benefits over costs are 'efficient' while policies with the opposite effect are 'inefficient'.[7] Efficiency is typically measured using either the Pareto or Kaldor-Hicks criteria. A 'Pareto efficient' change in policy produces net benefits for at least one individual affected by it, and does not leave anyone worse off than before the policy change. However, such policies are rare in reality – there are very few policies that do not result in at least one loser – and accordingly economists tend to use the alternative 'Kaldor-Hicks' efficiency criterion. This asks whether the benefits to those who gain from a change in policy exceed the costs to those who lose from it; if so, the policy change is said to be Kaldor-Hicks efficient. In general, benefits are measured in monetary terms; the inquiry as to whether a policy is efficient turns on whether the monetary value of the benefits to those who gain from the change in policy exceed the monetary costs to those who are hurt by it. In theory, the gainers could compensate the losers; however, this is not required to satisfy the test. The Kaldor-Hicks efficiency criterion thus forms the theoretical basis of standard cost-benefit analysis.[8]

There are several objections to the Kaldor-Hicks efficiency test.[9] First, it allows for the coercive imposition of losses on individuals. Second, the test assumes that one unit of currency has the same value regardless of who gains or loses it. This leads to the third criticism, that the test takes no account of distributional justice. Sykes illustrates these last two criticisms with the following example: a policy change that affects only two people – it will provide Donald Trump with an extra $10 and it will take $9.50 away from a homeless person. Sykes notes that this policy would be Kaldor-Hicks efficient, but would also be unattractive to most people. One reason for this is that the

---

[7] Alan O. Sykes, 'Comparative Advantage and the Normative Economics of International Trade Policy' (1998) 1 J. Int'l Econ. L. 49 at 57.

[8] Anthony I. Ogus, *Regulation: Legal Form and Economic Theory* (New York: Oxford University Press, 1994) at 25.

[9] *Ibid.*

'marginal utility' of money will be much greater for the homeless person than it is for Donald Trump. As a result, even if the policy creates a net gain measured in money, on balance it seems to contribute to human misery. Despite these tensions, economists find that 'efficiency' is still of interest as part of the normative case for trade. Sykes suggests four reasons for this:[10] (i) not all policy changes have the perverse quality of the Trump example; (ii) efficient policy decisions should result in a net increase in societal wealth, making it less likely that income distribution will be affected in any important, adverse way; (iii) unequal and/or unfair distribution may be fixed by a further policy change without any sacrifice of the efficiency gains from the original policy change; (iv) it is arguable that societal wealth is a worthy goal in itself that ought to be factored into decision-making.

Questions of efficiency can be particularly controversial in the context of international trade as governments have to consider whose welfare should count in the policy-making process.[11] Should the welfare consequences of national trade policies be evaluated from the national perspective, with little or no weight given to the welfare of persons outside the country, on the premise that governments ought to promote the welfare of their own citizens? Or should national policy pay regard to the interests of those from other countries?[12] Sykes argues that trade liberalization is likely to be Kaldor-Hicks efficient, but not in general Pareto efficient. For example, certain firms and workers may be better at producing certain goods than others, and cannot shift to the production of other things without suffering a reduction in their profits or wages. They might have invested in capital equipment that can be used to produce some things but not others, or workers might have learned skills that are valuable in one industry but do not transfer well to others. It can be said that firms and workers have investments in 'specific capital' (physical or human) – that is, investments that are specific to the industry in which they are working. While trade may benefit the individuals concerned by lowering the prices of other things that are imported, the net impact on them may be adverse. Thus trade has both gainers and losers, which means that the efficiency argument for trade must rest on Kaldor-Hicks efficiency.[13]

In terms of the welfare economics of trade policy, it is argued first that opening of trade is efficient in comparison to autarky;[14] and second, that

[10] Sykes, *supra* note 7 at 58.

[11] *Ibid.* at 59.

[12] *Ibid.* at 60.

[13] *Ibid.* at 62. Trade may also have an adverse effect on certain consumers by raising prices of exported goods.

[14] Autarky means an economy with no external trade. John Black, *Oxford Dictionary of Economics* (New York: Oxford University Press, 2002) at 17.

restrictions on trade are inefficient relative to allowing the market to achieve its own equilibrium.[15] Economists refer to the 'gains from trade', being improvements in national welfare as a result of trade. Welfare is improved because countries can both consume more, and earn more income, by trading more freely than they can in a state of autarky or by maintaining strong trade restrictions. This notion underpins the argument that further trade liberalization will promote growth and development.

Gains from trade are measured in various ways, including what is known as the *'consumption possibility frontier'* and the notion of *'consumer surplus'*. The consumption possibility frontier suggests that by trading a country can consume more of all commodities than it could in the absence of trade. It does so by specializing in production of products in which it has a comparative advantage and exporting the surpluses of its specialties in exchange for imports.

Dunkley provides a twofold critique of the *consumption possibility frontier* argument. First, he notes that the gains from trade take the form of increased import consumption and only *indirectly* the form of income. This assumes that higher import consumption is truly an increase in welfare and that people always prefer more to less.[16] Dunkley argues that this assumption is open to question. Second, he argues that the case for free trade underestimates costs that arise when a country restructures its economy in order to benefit from specialization. He finds that this process can have many costs, including relocation of employment, long-term unemployment, family disruption, devastation of certain towns or regions, loss of some industry-specific skills, and changes in the nature of society. That is, a range of personal and social costs need to be in some way subtracted from the quantitative gains from trade.[17]

Consumer surplus refers to the difference between what a person is willing to pay for a product, rather than go without it, and the market price which has to be paid. It can be considered as 'bonus utility', satisfaction people feel when prices are less than they are willing to pay. The concept has been a controversial one in the history of economic thought because of the difficulties in comparing or aggregating the respective utility or satisfaction of individuals.[18] As Pfouts wrote in 1953: 'Probably no single concept in the annals of economic theory has aroused so many emphatic expressions of opinion as has the consumer's surplus; indeed even today the biting winds of scholarly

---

[15] Sykes, *supra* note 7 at 60.

[16] Graham Dunkley, *Free Trade: Myth, Reality and Alternatives* (London: Zed Books, 2004) at 27.

[17] *Ibid.*

[18] *Ibid.* at 32.

sarcasm howl around this venerable storm centre'.[19] Dunkley argues that a key problem with using the notion of consumer surplus to illustrate the gains from trade is that there is no direct cash income to consumers, but rather, an indirect redistribution which people receive as 'consumer surplus' or a 'psychic bonus' as a result of prices being lower than they expect.[20] Other economists have argued that consumer surplus is a meaningful income equivalent due to its 'income effect' – that is, lower prices raise disposable income – but is still a limited notion due to conceptual uncertainties, measurement difficulties, the impossibility of knowing or meaningfully aggregating individual consumer preferences, the possibility of heterogeneous preferences, and the likelihood that people's desire for income will differ considerably and unpredictably.[21]

Protectionism is the antithesis of trade liberalization as it encourages high cost domestic firms to remain in the market while excluding low cost foreign firms, and prices consumers out of the market who would otherwise be willing to purchase goods at a price exceeding the marginal cost of production of efficient suppliers. Sykes argues in addition that regulatory protectionism (as opposed to genuine social regulation) tends to cause additional deadweight losses that make it more inefficient than other instruments of protection such as tariffs, quotas, and subsidies. This is because it raises the costs of foreign suppliers by inducing an expenditure of resources for no purpose (or with no effect) other than to raise the price of imported goods.[22]

5.2.3 Objections to Free Trade

While the case for trade liberalization is widely accepted in economic theory (albeit with some controversy regarding measurement of the gains from trade), the subject is more controversial in public discourse where a number of objections to trade liberalization have emerged. Of particular interest here is the concern that trade trumps environmental, health, and safety concerns.[23]

---

19 R.W. Pfouts, 'A Critique of Some Recent Contributions to the Theory of Consumer's Surplus' (1953) 19 Southern Economic Journal. Cited in John Martin Currie, John A. Murphy and Andrew Schmitz, 'The Concept of Economic Surplus and its Use in Economic Analysis' (1971) 81:324 Economic Journal 741 at 741.

20 Dunkley, *supra* note 16 at 30.

21 *Ibid.* at 33. Dunkley also cites P. Samuelson, *Foundations of Economic Analysis* (New York: Atheneum, 1974).

22 Alan O. Sykes, 'Regulatory Protectionism and the Law of International Trade' (1999) 66:1 U. Chicago L. Rev. 1 at 5–13.

23 Trebilcock and Howse, *supra* note 6 at 13. Trebilcock and Howse identify and discuss a number of further objections, including that (i) globalization is leading to a global monoculture; (ii) trade exacerbates inequalities of wealth within and among nations; (iii) trade adversely impacts labour standards and human rights; and (iv) self-sufficiency is preferable to dependency.

While this author rejects the charge that trade trumps health and safety concerns, it must be acknowledged that panels and the Appellate Body are sometimes required to engage in a balancing exercise with respect to conflicting health protection and trade liberalization objectives. The SPS Agreement does not define when measures are *necessary* to protect health and it is in cases where it is not clear that a measure is *necessary* that a balancing exercise is required. However, this does not equate to a trumping of health interests. Thus, while the objection raised is exaggerated, there are legitimate questions remaining as to how health objectives can best be balanced with those of trade liberalization.

From a theoretical perspective, some less mainstream strands of economic thought provide support for this particular objection to trade liberalization, or at least, to the argument that health ought to be prioritized in any balancing exercise between the objectives of health regulation and trade liberalization. The concept of 'gains from trade' is essentially based on monetary values and the human desire to consume more; however, some theorists argue that this is not the only value that counts. Dunkley speaks of a 'Gandhian Propensity' whereby 'people seek reasonable social justice, protection of cultural-spiritual traditions, or at the least the integrity rather than exact continuity of these, and maintenance of the community's natural environment, all being pursued partially at the expense of production and income maximization if necessary'.[24] This propensity is discussed by E.F. Schumacher who wrote *Small is Beautiful* in 1973. Schumacher's economic framework was labelled 'Buddhist economics' and exhibited a strong Gandhian influence. He was strongly critical of the concept of reducing everything to a value within a market-based framework, writing for example that 'to the extent that economic thinking is based on the market, it takes the sacredness out of life, because there can be nothing sacred in something that has a price. Not surprisingly, therefore, if economic thinking pervades the whole of society, even simple non-economic values like beauty, *health*, or cleanliness can survive only if they prove to be economic (emphasis added)'.[25] Schumacher was also critical of the notion of cost-benefit analysis which he considered to be simply a procedure by which the higher is reduced to the level of the lower and the priceless is given a price. As such, he wrote that:

> [I]t can never serve to clarify the situation and lead to an enlightened decision. All it can do is lead to self-deception or to the deception of others; for to undertake the immeasurable is absurd and constitutes but an elaborate method of moving from

---

24 Dunkley, *supra* note 16 at 65.

25 E.F. Schumacher, *Small Is Beautiful: Economics as if People Mattered – 25 Years Later . . . With Commentaries* (Vancouver: Hartley & Marks, 1999) at 29.

> preconceived notions to foregone conclusions; all one has to do to obtain the desired results is to impute suitable values to the immeasurable costs and benefits. The logical absurdity, however, is not the greatest fault of the undertaking: what is worse, and destructive of civilization, is the pretence that everything has a price or, in other words, that money is the highest of all values.[26]

Schumacher argued that it would be 'impossible to develop any economic theory at all, unless one were prepared to disregard a vast array of qualitative distinctions. But it should be just as obvious that the total suppression of qualitative distinctions, while it makes theorizing easy, at the same time makes it totally sterile.'[27]

These perspectives arguably provide more space for consideration of non-economic health concerns than orthodox economic theory. It is not suggested here that these alternative theories undermine the validity of the case for trade or even the role of cost-benefit analysis in welfare economics. To the contrary, economic analysis has shown itself capable of taking account of health concerns. However, it is suggested that they provide a complementary perspective that is capable of adding richness to a consideration of tensions between trade and health objectives. Within the WTO framework, consideration of these perspectives could be used to inform the balancing of interests required where domestic health regulations conflict with trade liberalization objectives. That is, even where the benefits of a health regulation are not easily quantifiable, this ought not necessarily to reduce its importance as something of value to be protected.

5.3 HEALTH REGULATION

There are many definitions of the term 'regulation', but the following is helpful for its broad coverage of potential types of regulation, including that related to health.

> . . . we can think of regulation as any process or set of processes by which norms are established, the behaviour of those subject to the norms monitored or fed back into the regime, and for which there are mechanisms for holding the behaviour of regulated actors within the acceptable limits of the regime (whether by enforcement action or by some other mechanism).[28]

---

[26] *Ibid.* at 31.

[27] *Ibid.* at 32.

[28] C. Scott, 'Analyzing Regulatory Space: Fragmented Resources and Institutional Design' (2001) Public Law 329 at 331. Cited in Bettina Lange, 'Regulatory Spaces and Interactions: An Introduction' (2003) 12:4 Social and Legal Studies 411 at 411.

Balancing trade and health objectives under the SPS Agreement requires consideration of the welfare consequences if the regulating country was prevented from enacting the measure in question. Would the welfare gains from liberalized trade outweigh the benefits of the disallowed measure? On the other hand, if countries could enact whatever health regulations they wished, regardless of the trade impact, what would be the welfare consequences? These questions raise the issue of how welfare is defined and measured. In this regard, the preceding discussion noted orthodox trade theory that trade-restricting regulations reduce welfare as defined in a relatively narrow economic sense. However, alternative approaches suggest a broader, less determinate notion of welfare. To a point, the SPS Agreement also recognizes non-monetary values; however, the scope of this recognition is not clear. Those charged with interpretation of the SPS Agreement should consider the basis upon which domestic regulation is made if they are to find a balance that advances welfare by best serving both trade liberalization goals and countries' domestic regulatory interests.

With these questions in mind, this section examines the purpose and benefits of domestic regulation and the forces that drive regulatory decision-making. First, however, it should be noted that the theoretical perspectives discussed in this section are those of scholars from Western, industrialized nations, for the most part the United Kingdom and the US. Yet this book deals with the right of over 150 nations to regulate in the light of their international trade obligations.[29] It might be questioned whether it is even possible to make any generalizations about the topic of domestic regulation when so many nations are involved, particularly given the comments about regulatory diversity in Chapter 4. The approach taken here is that there are enough commonalities to make such a discussion worthwhile. First, WTO Members have agreed to be bound by rules that seek to further liberalize trade; thus, there is a common bond, acceptance of the merits of trade liberalization. Second, Members have recognized the importance of protecting health. There must be at least some common elements in such an effort, including a recognition of the importance of health. Third, while it would be impossible, within the scope of this book, to examine the applicability of the regulatory theories identified to every WTO Member, it is suggested that the theories discussed have a potentially broad application across countries. Finally, Kagan argues that in an increasingly integrated global economy, national regulatory systems – at least in economically advanced democracies – have similar goals and standards.[30]

---

[29] There are 152 WTO Members.

[30] Robert A. Kagan, 'Introduction: Comparing National Styles of Regulation in Japan and the United States' (2000) 22:3&4 L. & Pol'y 225 at 226.

rding health, this is partly due to the fact that national environmental and umer groups, and other activist organizations, are linked by international networks and make similar demands on governments. Meanwhile, multinational corporations tend to lobby for harmonization of national laws on matters such as food and product safety.

The outcomes postulated by the various theories discussed will thus hold to a greater or lesser extent in different geographical, political, economic, and cultural settings. In each country, regulatory decision-making and implementation will be subject to the particular domestic context. There will be both differences and similarities: decision-makers in any system must have some reason for enacting regulations; and in any country, there will be instances of regulatory failure. It is these similarities that make the theories useful in providing a deeper understanding of domestic regulation. However, differences in regulatory preferences, processes, and capacities must also be acknowledged.

5.3.1 Normative Theories of Regulation

Normative economic analysis is also referred to as 'welfare economics' and asks whether a particular policy will make individuals affected by it better off in terms of how they judge their own welfare. Welfare economics judges the efficiency of a policy, with the normative goal being 'allocative efficiency', determined according to whether the policy is either Pareto or Kaldor-Hicks efficient.[31] Decisions made by regulators are 'collective' (as opposed to decisions which are the result of voluntary agreement among affected parties) which raises the question of whether the net effect of the decision is to increase social welfare, as judged by all affected individuals in terms of its impact on their level of present or prospective utility.[32]

Regulation can be broadly categorized as one of two types, *social* or *economic regulation*.[33] *Economic regulation* covers a relatively narrow range of activities and will not be discussed here. Essentially, it applies to industries with monopolistic tendencies and seeks to provide a substitute for competition in relation to natural monopolies.[34] *Social regulation*, on the other hand, deals

---

[31] Michael J. Trebilcock, 'An Introduction to Law and Economics' (1997) 23:1 Monash U.L. Rev. 123 at 132.

[32] *Ibid.* at 133.

[33] Throughout this book, the term 'regulation' is used to denote legislation (that is, statutes or Acts of Parliament) and regulation (that is, rules promulgated pursuant to legislation). WTO agreements refer to 'measures' and this term includes both legislation and regulations.

[34] Ogus, *supra* note 8 at 5.

with a wide range of matters including human, animal, and plant health. Grounds for social regulation are usually categorized into economic and non-economic goals. The economic goals of correcting market failure and information deficits are discussed, followed by the non-economic goals of distributive justice, paternalism, community values, and values specific to the protection of human, animal, and plant health. In Western democracies, regulation must be justified in light of its tendency to interfere with the freedom and/or property rights of individuals and businesses.

5.3.1.1 Economic goals

Market failure Market failure occurs when markets do not lead to an efficient outcome; that is, when an action leads to a result which is neither Pareto nor Kaldor-Hicks efficient. In some cases, market failure will justify government intervention to correct the failure. Forms of market failure commonly implicated in regulatory intervention to protect health are negative externalities, public goods, and information deficits.

Negative externalities Externalities occur when one individual's actions affect another's well-being and the relevant costs or benefits are not borne by the first individual nor reflected (or 'internalized') in market prices.[35] This gives rise to a disconnect between the social cost or benefit[36] of an action and its private cost or benefit.[37] Externalities may be either positive or negative: positive if the behaviour in question makes someone else better off, and negative if the behaviour makes someone else worse off. Health regulations are concerned with negative externalities.

The classic example of a negative externality is pollution caused by industry. For example, where a factory discharges pollutants into a lake, the owner does not bear the resulting costs, which may include negative impacts on the lake's ecosystem as well as on the public who swim in and boat on the lake and rely on it as a source of drinking water. This results in market failure because the factory owner does not have to pay for the true social cost of his actions (that is, the harm to the ecosystem and to users of the lake) and, as

---

[35] Tyler Cowen, Public Goods and Externalities (2005), online: The Library of Economics and Liberty: The Concise Encyclopaedia of Economics <http://www.econlib.org/library/Enc/PublicGoodsandExternalities.html> (Date of Access: 2005).

[36] The social cost or benefit is the sum of the private cost or benefit and the external cost or benefit.

[37] Julie Soloway, 'Institutional Capacity to Constrain Suboptimal Welfare Outcomes from Trade–Restricting Environmental, Health and Safety Regulation under NAFTA' (University of Toronto, 2000, unpublished) at 71.

such, has an incentive to produce too much of the product causing harm. Consumers contribute to the pollution by demanding too much of the factory's output and, like the factory owner, do not bear the costs in the price of the goods.

Problems caused by negative externalities may be solved through 'internalization' whereby those producing externalities are required to take account of them in their decision-making. Where there are negative externalities, such motivation may come from either private law (for example, an action in nuisance) or regulation (for example, imposing environmental standards or taxing discharges).[38] Yet internalization is not always a simple task given that the underlying problem is that there are competing and conflicting claims by two or more parties (the factory owner and users of the lake) for use of a single resource – in this example, the lake. A statement made by Coase in 1960 in his seminal article on social costs still has significant force in this regard: 'The problem which we face in dealing with actions which have harmful effects is not simply one of restraining those responsible for them. What has to be decided is whether the gain from preventing the harm is greater than the loss which would be suffered elsewhere as a result of stopping the action which produces the harm.'[39] Applied to the example given above, he explained further: 'if we assume that the harmful effect of the pollution is that it kills the fish, the question to be decided is: is the value of the fish lost greater or less than the value of the produce which the contamination of the stream makes possible?'[40]

Taking Schumacher's approach, it could be argued that the fish should not be assigned a monetary value. Similarly, it might be argued that health cannot be quantified in monetary terms. Such an approach is reflected in an excerpt from an American public health law manual published in 1929 which argues for regulation to address negative externalities in the context of public health on the following grounds:[41]

> Sanitarians work toward the ideal that all people will in time know what healthful living is, and that they will in time reach that moral plane when they will practice what they know. However, law is still necessary. People have an inclination toward acts which are not for their neighbours' good. In our complicated civilization, many restrictions must be placed on individual conduct so that we may live happily and healthfully one with another.

---

38 Ogus, *supra* note 8 at 35.

39 R.H. Coase, 'The Problem of Social Cost' (1960) III J.L. & Econ. 1.

40 *Ibid.* at 2.

41 C.V. Chapin, Foreword in J.A. Tobey, *Public Health Law: A Manual of Law for Sanitarians* (Baltimore, Md: Williams & Wilkins Co, 1926). Cited in Lawrence O. Gostin, 'Public Health Law in a New Century: Law as a Tool to Advance the Community's Health' (2000) 283:21 Journal of the American Medical Association 2837 at 2841.

Countless examples could be given of health regulations justified by the presence of negative externalities. One is that of mercury emissions from power plants into lakes and rivers resulting in a negative externality in the form of increased mercury levels in fish supplies, posing a health risk to children and pregnant women, and justifying regulation to control emission levels. In the context of animal and plant health, failure by one individual or entity to prevent or contain pests or diseases can result in costs that they do not have to bear such as illness caused by contamination of the food chain, or infection of another's herds or crops. Where market mechanisms alone fail to prevent or correct such negative externalities, then governments might regulate to prevent entry or reduce the risks of the threat.

Public goods Most goods are private in nature. That is, their consumption can be withheld until a payment is made in exchange, and once consumed, they cannot be consumed again. For example, cheese can be withheld from a consumer until they pay the cheesemonger their asking price, and once the consumer has eaten the cheese, it cannot be eaten again. These characteristics are known as *excludability* and *rivalrous consumption*.[42]

A public good, on the other hand, is one whose benefit is shared by either the public as a whole or a sub-group thereof. It has two characteristics which are the opposite of those which define private goods: first, it is impossible or too expensive for the supplier to exclude those who do not pay from the benefit (*non-excludability*); and second, consumption by one person does not leave less for others to consume (*non-rivalrous competition*).[43]

Consumable goods occupy a continuum with 'pure' public goods at one extreme and 'private' goods at the other. Pure public goods are both non-rivalrous and non-excludable. Lying on the continuum between the extremes of public goods and private goods are goods with varying degrees of excludability and/or extent of non-rivalrous consumption. Goods can, for example, be *non-excludable but rivalrous in consumption* (for example, forests: the environmental benefits of forests are not excludable, but if they are used for logging these benefits are foregone), or *excludable but non-rivalrous* (for example, cable television which is excludable through subscription requirements, but can be consumed by many consumers).[44] Many of these kinds of goods are treated as if they were pure public goods. In some cases, public

---

[42] Richard D. Smith, *et al.*, eds, *Global Public Goods for Health: Health Economic and Public Health Perspectives* (New York: Oxford University Press, 2003) at 4.

[43] Ogus, *supra* note 8 at 33.

[44] Smith, *supra* note 42 at 5.

goods problems can be solved through the use of market mechanisms,[45] while in other cases, governments step in to regulate and/or provide goods or services where required.

Public health is arguably a public good. Gostin emphasizes its public good characteristics:[46]

> [P]ublic health can be achieved only through collective action, not individual endeavour. Acting alone, individuals cannot ensure even minimal levels of health . . . no single individual, or group of individuals, can ensure the health of the community. Meaningful protection and assurance of the population's health require communal effort. The community as a whole has a stake in environmental protection, sanitation, clean air and surface water, uncontaminated food and drinking water, safe roads and products, and control of infectious disease. Each of these collective goods, and many more, are essential conditions for health. Yet, these goods can be secured only through organized action on behalf of the public.
>
> Moreover, the population, or electorate, legitimizes systematic community activity for the public's health. Public health activities in a democracy cannot be organized, funded, or implemented without the assent of the people. It is the public that bands together to achieve social goods that could not be secured absent collective action. And it is the public that legitimizes government action for the common welfare. Elected officials are at least putatively committed to securing the public's health; and constituents are committed to bear the necessary burdens.

Furthering the notion of public health as a public good, Gostin states that 'public health takes on a special meaning and importance in political communities. Health is indispensable not only to individuals, but to the community as a whole. The benefits of health to each individual are indisputable . . . perhaps not as obvious, however, health is also essential for communities. Without minimum levels of health, populations cannot fully engage in the social interactions of a community, participate in the political process, generate wealth and ensure economic prosperity, and provide for common defense and security.'[47]

Plant and animal health also have public good characteristics. In the case of crops or livestock, one farmer acting alone is unlikely to be able to ensure

---

[45] Companies may *exclude non-payers* from enjoying the benefits of a good or service. For example, cable television companies scramble their signals so that broadcasts can only be received by subscribers. Other public goods problems can be solved by *defining individual property rights* in the appropriate economic resource. Cleaning up a polluted lake, for instance, involves a free-rider problem if no one owns the lake. The benefits of a clean lake are enjoyed by many people, and no one can be charged for these benefits. Once there is an owner, however, that person can charge higher prices to users who benefit from the lake. Cowen, *supra* note 35.

[46] Gostin, *supra* note 41 at 2883.

[47] Gostin, *ibid.* at 2838.

minimum levels of health. Further, the concept has particular salience in the case of endangered flora and fauna where enjoyment by the public may not be able to be restricted without some kind of regulatory incentives.

Information deficits and bounded rationality A market economy where resources flow to where they are most valued is based on consumer choice, with preferences being reflected in demand. This market system of allocation is based on two fundamental assumptions.[48] First, it is assumed that consumers have adequate information on the set of alternatives available to them, including the consequences to them of making different choices; and second, that they are capable of processing that information and of 'rationally' behaving in a manner that maximizes their expected 'utility'. Economic analysis finds that where there is a failure of either assumption, there may be a case for regulatory intervention on the grounds of market failure.

Economic analysis tends to assume the existence of 'perfect information' in markets, but this is rarely the case in reality. Rather, information asymmetries between manufacturers and sellers of goods on the one hand, and consumers on the other, are common and provide a strong rationale for regulation to protect consumers.[49] Yet as Ogus argues, from a public interest perspective, the absence of 'perfect' information cannot by itself justify regulatory intervention.[50] The relevant question for policy-makers is whether the unregulated market generates 'optimal' information in relation to a particular area of decision-making. In other words, are the marginal costs of supplying and processing the level and quality of information in question approximately equal to the marginal benefits that result? Consumers often face high costs in obtaining adequate information with which to make a purchasing decision, which means that they often must choose at what point they should remain rationally ignorant. [51] Suppliers, on the other hand, can generally provide information more cheaply through advertising because of economies of scale. In a competitive market, they have an incentive to provide information in order to distinguish their products from those of their competitors.[52] Ogus identifies several factors which may serve to lessen this incentive or which may result in countervailing inefficiencies. First, information has a public good aspect to it – it is difficult to restrict it to those who directly or indirectly pay for it and its use

---

[48] Ogus, *supra* note 8 at 37.

[49] Gillian Hadfield, Robert Howse and Michael J. Trebilcock, *Rethinking Consumer Protection Policy* (Prepared for and revised based on the University of Toronto Roundtable on New Approaches to Consumer Law, 1996) at 6.

[50] Ogus, *supra* note 8 at 38–9.

[51] Hadfield, Howse and Trebilcock, *supra* note 49 at 6.

[52] Ogus, *supra* note 8 at 40.

by one consumer does not lower the value it has to others. This suggests that in an unregulated market, information will be under-produced. Second, advertising can lead consumers to believe that one supplier's product is different from another's when in fact there are either no or insignificant differences. This results in the supplier gaining some degree of (undesirable) market power. Third, suppliers may have an incentive to supply false or misleading information in order to boost profits. Fourth, information may be incomplete where suppliers fail to provide any information about the product's negative qualities.[53]

The second assumption of economic analysis – that consumers make rational choices – is also questionable. The term 'bounded rationality' refers to the idea that individuals have limited capacity to receive, store, and process information.[54] Work that examines this proposition has been undertaken by Sunstein and others under the rubric of 'behavioural law and economics', which will be discussed later.[55] Food quality provides an example of bounded rationality. Food science is so complex that many people will not be able to judge for themselves whether a processed food product is a good choice or not.

The problems of information deficits and bounded rationality are of particular concern where human health may be adversely affected. Policy responses to the problems include: mandatory disclosure requirements on providers of goods (for example, requiring hazard labels on certain household chemicals); minimum standard-setting for products (for example, placing a ceiling on the level of pesticide residue allowed on vegetables); and product bans (for example, banning the production of hormone-treated beef, or the use of trans-fatty acids in processed foods). In each of these cases, the basic rationale is that due to costs which mean producers are unlikely to provide the information of their own accord and/or consumers' bounded rationality, it is justifiable to regulate in the interests of consumer protection. As we have seen, such regulation may in some cases be trade-restrictive and the cause of trade disputes.

5.3.1.2 Non-economic goals

Distributive justice The economic goal of allocative efficiency is concerned with maximizing welfare which is coterminous with wealth maximization. It is not concerned with the distribution of wealth between people or groups in society. Thus, regulatory intervention may be motivated by a desire to achieve a 'fair' or 'just' distribution of resources so that all individuals in the society have a fair share of at least some goods and services.[56]

---

53 *Ibid.*
54 *Ibid.* at 41.
55 See discussion, *infra* at section 5.3.2.3.
56 Ogus, *supra* note 8 at 46.

Ogus suggests that theories of distributive justice influence regulatory policy both directly and indirectly, with the indirect influence occurring where intervention is primarily justified on another ground such as market failure. For example, where the meat-packing industry has the potential to cause illness through poor hygiene or diseased livestock, the government might seek to deal with the externality by using public funds to pay for the installation of a food inspection system or the industry might be taxed more to pay for the inspection system. In the first scenario, shareholders and consumers of the firm's products would benefit, while the second would benefit members of the non meat-eating public at the expense of shareholders and consumers.[57] In such cases, regulatory theory suggests that policies should attempt to predict the distributional consequences of proposed regulation and ensure that measures are adopted which will lead to fair and just outcomes.[58] The direct influence is seen in regulatory measures designed specifically to achieve redistributional goals such as taxation systems that fund health-care systems that provide free services such as flu shots or vaccinations.[59] Ogus notes that redistributive policies may operate within a 'temporal' dimension, for example, where the use or consumption of some resources may affect what is available for future generations (for example, allowing exotic animal species to be destroyed).[60]

Paternalism The concept of paternalism is in tension with liberal theory, which plays a fundamental role in a market-based economy. In the public policy context, paternalism supports the introduction of regulatory measures that limit people's choice. This contrasts with a liberal approach which assumes that individuals will make rational choices to further their own utility, and leaves them free to make those choices. Paternalism has been defined as 'the interference with a person's liberty of action justified by reasons referring exclusively to the welfare, good, happiness, needs, interests or values of the person being coerced'.[61]

Numerous theorists have attempted to reconcile the relationship between liberal notions of personal autonomy and personal good.[62] Feinberg suggests

---

[57] *Ibid.* at 48.

[58] *Ibid.*

[59] Other common forms of direct redistribution include those designed to reduce inequality of income and wealth by providing various forms of welfare assistance, to equalize resources between regions, to transfer resources to those who have been the victim of a misfortune such as injury, illness, or unemployment. Ogus, *ibid.* at 49.

[60] *Ibid.*

[61] R. Dworkin, 'Paternalism' in R. Wasserstrom, ed., *Morality and the Law* (1971), cited in Ogus, *ibid.* at 51.

[62] See for example Michael J. Trebilcock, *The Limits of Freedom of Contract* (Cambridge, MA: Harvard University Press, 1993) at 149.

that in most cases, a person's own good will be most reliably furthered if they are allowed to make their own choices in self-regarding matters, but when self-interest and self-determination do not coincide, decision-makers must simply do their best to balance autonomy against personal well-being and decide between them intuitively, because neither has automatic priority over the other.[63]

Paternalistic public policies regarding health may usefully be viewed through the lens of Feinberg's 'need to balance' construct. Policies may be justified on a number of grounds, including where it is considered that: individuals have a limited capacity to process complex information about a risk; an individual's choice would likely result from ignorance of factual circumstances or the consequences of an action; and/or in the situation at hand, individuals have limited willpower.[64] As Gostin writes, 'they may know, objectively, what is in their best interests, but find it difficult to behave accordingly'.[65] This may be the case in relation to addictive habits such as smoking and drinking, but also with respect to eating unhealthy foods such as fast food.[66] Ogus argues that in these kinds of situations, the likelihood of people surrendering to temptation and its accompanying short-term pleasures, combined with the likely long-term health costs, means that individual utility will be maximized if they consent to being deprived of the temptation.[67] In this regard, he finds that paternalism distinguishes between an individual's 'real' and 'apparent' desires, finding that paternalism gives effect to the former.[68] Regarding this distinction, Trebilcock speaks of paternalism as having the function of enhancing welfare by facilitating the realization of preferences that reflect underlying utility functions.[69] In this regard, it is possible to argue that

---

63 Joel Feinberg, *Harm to Self* (1986) cited in Trebilcock, *ibid.*

64 The notion of paternalism is connected to the concept of 'bounded rationality' because justification relies on the fact that people face constraints on their capacity to make rational choices. See discussion of bounded rationality, *infra* at section 5.3.2.3.

65 Lawrence O. Gostin, 'Public Health Law in a New Century — Part III: Public Health Regulation: A Systematic Evaluation' (2000) 283:23 Journal of the American Medical Association 3118 at 3119.

66 Another ground for paternalistic regulations is the importance of social, economic, and environmental constraints on people's behaviour, such as parents and family, peers and community, media and commercial advertising. Gostin, *supra* note 65 at 3119.

67 Ogus cites G. Loomes and R. Sugden, 'Regret Theory: An Alternative Theory of Rational Choice under Uncertainty' (1982) 92 Econ. J. 805. Ogus, *supra* note 8 at 52.

68 Ogus, *supra* note 8 at 53. It is arguable, however, that this is where paternalism may prove to be a dangerous notion as it leaves open the possibility of justifying regulations on the basis of 'apparent' desires that do not in fact exist.

69 Trebilcock, *supra* note 62 at 147.

primacy should be accorded to the 'good' of the individual rather than to his or her preferences.[70]

Paternalistic policies tend to be problematic because they do not allow for distinctions between individuals' decision-making ability, meaning that some individuals who could have made decisions in their own best interest are not given the choice.[71] Further, as Sunstein argues, laws may reflect the majority's preferences, or second-order preferences at the expense of first-order preferences.[72] Nevertheless, empirical observation shows that paternalism is an important concept in the health arena, with many regulations designed to control individual behaviour. Ogus argues that while paternalism is not often invoked in policy discussions, it is nevertheless a powerful motivation for regulation, even where other justifications, such as externalities, are also relied upon.[73] An example is smoking, an action which has an adverse impact on the individual smoker's health – as well as on others through the impacts of second-hand smoke – and may therefore give rise to health-care costs which (in a publicly funded health-care system) are borne by taxpayers. Taxpayers therefore have an interest in the introduction of non-smoking laws.[74]

However, a paternalistic motivation is also apparent, given the indisputable adverse health effects of smoking but many people's apparent lack of willpower to quit. From a trade perspective, paternalism may co-exist with protectionist intentions.

Community values Community values are a set of public interest goals that reflect the 'desire to expand the social, intellectual, and physical environment; rather than the desires of individuals to consume or use certain goods'.[75] The emphasis is on providing the opportunity for members of the community to 'experience and test different conceptions of the good', and on fostering this through a mutuality of concern and respect, as well as by increased participation

---

70 *Ibid.* at 149.

71 Ogus, *supra* note 8 at 53.

72 Although, as Trebilcock notes, Sunstein does not make it clear how to go about determining first-order and second-order preferences. Trebilcock, *supra* note 62 at 153.

73 Ogus, *supra* note 8 at 52.

74 Some have argued, however, that smokers may actually save health-care costs by virtue of the fact that they tend to die at a younger age than non-smokers and therefore incur lower health-care costs in the long run. Jan J. Barendregt, Luc Bonneux and Paul J. Van der Maas, 'The Health Care Costs of Smoking' (1997) 337:15 New England Journal of Medicine 1050.

75 Ogus, *supra* note 8 at 54.

in the decision-making processes of collective affairs.[76] As Ogus notes, an unregulated market only has a limited capacity to achieve these goals – given its failure to take into account the demands likely to be made by future generations and the 'free-rider' problem – and accordingly regulatory intervention may be required. Examples of a community value in public health are regulations to ensure food quality which are in part motivated by a desire to have a healthy society. Similarly, regulations to protect endangered flora and fauna stem in part from a desire to have the kind of community which values and protects its natural resources.

Obligation to protect human health Regarding human health, a further non-economic imperative to regulate exists, namely, an obligation on the part of governments to protect health which derives from the concept of a 'social contract' between governments and those whom they govern.[77] Reynolds suggests that even on a minimal view, it is the duty of government to preserve life, liberty and property: 'The protection of the public's health is one element of the protection of life, and the long legislative involvement of governments in many countries in the operation of public health systems, suggests that protecting public health is a legitimate responsibility of government'.[78]

Writing about the US, Gostin cites a 1927 article which states that government is 'organized for the express purpose, among others, of conserving the public health and cannot divest itself of this important duty'.[79] More generally, Gostin argues that theories of democracy and political communities help to explain the primacy of government in matters of public health. He cites Walzer who said about political communities that 'membership is important because of what the members of a political community owe to one another . . . and the first thing they owe is the communal provision of security and welfare'.[80]

The importance of health is internationally recognized and the right to health is now recognized as a fundamental human right. The WHO made the first declaration of health as a human right in the Preamble to its *1946*

---

[76] R. Stewart, 'Regulation in a Liberal State: The Role of Non-Commodity Values' (1983) 92 Yale L.J. 1537 at 1566.

[77] Reynolds refers to John Locke (1632–1704) who expressed the idea that governments were not an expression of a divine right of rulers but the result of a tacit agreement of rulers and those who were ruled through a 'social contract'. Christopher Reynolds and Genevieve Howse, *Public Health: Law and Regulation* (Sydney: The Federation Press, 2004) at 22.

[78] Reynolds and Howse, *ibid.* at 23.

[79] J.A. Tobey, 'Public Health and the Police Power' (1927) 4 N.Y.U. Univ. L. Rev.; 126–33. Cited in Gostin, *supra* note 41 at 2838.

[80] M. Walzer, *Spheres of Justice: A Defense of Pluralism and Equality* (New York: Basic Books, 1983). Cited in Gostin, *ibid.*

Constitution which states that: 'the enjoyment of the highest attainable state of health is one of the fundamental rights of every human being' and that 'the health of all peoples is fundamental to the attainment of peace and security and is dependent upon the fullest co-operation of individuals and States'.[81] The WHO has since reaffirmed the right to health as a fundamental human right on numerous occasions. For example, the *Declaration of Alma Alta* in 1978 stated that health is 'a fundamental human right and that the attainment of the highest possible level of health is a most important world-wide social goal whose realization requires the action of many other social and economic sectors in addition to the health sector'.[82]

A number of international and regional human rights instruments also recognize the right to health. Most importantly, Article 12 of the *International Covenant on Economic, Social and Cultural Rights (1966) (ICESCR)*[83] states that '. . . the States Parties to the present Covenant recognize the right of everyone to the enjoyment of the highest attainable standard of physical and mental health'. The Covenant, which is binding under international law, seeks to ensure that states provide citizens with the conditions essential for achieving good health. It prescribes steps to be taken by governments in order to achieve the full realization of this right, including those necessary for the improvement of all aspects of environmental and industrial hygiene, as well as the prevention, treatment, and control of epidemic, endemic, occupational, and other diseases.[84]

In the Americas, the (non-binding) *American Declaration on the Rights and Duties of Man* (1948) provides that '[e]very person has the right to the preservation of his health through sanitary and social measures relating to food, clothing, housing and medical care, to the extent permitted by public and

---

[81] Constitution of the WHO (1946), available online at www.who.org (date accessed: 25 April 2005). Currently 192 nations are Members of the WHO.

[82] *Declaration of Alma Alta*, Alma Alta, 1978, available online at the Pan American Health Organization website, http://www.paho.org/English/DD/PIN/alma-ata_declaration.htm.

[83] Adopted and opened for signature, ratification, and accession by General Assembly Resolution 2200A (XXI) of 16 December 1966, available online: http://www.hri.ca/uninfo/treaties/2.shtml (date accessed: 25 April 2005).

[84] *International Covenant on Economic, Social and Cultural Rights*, UN General Assembly Resolution 2200A(XXI) A/6316 (1966). Key weaknesses in the ICESCR should be noted, notably, it is questionable whether states are required to incorporate the Covenant into domestic law, and there is no individual or inter-State complaint mechanism. Yutaka Arai-Takahashi, 'The Right to Health in International Law – A Critical Analysis' in Robyn Martin and Linda Johnson, eds, *Law and the Public Dimension of Health* (London: Cavendish Publishing Limited, 2001) 143 at 146.

community resources'.[85] The *Additional Protocol to the American Convention on Human Rights in the Area of Economic, Social and Cultural Rights (Protocol of San Salvador)*, adopted by the Organization of American States (OAS) in 1988 characterizes health as a public good and, among other things, requires States to take measures to prevent diseases.[86] In Europe, the *European Social Charter* stipulates that measures must be taken to ensure the right to protection of health, with objectives including the removal of the causes of ill health and the prevention of diseases.[87] Other regional instruments which speak of health as a human right include the *African Charter on Human and Peoples Rights* (1981).[88]

The right to health as articulated in these international and regional human rights instruments has two aspects, namely, a *right to health care* and *a right to healthy conditions.* It is the latter which is of interest here. In this regard, *General Comment 14 on the right to the highest attainable standard of health* issued by the Committee on Economic, Social and Cultural Rights (CESCR) in July 2000 is of interest.[89] First, General Comment 14 includes the right to the 'underlying determinants of health' which would require governments to ensure – among other things – an adequate supply of safe food, healthy occupational and environmental conditions.[90] General Comment 14 also elaborates on the meaning of the right to a healthy natural environment as provided in Article 12.2(b) of the ICESCR by providing a non-exclusive list of what the right comprises, including the prevention and reduction of the population's exposure to harmful substances such as radiation and harmful chemicals or other detrimental environmental conditions that directly or indirectly impact

---

[85] *American Declaration on the Rights and Duties of Man*, Adopted by the Ninth International Conference of American States, 1948, OAS.

[86] *Additional Protocol to the American Convention on Human Rights in the Area of Economic, Social and Cultural Rights, Protocol of San Salvador*, 17 November 1988 (Ratified 16 November 1999), OAS, Treaty Series, No. 69.

[87] See Article 10 (Right to Health, and also Article 11 *European Social Charter*, Turin, 18 October 1961 (entered into force 26 February 1965), European Treaty Series, No. 35. Since its implementation, the European Social Charter has been amended through the addition of a Protocol which introduced a collective complaint system. See *Additional Protocol to the European Social Charter Providing for a System of Collective Complaints*, Strasbourg, 9 November 1995, European Treaty Series, No. 158.

[88] Article 16, *African Charter on Human and Peoples Rights*, 1981 (entered into force 1986), OAU Doc. CAB/LEG/67/3 rev. 5, 21 I.L.M. 58 (1982), online at http://www1.umn.edu/humanrts/instree/z1afchar.htm.

[89] *Committee on Economic, Social and Cultural Rights (CESCR) General Comment 14, The right to the highest attainable standard of health*, E/C.12/2000/4.

[90] *Ibid.*

upon human health.[91] The right to a healthy natural environment arguably also includes the right to safe products where those products may otherwise have adverse consequences for health.[92]

General Comment 14 categorizes government obligations as *respecting*, *protecting*, and *fulfilling* the right to health.[93] The obligation to *protect* the right to health refers mainly to governmental obligations to make efforts to minimize risks to health and to take all necessary measures to safeguard the population from infringements of the right to health by third parties. It may require governments to adopt appropriate legislative, regulatory, administrative, budgetary, judicial, promotional, and other measures.[94] Examples include food safety laws that protect the public from unscrupulous or unsafe food manufacturers, or import restrictions that protect people from diseases or contaminants in imported food.[95] General Comment 14 states that violations of the obligation to protect health follow from the failure of a State to take all necessary measures to safeguard persons within their jurisdiction from infringements of the right to health by third parties. This includes failure to protect consumers from practices detrimental to health such as by manufacturers of medicines or food and to discourage the production, marketing, and abusive consumption of tobacco, narcotics, and other harmful substances.[96]

General Comment 14 recognizes that there are significant differences between countries in their state of development, financial resources, health status, and social conditions and that, as such, the process of fully realizing the right to health will require variable amounts of time and resources. It also

---

91 *Ibid.*

92 For further discussion of the content of the right to health, see Judith Asher, *The Right to Health: A Resource Manual for NGOs* (Washington DC: American Association for the Advancement of Science, 2005) at chapters 2 and 3.

93 *Comment on Economic, Social and Cultural Rights (CESCR) General Comment 14, The Right to the Highest Attainable Standard of Health*, E/C.12/2000/4 [*General Comment 14*].

94 Asher, *supra* note 92 at 15.

95 *Respecting* the right to health applies mainly to government laws and policies and requires that states refrain from undertaking actions that inhibit or interfere (directly or indirectly) with people's ability to enjoy the right to health, such as by introducing actions, programmes, policies, or laws that are likely to result in bodily harm, unnecessary morbidity, and preventable mortality. *Fulfilling* the right to health refers to positive measures that governments are required to take, such as by providing relevant services, to enable people to enjoy the right to health in practice. The obligation is often divided into the obligations to facilitate, provide, and promote the right to health. Definitions are based on General Comment 14 as noted in Asher, *supra* note 92 at para 3.2.

96 General Comment 14, *supra* note 93.

recognizes that the most appropriate measures to implement the right to health will vary significantly from one state to another.

In addition to the right to health, international human rights law emphasizes the *right to participate in decision-making*. That is, individuals and groups have the right to participate in decision-making processes that might affect their health and development. The International Covenant on Civil and Political Rights (ICCPR) states that: 'every citizen shall have the right and the opportunity, without . . . [discrimination] . . . and without unreasonable restriction: . . . to take part in the conduct of public affairs, directly or through freely chosen representatives'.[97] More specifically, General Comment 14 requires governments to ensure that people can participate in decision-making processes which may affect their health and development. Essentially, a human rights approach to health emphasizes that the 'effective and sustainable provision of health-related services can only be achieved if people participate in the design of policies, programmes and strategies that are meant for their protection and benefit'.[98]

In sum, a human rights approach to health re-frames health needs as health rights.[99] Becoming and remaining healthy is not merely a medical, technical, or economic problem, but is a question of social justice and concrete governmental obligations.[100] Thus the balancing act required under the SPS Agreement is not merely one of a nation's *right* to regulate but of a nation's *obligation* to regulate to protect health and its obligation not to restrict trade. In some cases, trade-restrictive health regulations will be necessary for a government to fulfil its obligations to protect its citizens' right to health.

Protection of animal and plant health Efforts to protect the life and health of livestock, fisheries, and commercial crops are typically primarily motivated by economic interests (although human health may also be at issue). There are also, however, non-economic goals and these lie primarily behind measures to protect wild flora and fauna.

The values behind protection of wild flora and fauna may be either *instrumental* or *intrinsic*. Instrumental values in turn may be either *economic* or *non-economic*. As with livestock and commercial crops, wild fauna and flora may demand protection on an economic basis, for example, commercially valuable fishing resources. Non-economic values may also be considered sufficiently important to require protection. A number of international conventions for the

---

[97] *International Covenant on Civil and Political Rights*, UN General Assembly Resolution 217A (III). A/8180 at 71 (1948).

[98] Asher, supra note 92 at chapter 2.

[99] *Ibid.*

[100] *Ibid.*

protection of wild fauna and flora cite both aesthetic and scientific value as primary grounds for protection.[101] Conventions also cite historic,[102] symbolic,[103] social,[104] utilitarian,[105] and ethical[106] value for protection of wild fauna and flora. The Western Hemisphere Convention is the main convention based on non-economic values alone; most conventions rely on both economic and non-economic values.

Intrinsic values are those values which are considered to be inherent in the object itself, rather than being inferred from human values.[107] Hayden wrote in 1942 that it is possible to 'assert openly the right of animals and plants to exist for their own sake'.[108] Although controversial, the notion has been gaining currency in international and regional conventions. The 1979 Bern Convention on the Conservation of European Wildlife and Natural Habitats was one of the first conventions to name intrinsic value as one of the grounds for the preservation of wild fauna and flora.[109] It states in the Preamble that 'wild flora and fauna constitute a natural heritage of aesthetic, scientific, cultural, recreational, economic and intrinsic value that needs to be preserved and handed on to future generations'.[110] In 1982, the General Assembly of the United Nations passed Resolution 37/7 (the 'World Charter for Nature').[111] The Charter is based on the premise that 'every form of life is unique, warranting respect regardless of its worth to man'. More recently, the Preamble of the Biodiversity Convention of 1992 refers to the 'intrinsic value of biological diversity and of the ecological, genetic, social, economic, scientific, educational, cultural, recreational and aesthetic values of biological diversity and its components'.[112]

---

[101] See for example the founding declaration of the International Council for Bird Protection (ICBP) which cites aesthetic value as well as scientific value as one of the primary grounds for the protection of birds before economic values. See also the Western Hemisphere Convention of 1940, the African Convention of 1968, the World Heritage Convention of 1972, CITES of 1973, and the Bonn Convention of 1979.

[102] See for example the Western Hemisphere Convention of 1940.

[103] See for example the World Conservation Strategy.

[104] See for example the Bonn Convention of 1979.

[105] See for example the World Conservation Strategy.

[106] *Ibid.*

[107] P. van Heijnsbergen, *International Legal Protection of Wild Fauna and Flora* (Amsterdam: IOS Press, 1997) at 53.

[108] S.S. Hayden, *The International Protection of Wild Life* (New York: Columbia University Press, 1942) at 31. Cited in van Heijnsbergen, *ibid.* at 57.

[109] Council of Europe, Bern, 19.IX.1979.

[110] *Ibid.*

[111] United Nations, General Assembly, A/RES/37/7, 28 October 1982.

[112] Convention on Biological Diversity, 22 EPL 4, 1992, p. 251, Preamble.

There are also two general international law principles that speak to a duty to protect wild fauna and flora.[113] First, the preservation principle may be deduced from various treaties and conventions. The Western Hemisphere Convention of 1940 refers to the desire of the signatory nations to 'protect and preserve . . . representatives of all species and genera of their native flora and fauna'.[114] In 1972, the Stockholm Declaration adopted by the UN Conference on the Human Environment stated that: 'The natural resources of the earth, including . . . flora and fauna, and especially representative samples of natural ecosystems, must be safeguarded'.[115] The Convention on International Trade in Endangered Species (CITES) declares in the Preamble that 'wild fauna and flora . . . are an irreplaceable part of the natural systems of the earth, which must be protected for this and the generations to come'.[116] Similar statements are found in other conventions, including the Bonn and Bern Conventions,[117] and the Biodiversity Convention of 1992.[118] Van Heijnsbergen argues that the preservation principle may be regarded not only as a principle of international law, but also as a part of international customary law.[119]

The second identifiable principle is the 'common heritage' principle, pursuant to which states are responsible for preserving fauna and flora found within their territory.[120] The notion of flora and fauna as 'common heritage' has been incorporated in a number of international conventions and declarations and is seen as a principle of international law. For example, the Stockholm Declaration of 1972 refers to a 'responsibility to safeguard the heritage of wildlife',[121] while the 1992 Biodiversity Convention refers to humanity's interest in ensuring the conservation of biological diversity.

As with human health, where an SPS measure is enacted in order to protect the health and life of wild flora and fauna, the issue under the SPS Agreement

[113] General principles of international law may be inferred from legal instruments such as treaties or resolutions of international organizations. They are different from customary international law which derives from general practice recognized as law. Van Heijnsbergen, *supra* note 107 at 63.

[114] 161 UNTS, p. 193.

[115] At Principle 2. A/CONF.48/14/Rev.1, p. 4.

[116] 993 UNTS, p. 243.

[117] The Bonn Convention on the Conservation of Migratory Species of Wild Animals states that 'States are and must be the protectors of the migratory species of wild animals' (19 ILM, p. 11); while the Bern Convention states that 'wild flora and fauna constitute a heritage . . . that needs to be preserved' (IEL 979:70).

[118] Providing that 'States are responsible for conserving their biological diversity' (EPL4, 1992, p. 251).

[119] Van Heijnsbergen, *supra* note 107 at 66.

[120] *Ibid.* at 70.

[121] A/CONF.48/14/Rev.1, p. 4.

becomes one of balancing a nation's *obligation* to protect health with its commitments not to restrict trade.

5.3.2 Positive Theories of Regulation

Positive economics seeks to explain the law as it is. It provides a tool by which to explain regulation through the way that politicians, bureaucrats, and regulators respond to incentives other than welfare impacts.[122] Among positive theories of regulation, a broad distinction can be made between 'public' interest and 'private' interest theory. Both theories rely on the fundamental assumptions of economics to explain who will receive the benefits or burdens of regulation, what form regulation will take, and the effects of regulation upon the allocation of resources.[123] The following sections provide an overview of the key ideas within both of these strands of positive economic analysis.

5.3.2.1 Public interest theory

The public interest theory provided a long-standing and well-accepted explanation for regulation from the late nineteenth century until it was seriously challenged in the 1970s. Despite its place in the history of economic theory, it is sometimes suggested that the theory has been more often assumed than articulated.[124] Central to the theory is the notion that regulatory decision-makers act in the pursuit of 'public interest'-related objectives as opposed to group, sector, or individual self-interest.[125] As such, where market failure occurs, regulatory intervention may occur in the name of the public interest.

The public interest theory has been criticized on a number of grounds. A fundamental challenge was launched in the early 1970s on the grounds that there is no agreed notion of what constitutes the public interest. In a comprehensive examination of the role of the 'public interest' in regulation, Feintuck argues that the notion of the public interest is extremely nebulous and insubstantial, with no identifiable normative content.[126] This results, he argues, in the notion being 'hopelessly vulnerable to annexation or colonization by those

---

[122] Soloway, *supra* note 37 at 61.

[123] George J. Stigler, 'The Theory of Economic Regulation' (1971) 1 Bell Journal of Economics 3 at 3.

[124] See Ogus, *supra* note 8 at 55; and Richard A. Posner, 'Theories of Economic Regulation' (1974) 5:2 The Bell Journal of Economics and Management Science 335 at 335.

[125] Robert Baldwin and Martin Cave, *Understanding Regulation: Theory, Strategy, and Practice* (New York: Oxford University Press, 1999) at 21.

[126] Mike Feintuck, *'The Public Interest' in Regulation* (New York: Oxford University Press, 2004) at 33.

who exercise power in society'.[127] He finds, however, that despite significant problems in articulating the concept's meaning, its use has been remarkably persistent, and even that it has an almost mythical or folkloric quality.[128]

The previous sections have set out a case for regulatory intervention to protect human, animal, and plant health. Based on these grounds, it seems reasonable to suggest that in the case of health protection, the 'public interest' is indeed normatively identifiable. That is, health protection exists within a legitimate public sphere of activity.[129] Yet, as Feintuck argues, in many cases the original justification or combination of justifications for regulatory intervention may only be a hazy memory by the time regulatory objectives and strategies are determined and implemented. Even where health protection is based on accepted norms of what it means to be healthy, uncertainties and differing opinions will exist regarding what is required to achieve the stated objectives. Accordingly, regulatory intervention risks becoming subjective and unpredictable.[130]

Other grounds of criticism of the public interest theory of regulation include doubts concerning the expertise, objectiveness, efficiency, and competency of regulators;[131] arguments that the theory understates the degree to which economic and political influence has an impact on regulation;[132] as well as a lack of empirical evidence of regulation being enacted in the public interest and observed cases of regulatory failure.[133] With regard to the latter, a number of theorists have argued that much regulatory policy is ill-advised and socially counterproductive.[134] They have identified various government programmes that achieved publicly beneficial outcomes, but at an unnecessarily inflated cost; aim for one outcome but achieve another; pursue trivial problems while ignoring more substantial hazards; and those that serve primarily to benefit discrete groups with no credible distributive-justice claim to publicly funded beneficence.[135]

---

[127] *Ibid.*

[128] *Ibid.* at 25.

[129] *Ibid.* at 18.

[130] *Ibid.* at 24.

[131] Baldwin and Cave, *supra* note 125 at 20. See also M.J. Trebilcock, *et al.*, *The Choice of Governing Instrument* (A study prepared for the Economic Council of Canada, Toronto, 1982).

[132] Baldwin and Cave, *ibid.*

[133] See generally Steven P. Croley, 'Theories of Regulation: Incorporating the Administrative Process' (1998) 98:1 Colum. L. Rev. 1 at 70.

[134] See discussion in Jeffrey J. Rachlinski and Cynthia R. Farina, 'Getting Beyond Cynicism: New Theories of the Regulatory State Cognitive Psychology and Optimal Government Design' (2002) 87 Cornell L. Rev. 549.

[135] *Ibid.* at 550. For another American commentator, see Cass R. Sunstein, *After*

Observations of regulatory failure have prompted a search for the causes of the failure. As Ogus asks, 'what explanations could be offered for the defects in the design of institutional structures and for the divergence between the ostensible public interest goals and achievement (or lack of it)?'[136] Various explanations have been offered. Ogus argues that sometimes regulations fail for 'technical' reasons; that is, 'inadequate expertise and forethought was brought to bear on the methods of achieving the public interest goals'.[137] Sunstein notes that sometimes statutes fail because they are based on a misdiagnosis of the problem, on poor policy analysis, or on inadequate information.[138] In other cases, the original regulatory design may have been appropriate yet rendered ineffective by inadequate enforcement.[139] These explanations can be accommodated within public interest theory. An alternative explanation is provided by private interest theories.

5.3.2.2 Private interest theory: public choice

In rejecting the public interest theory, private interest theorists find that regulation is not driven by pursuit of the public interest, but by the demands of private interests. In a seminal article in 1971, Stigler wrote that as a rule, regulation is acquired by industry and is designed and operated primarily for its benefit.[140] In other words, industry 'captures' the regulatory agency and uses regulation to prevent competition. This outcome is facilitated by the power of governments to coerce and thus to implement regulatory policies that will benefit industry.[141] Industry thus has an incentive to influence regulators so as to benefit from a 'regulatory rent'. A market for regulation is thereby formed.

Much attention has been paid to a particular branch of private interest theory known as public choice theory. In the context of representative democracy, public choice theorists examine the voting patterns of campaigning or elected officials. Using the concept of *homo economicus* (the utility-maximizing

---

the Rights Revolution: Reconceiving the Regulatory State (Cambridge, MA: Harvard University Press, 1990). Ogus cites various commentators who have made similar findings in the UK, including: C.K. Rowley and G.K. Yarrow, 'Property Rights, Regulation and Public Enterprise: The Case of the British Steel Industry 1957–1975' (1981) 1 Int'l Rev. L. & Econ. 63; A. Peacock, ed., *The Regulation Game: How British and West German Companies Bargain with Government* (1984); W. Carson, *The Other Price of Britain's Oil – Safety and Control in the North Sea* (1982). Ogus, *supra* note 8 at 56.

136 Ogus, *ibid.* at 56.

137 *Ibid.*

138 Sunstein, *supra* note 135 at 86.

139 Ogus, *supra* note 8 at 56.

140 Stigler, *supra* note 123 at 3.

141 Stigler finds that there are four policies that industry may seek of the state: subsidies; control over entry by new rivals; those which affect substitutes and complements; and price-fixing. *Ibid.* at 4.

rational person) theorists have produced models finding that legislators or politicians make decisions that 'maximize their own self interest based on such factors as votes, power, and political income'.[142] In doing so they act without regard to collective cost.[143] In other words, the theory finds that government officials are self-regarding and do not reliably serve the public interest. Public choice theorists thus make a radical departure from the public interest theorists who see government officials as acting in the public interest.

Two key elements used to explain self-interested behaviour of politicians are (i) the self-interested legislator looking to maximize the likelihood of election or re-election, and (ii) the rational ignorance of the voter.[144] The theory finds that legislators are motivated primarily by maximizing their chances of election or re-election. Given that few politicians can finance their election campaigns independently, they seek to attract resources from supporters.[145] To this end, they support programmes and vote for laws/regulations that are most responsive to active special groups who can provide those resources.[146] Large groups may be able to deliver votes while financial resources pay for advertising that in turn deliver votes. In return, these interest groups get to enjoy the benefits of their favoured regulatory policies.[147]

The concept of the rational ignorance of the voter refers to the notion that voters face high costs to inform themselves about issues. Further, because voters know there is very little chance that their vote will actually be decisive, they have little incentive to invest the time, money, and energy required to become well-informed. In other words, it makes sense (or is rational) for voters to remain ignorant on many issues. This lack of incentive to become well-informed – combined with the strong incentives for officials to look to short-term strategies or policies to boost their chances of election or re-election – means that regulatory outcomes are likely to be inefficient.[148]

Bureaucrats often play a critical role in regulatory decision-making, especially where legislation leaves room for discretion regarding implementation. With bureaucratic discretion arises the possibility of divergence between

---

[142] Nicholas Mercuro and Steven G. Medema, *Economics and the Law: From Posner to Post-Modernism* (Princeton, NJ: Princeton University Press, 1997) at 92.

[143] Daniel A. Farber and Philip P. Frickey, *Law and Public Choice: A Critical Introduction* (1991) 21–33 cited in Rachlinski and Farina, *supra* note 134 at 549.

[144] Mercurio and Medema, *supra* note 142 at 90. This account accords with the interest group theory put forward by Stigler.

[145] Steven P. Croley, 'Public Interested Regulation' (2000) 28:7 Florida State University Law Review 7 at 9.

[146] Mercurio and Medema, *supra* note 142 at 92.

[147] Croley, *supra* note 145 at 9.

[148] Explanation taken from Mercurio & Medema, *supra* note 142 at 92.

legislative intent and its final impact.[149] In order to examine how legislation is implemented, public choice theory therefore looks at the role of the incentives placed upon, and the resulting actions of, bureaucrats.[150] Through such analysis, public choice theory finds that the outcomes of bureaucratic choice are often wasteful and inefficient. As with legislators and regulators, the theory finds that utility-maximizing bureaucrats have interests that do not always coincide with the maximization of social welfare. Various models of the theory suggest that bureaucrats are influenced by a number of factors that conflict with social welfare, such as power, prestige, job security, future salary and employment opportunities, working conditions, and the size of their budget.[151]

Criticisms of private interest theory Private interest theorists have been criticized on several grounds. A key criticism – which has led to a search for alternative theories as discussed in the next section – is that private interest theories do not always seem to explain regulatory outcomes. That is, sometimes regulatory institutions appear to advance broad, diffuse interests, even at the expense of more powerful, concentrated interests. This raises some intriguing questions, such as that raised by Croley who asks 'under what set of conditions, politically and especially legally, can regulatory bodies, federal administrative agencies in particular, deliver broad-based benefits – "public interest" or, better, "public interest*ed*" rather than "special interest" regulation? How, and perhaps more to the point, why do they at times seem to deliver broad-based benefits even over the strong opposition of well-organized and well-funded interests?'

Croley finds that private interest theory, while compelling at one level of generality, has flaws. He questions each of its key claims which he summarizes as follows:[152]

1. Interest groups seek regulatory decisions that advance the selfish interests of their members – the 'interest group motivation claim'.
2. Small, narrowly focused interest groups, whose members individually have much at stake, are unusually well able to overcome the collective

---

[149] Mercurio and Medema, *ibid.* at 93.

[150] *Ibid.*

[151] *Ibid.* Mercuro and Medema also note that legislators often lack the incentive to curb the inefficient and wasteful performance of bureaucrats, with reasons cited in the public choice literature including lack of accountability, biased information regarding performance, and the fact that bureaucrats and their clients can form powerful interest group coalitions capable of effectuating political decisions.

[152] Croley, *supra* note 145 at 16.

action problem that generally impedes interest group mobilization. This ability creates a strong bias in the demand for regulation in favour of narrow interests – the 'collective action claim'.

3. Legislators seek to trade favourable regulatory treatment for needed political resources from the interest groups best able to provide them – the 'legislator motivation claim'.
4. Legislative control over administrative agencies is sufficient to allow legislators to deliver the regulatory treatment interest groups seek from agencies – the 'legislative dominance claim'.

Croley argues that the *'interest group motivation claim'* overstates matters because (i) many interest groups actually purport (and in some cases, actually appear) to represent general interests; and (ii) some organized interest groups appear primarily to promote interests extending beyond those of their own membership, for example, environmental groups, consumer groups, and other public interest groups.[153] Croley finds that the existence of these groups complicates the interest group motivation claim.[154]

Regarding the *'collective action'* claim, Croley takes issue with the emphasis on the collective action barriers facing the individual voter who is unable to vindicate his or her regulatory interests, while at the same time assuming that regulation-demanding interest groups can overcome their collective action problems.[155] He questions why, if some interest groups can overcome these problems, those representing the interests of the average voter cannot do the same. He argues that this problem introduces considerable conceptual indeterminacy into the theory as its logic implies that groups (as opposed to individual group members) are not as central to understanding political behaviour as is commonly thought.[156]

Croley finds that the *'legislator motivation claim'* which implies that re-election-minded legislators inevitably satisfy interest group demands is misplaced. He argues that not all legislators are electorally vulnerable (they may, for example, be from 'safe' districts with no strong incentive to satisfy

---

[153] *Ibid.* at 17.

[154] This argument may to some extent be countered by the argument that interest groups also have an incentive to act to increase their membership.

[155] Croley, *supra* note 145 at 45.

[156] *Ibid.* at 20. One explanation for this is that put forward by Olsen who finds that producer and industry groups are able to organize and exert influence for several reasons. First, they are typically homogenous – unlike groups representing consumers and other public interests. Second, they tend to have a small number of members. Third, they are largely effective at controlling their members, for example, by requiring financial contributions as a condition of membership. M. Olsen, *The Logic of Collective Action* (1965) 141–3. Cited by Ogus, *supra* note 8 at 71.

interest group regulatory demands); and that while re-election may be an important goal for legislators, it may be traded off against other important goals such as ideology.[157] He finds it implausible that legislators would have no goal other than job security; and suggests that ideology is a better predictor of legislative behaviour than the re-election incentive.[158] If legislators had no other goals than to trade regulatory goods for re-election – why would they seek electoral office in the first place? Similarly, Rachlinski argues that those who serve in administrative agencies are often motivated by a sincere desire to pursue some conception of the public good; and cites empirical evidence that public officials often make choices that are more consistent with their ideological beliefs than their self-interest.[159] In other words, public choice theory is seen as being too cynical and simplistic. Finally, in terms of the *'legislative dominance claim'*, Croley argues that agencies are not solely answerable to legislators but are also influenced by other bodies such as the courts.[160]

Another critic of the public choice theory is Trebilcock, who argues that it obscures the importance of a range of non-economic and non-self-interested values that commonly motivate various participants in collective decision-making processes, including notions of distributive justice, corrective justice, due process, and communitarianism. Trebilcock raises other issues with public choice theory as well, arguing first that the theory's framework does not pay sufficient attention to the long-run, independent impact of incremental changes to institutional design and operation on subsequent policy outcomes; and second, that public choice theory does not provide a well-developed, dynamic account of what sorts of forces disrupt existing political equilibria and lead over time to non-incremental policy change.[161]

Croley also criticizes public choice theorists' reliance on the principal–agent model whereby it is argued that administrators and bureaucrats must follow legislators' wishes at least often enough to ensure their own survival because legislators are in a position to punish errant agencies. He argues that it is empirically not clear that legislators are able to monitor bureaucrats as successfully as public choice theorists believe.[162] Given that

---

157 Croley, *ibid.* at 22.

158 *Ibid.* at 43.

159 Rachlinski and Farina, *supra* note 134 at 553.

160 Croley, *supra* note 145 at 23.

161 Michael J. Trebilcock, 'The Choice of Governing Instrument: A Retrospective' in Pearl Eliadis, Margaret M. Hill and Michael Howlett, eds, *Designing Government* (Montreal and Kingston: McGill-Queen's University Press, 2005) 51 at 54.

162 Croley, *supra* note 133 at 44.

administrators are more insulated from political pressures than those holding elected office, it is not clear that they will always pursue policies demanded by interest groups.

Finally, Croley argues that the empirical evidence for public choice theory in practice is not very compelling. He suggests that the theory is outdated and refers to examples of deregulation in various industries in the US such as airlines, securities, telecommunications, and trucking which he considers to be evidence of the declining validity of the theory. He argues that deregulation of these industries, which opened them up to entry and competition, was widely considered to be in the general public interest, as regulation had tended to benefit only the industries themselves.[163]

Despite these criticisms, the public choice theory arguably still has some salience. Soloway finds relevance by considering how public and private interests may complement each other.[164] She suggests that an explanation may in some cases be found in Yandle's 'Baptist–Bootlegger' theory which holds that special interests cannot simply be the subject of a transfer of power but that instead 'politicians in a democracy must find ways to dress their actions in public interest clothing'.[165] Yandle suggests that this is achieved by formation of so-called 'Baptist–Bootlegger' coalitions where different groups collaborate who want the same goal but for different reasons. The term originates in studies of the prohibition era in the US where political support for keeping certain counties 'dry' came from both Baptists, who favoured prohibition on moral grounds; and bootleggers, whose business depended on the continuing illegality of liquor sales.[166] Vogel finds that such coalitions have been a long-standing factor in domestic regulatory policy-making in the US. He cites a number of examples, including the passage of the *Federal Meat Inspection Act* of 1906 which was due to the combined efforts of American consumers who were outraged following publication of Upton Sinclair's *The Jungle* – a nightmarish account of the horrendous conditions in the American meat-packing industry – and large American meat-processing firms, who wanted stricter federal meat inspection in order to convince European governments to provide market access for their products.[167] Thus, the Baptist–Bootlegger theory supports public choice theory to the extent that public interest regulation may also satisfy industry groups.

---

[163] *Ibid.* at 54.

[164] Soloway, *supra* note 37 at 86.

[165] *Ibid.* at 89. Citing B. Yandle, 'Bootleggers and Baptists in the Market for Regulation' in J.F. Shogren, ed., *The Political Economy of Government Regulation* (Boston, MA: Kluwer Academic Publishers, 1989). Cited in Soloway, *ibid.* at 90.

[166] David Vogel, *Trading Up: Consumer and Environmental Regulation in a Global Economy* (Cambridge, MA: Harvard University Press, 1995) at 20.

[167] *Ibid.*

Public choice theory has particular salience in the context of trade disputes over health regulations. The scientific evidence requirements of the SPS Agreement are often described as a means of preventing domestic industries from generating risk regulation that serves as a disguised form of protectionism.[168] If one accepts that the public choice theory has some legitimacy, then there is a stronger imperative for strict implementation of the SPS Agreement on the grounds that regulators will respond to domestic producer demands for trade-restrictive regulations. Even if one does not accept the public choice theory, evidence of failed regulation in the public interest should also support strict implementation. Howse argues that because '"democratic" outcomes typically reflect capture of the regulatory process by concentrated interests . . . hand-tying of the political process by international rules, or by an apolitical authority such as "science", may actually enhance domestic welfare'.[169]

5.3.2.3 Alternative explanations of regulation

This overview of regulatory theory concludes with a brief consideration of two alternative perspectives which may play a useful complementary role in examining health regulatory decision-making, namely, administrative process theory, and the contributions of behavioural law and economics.

Administrative process theory The administrative process theory has been articulated by Croley whose conclusions seem to concur with those of the so-called 'institutional theorists' who are critical of the use of the *homo economicus* concept in private interest theory. These theorists find that there is more to regulatory decision-making than the aggregation of individual preferences.[170] Instead, they find that individuals do not simply act to further their own self-interest but are influenced by rules as well as organizational and social settings, and their preferences are shaped by institutional procedures, principles, expectations, and norms that are encountered in cultural and historical frameworks. This approach differs from both the public interest and

---

168 Howard F. Chang, 'Risk Regulation, Endogenous Public Concerns, and the Hormones Dispute: Nothing to Fear But Fear Itself?' (2004) 77 S. Cal. L. Rev. 743 at 27.

169 Robert Howse, 'Democracy, Science, and Free Trade: Risk Regulation on Trial at the World Trade Organization' (2000) 98 Mich. L. Rev. 2329 at 2333.

170 Baldwin and Cave, *supra* note 125 at 27. The authors cite a number of sources for these propositions, including J. March and J. Olsen, 'The New Institutionalism: Organisational Factors in Political Life' (1984) 78 Am. Pol. Sci. Rev. 734; W. Powell and P. Di Maggio, eds, *The New Institutionalism in Organizational Analysis* (Chicago, 1991); and J. Meyer and B. Rowan, 'Institutionalised Organisations: Formal Structure as Myth and Ceremony' (1977) Am. J. Sociol. 340.

private interest theories of regulation. Rather than being shaped by the public interest or by competition between different private interests, regulation is shaped by institutional arrangements and rules as well as social pressures.

Croley suggests that regulatory outcomes are only explicable with due emphasis on the full legal-procedural context in which administrative agencies operate and argues that the administrative process can be understood as a source of control of interest groups. What he terms the 'administrative process theory' holds first that administrative regulators are motivated by more than self-interest, and in particular, are often motivated by concerns for general, public-oriented interests. Second, administrative process rules governing regulatory decision-making by agencies promote agency autonomy more than they advance legislative control. Third, by fostering procedural independence, extra-legislative influences on agency decision-making further promote agency autonomy more than they advance legislative control of agencies.[171]

Croley argues that 'by delegating substantial lawmaking authority to agencies, the legislature insulates both itself and administrative regulators from interest group pressure. It does so, specifically, by equipping agencies with legal authority and political autonomy sufficient to advance broad-based interests'.[172] He predicts that where administrative regulators are motivated to advance general interest regulation,[173] certain rules of the administrative process, as well as the legal-institutional context in which they operate, will equip them to realize their general-interest regulatory aims, and that regulatory outcomes may therefore reflect those aims.[174]

For the theory to be plausible, Croley notes that several things must be true. First, administrators themselves must be in part committed to ideology or principle (that is, some vision of the public interest). Second, administrative decision-makers must truly enjoy some substantial degree of legal and political decision-making autonomy.[175] The extent to which these factors are present will clearly vary by jurisdiction.

---

171 Croley, *supra* note 145 at 27.

172 *Ibid.* at 26.

173 *Ibid.* at 106. Croley argues that empirical evidence in the US finds that regulatory activity in the environmental, consumer protection, and financial services arenas suggests that administrators are at least sometimes motivated by general interest goals.

174 *Ibid.* at 26.

175 In reaching his conclusions, Croley disagrees with Mathew McCubbins, Roger Noll, and Barry Weingast ('McNollgast') who argue that most of administrative law is 'written for the purpose of helping elected politicians retain control of policymaking'. Mathew McCubbins, Roger Noll and Barry Weingast, 'Administrative Procedures as Instruments of Political Control' (1989) 3:2 J. Law Econ. Org. 243. Croley finds that it is equally arguable that administrative decision-making procedures actually foster agency autonomy and independence from the legislators. Croley, *ibid.* at 33.

In his analysis, Croley seems to imply that the public interest theory which has for so long been largely disregarded may have some salience after all. Indeed, as noted above, in the context of health regulation, it is arguable that some normative content can plausibly be given to the notion of the 'public interest'. Croley's theory might be used to predict procedures that are more likely to result in outcomes that genuinely further health interests particularly in agencies involved in setting SPS measures.

Behavioural law and economics Another explanation for regulatory failure is that poor decisions are the result of fallibility rather than culpability.[176] A number of scholars have relied on insights from cognitive psychology[177] to identify categories of errors that regularly play a part in human decision-making. In doing so, they have argued that bad public policy can often be traced to flaws in human judgment and choice among governmental actors.[178] Even where decision-makers have the public good as their motive, this may not be enough to result in good policy if they exercise bad judgment. As a result, it is argued that aligning and channelling self-interest toward pursuing the public interest (as normative public choice theory would advocate) will not guarantee good policy outcomes. Likewise, even publicly motivated agencies operating in an autonomous legal-institutional context will not always make good policy.

Jolls, Sunstein, and Thaler find that 'real' people differ from *homo economicus* in that they are subject to three important 'bounds' on their behaviour that raise questions about the central economic ideas of utility maximization, stable preferences, rational expectations, and optimal processing of information. The three bounds they suggest are bounded rationality, bounded willpower, and bounded self-interest.[179]

Bounded rationality refers to the fact that human cognitive abilities are not infinite. People have limited computational skills and flawed memories. Psychologists have found that people rely on two primary strategies to make the most of their cognitive abilities and overcome their bounded rationality.

---

176 Rachlinski and Farina, *supra* note 134 at 554.

177 The core premise of cognitive psychological theory is an understanding that the 'human brain is a limited information processor that cannot possibly manage successfully all of the stimuli crossing its perceptual threshold'. Rachlinski and Farina cite Daniel Kahneman and Amos Tversky, 'On the Reality of Cognitive Illusions' (1996) 103 Psychol. Rev. 582, 583. Rachlinski and Farina, *ibid.* at 555.

178 Cass Sunstein, 'Cognition and Cost-Benefit Analysis' 29 J. Legal Stud. (2000) 1059 at 1060–61.

179 The following discussion of the three bounds is taken from Christine Jolls, Cass R. Sunstein and Richard Thaler, 'A Behavioural Approach to Law and Economics' (1998) 50 Stan. L. Rev. 1471 at 1477.

First, they rely on mental shortcuts, which psychologists call 'heuristics'.[180] Heuristics are rules of thumb that allow quick information processing by substituting an easy question for a harder one.[181] Second, people rely on organizing principles, which psychologists call schema, to process information.[182] These schema involve a scripted set of default information and organization themes that help people focus on the information most likely to be relevant, thus allowing them to ignore information likely to be irrelevant.

Key heuristic devices include: *availability*; *anchoring*; and *case-based decisions*. The *availability* device finds that people tend to think that risks are more serious when an incident is readily called to mind or 'available'. That is, if something has happened in the immediate past, such as contamination of spinach originating from a Californian farm,[183] people are more likely to fear that it will happen again. The availability heuristic can produce systematic errors because risk assessments may be biased in the sense that people will think that some risks are high, whereas others are low. Sunstein notes that the availability heuristic appears to affect the demand for law, and can result in under- as well as overregulation.[184] For example, in the spinach contamination example, people may demand regulation following an outbreak of E. coli poisoning, but may continue to ignore health risks such as poor diet and lack of exercise.[185]

---

180 Rachlinski and Farina cite Amos Tversky and Daniel Kahneman, 'Judgment under Uncertainty: Heuristics and Biases' (1974) 185 Science 1124. Rachlinski and Farina, *supra* note 134 at 555.

181 Cass R. Sunstein, 'Book Review – The Laws of Fear' (2001) 115 Harv. L. Rev. 1119 at 1143. Sunstein gives the example of a person being asked about the probability of getting caught in a major traffic jam on a particular route. Rather than looking at the statistical probabilities, the person is likely to ask whether they, or people they know, have been stuck in traffic jams on that route.

182 Rachlinski and Farina cite Susan T. Fiske and Shelley E. Taylor, *Social Cognition*, 2nd ed. (1991) 97–9, 136–9. Rachlinski and Farina, *supra* note 134 at 549.

183 On 14 September 2006, the US Centres for Disease Control and Prevention issued an alert concerning an outbreak of E. coli 0157:H7, a potentially lethal pathogen typically associated with adulterated beef. The outbreak had been traced to consumption of raw spinach sold in bags. See 'Sickening Spinach' (*The Economist*, 21 September 2006).

184 Cass R. Sunstein, ed., *Behavioural Law and Economics* (Cambridge, UK: University of Cambridge Press, 2000) at 5. See also Jolls, Sunstein and Thaler, *supra* note 179 at 1519.

185 Slovic writes that the availability heuristic is partly responsible for participants in a study significantly overestimating highly publicized causes of death, including tornadoes, cancer, botulism, and homicide. On the other hand, they underestimated the number of deaths from strokes, asthma, emphysema, and diabetes. As well, people mistakenly thought the number of deaths from accidents is higher than those from disease; and that more people die from homicides than from suicides. Paul Slovic, *The Perception of Risk* (London: Earthscan Publications Ltd, 2000) at 108.

The anchoring heuristic finds people making probability judgments on the basis of an initial value, or 'anchor', for which they make insufficient adjustments. The initial value may have an arbitrary or irrational source. When this is the case, the probability assessment may be clearly wrong. The *case-based decisions* heuristic involves people simplifying the task of calculating the expected costs and benefits of alternatives by reasoning based on past cases. The result of using heuristics is that while someone may seem to be behaving rationally, they will make forecasts that are different from those that would result from the standard rational choice model.[186]

Bounded willpower refers to the fact that people often take actions (such as smoking) that they know will have an adverse impact on their interests in the long run. Jolls, Sunstein, and Thaler recognize that many people know they have bounded willpower and take steps to address the consequences of this. For example, they join a pension plan to prevent under-saving and avoid keeping chocolate around the house when they are trying to diet.[187] They suggest that the demand for and supply of regulation may reflect people's understanding of their own (or others') bounded willpower.[188]

Bounded self-interest refers to the fact that self-interest is bounded in a much broader range of settings than suggested by conventional economics. That is, people care, or act as if they care, about others, even strangers, in some circumstances. Jolls, Sunstein, and Thaler suggest that, as a result of these concerns, agents in a behavioural economic model are both nicer and (when they are not treated fairly) more spiteful than the agents postulated by classical economic theory.[189]

Jolls, Sunstein, and Thaler seek to show how the incorporation of these understandings of human behaviour bears on the actual operation and improvement of the legal system. They suggest first that *bounded rationality* will come into play whenever actors in the legal system are called upon to assess the probability of an uncertain event. This will often be the case where decision-makers are called upon to enact measures in response to a perceived risk to health. Second, they find that bounded rationality will come into play whenever actors are valuing outcomes, for example, where risk assessments are being undertaken. The relevance of *bounded willpower* lies primarily with decisions that have consequences over time, such as anti-smoking legislation. *Bounded self-interest* is relevant mostly in situations where one party is seeking to punish 'unfair' behaviour on the part of another.[190]

---

186 Jolls, Sunstein and Thaler, *supra* note 179 at 1478.
187 *Ibid.* at 1479.
188 *Ibid.*
189 *Ibid.*
190 *Ibid.* at 1480.

Rachlinski and Farina argue that while heuristics and schema are essential to navigating the huge amount of information that confronts individuals every day, they come at the cost of systematic errors in judgment.[191] They find that such errors are not the preserve of lay persons but that even experts with knowledge, experience, and training in decision-making can fall prey to the illusions of judgment.[192] They argue in many situations that experts do not receive reliable feedback, which inhibits their opportunity to learn from experience. As well, they tend to be overconfident in their judgments, placing too much faith in their abilities, and failing to consider alternative policies beyond the boundaries of their expertise.

5.4 CONCLUSION

This chapter has examined regulatory theory in order to shed light on why nations regulate. Regulations not only have potential to restrict trade but also to interfere with the rights and freedoms of individuals and business. As such, representative democracies require that regulations be based on grounds that justify those restrictions. Looking at these grounds of justification (a normative theory of regulation) is helpful in enabling us to unpack the domestic (and global) welfare benefits from regulation that must be weighed against the benefits to be gained from trade liberalization. Similarly, it is important to conduct a positive analysis of regulatory decision-making to investigate how regulatory decisions are actually made as this will impact on how they ought to be reviewed by WTO panels and the Appellate Body.

None of the theories offered in this chapter can, by themselves, provide a conclusive explanation for regulatory intervention in every case. Baldwin and Cave note the folly of attempting to provide an exhaustive account of the potential theories available to explain regulation. As they state: 'it would be optimistic, even rash, to suggest that such theories can be synthesized so that reliable predictions can be made about all or most regulatory processes[193] . . . different theories exist at differing levels of generality and have varying applications and uses as explanatory tools.'[194] Similarly, Trebilcock finds that no single positive theory of the process of public policy decision making has

---

[191] Rachlinski and Farina, *supra* note 134 at 558.

[192] *Ibid*. at 560.

[193] Baldwin and Cave, *supra* note 125 at 32. Citing M.E. Levine and J.L. Forrence, 'Regulatory Capture, Public Interest and the Public Agenda: Towards Synthesis' (1990) J. Law. Econ. Org. 167.

[194] Baldwin and Cave, *supra* note 125 at 32.

broad explanatory power.[195] Further difficulties are presented by differing national political, economic, social, and cultural contexts which play an important role in regulatory divergences between countries. However, while none of them may be exhaustive, the theories discussed all have a valuable story to tell. The justification cited for much social regulation is the economic goal of correcting market failures. However, notions of distributive justice, paternalism, and community values also play a role. Where regulation aims to protect human health, another justification comes into play, that of an obligation on the part of governments to protect health. Such an obligation is supported by the declaration of health as a fundamental human right by the WHO and various other human rights instruments. Similarly, governments arguably have an obligation to protect wild fauna and flora pursuant to principles of international law.

In terms of how regulations are actually made and implemented, public and private interest theorists both provide insights into the decision-making process. Public interest theory finds that regulation aims to further the public interest. However, it is a trite observation that domestic regulation does not always result in the furthering of public interest, even where the 'public interest' is able to be clearly articulated and defined. Even where the goal is to protect health, there will always be cases where it is unclear what exactly is in the public interest, for example, where scientific uncertainty means it is unclear whether or not a particular regulation will really serve to protect health.

Private interest theories of regulation introduce the goals and motivations of both interest groups and decision-makers into the regulatory decision-making process and these will not always be consistent with the public interest. Thus, regulatory decision-making may be captured by those seeking to protect local producer interests and prevent entry by foreign producers. From a trade liberalization perspective, the fact that regulation purportedly designed to further the public interest may not only fail to do so, but may actually promote private interests, highlights the importance of close examination of regulatory decisions where they result in trade restrictions as they may in actuality be protectionist devices. On another note, the behavioural law and economics approach argues that poor decisions are often the result of fallibility on the part of regulators, rather than culpability. From the SPS Agreement's post-discriminatory perspective, regulations made due to poor judgment will be deemed problematic if they result in trade restrictions, even if there is no protectionist intent.

---

[195] Trebilcock, *supra* note 161. Trebilcock refers to a report written in 1982 in which he and Douglas Hartle, Robert Prichard, and Donald Dewees took a less hopeful view of the policy-making process: Trebilcock *et al.*, *supra* note 131.

6. Identifying tension: the 'difficult' (or 'amber') cases

6.1 INTRODUCTION

On the face of the SPS Agreement, any conflict or tension that might otherwise exist between health and trade objectives is avoided by giving recognition to a nation's right to enact trade-restrictive SPS measures when necessary to protect health. What then is meant by the statement that trade and health objectives sometimes conflict or are in tension with each other?

The SPS Agreement does indeed avoid conflict or tension by recognizing the importance of both trade and health, and giving priority to health in *some* cases, namely, when measures are *necessary* to protect health. Through adoption of requirements for measures to be based on scientific evidence, the SPS Agreement's negotiators appear to have assumed that it will be possible to determine on the basis of science *when* a measure is *necessary*. However, this will not always be the case. This chapter argues that there are a range of cases where it will be extremely difficult, if not impossible, to determine by any objective scientific standard whether or not a measure is *necessary*. In these (difficult) cases, conflict between trade and health objectives persists and WTO panels and the Appellate Body will face difficulties – both conceptual and factual – as they try to balance the competing objectives of health and trade.

6.2 PRELUDE TO THE 'DIFFICULT' CASES: WHAT IS A RISK TO HEALTH?

The identification of a risk to health lies at the core of the issues that arise in the difficult cases. As discussed in the next section, many of these cases involve uncertainty or disagreement about the existence of a risk and/or the seriousness of that risk. What then is a risk to health? The *Oxford English Dictionary* defines 'risk' as a 'hazard, danger; exposure to mischance or peril'.[1] Kraemer, Lowe, and Kupfer offer a more detailed definition for risks

[1] *Oxford English Dictionary Online*, *s.v.* 'risk'.

to human health, namely: 'the probability of an outcome within a population'.[2] They elaborate on this definition by looking more closely at the three elements it contains: first, risk is a *probability*; second, researchers measure risk within a well-specified *population*; and third, researchers measure risk for a precisely specified *outcome*.[3]

Probability refers to the likelihood that a given person within a specified population will experience the specified outcome. Risk is generally expressed as a number or fraction where the outcome is health-related such as contracting a disease or death (for example, 1 out of 100 people will contract the disease in question).[4] Risk is measured within a specified *population* in recognition of the fact that differences between populations may lead to significantly different results. Differences may be defined in many ways including gender, age, and geography. It is also necessary to specify *outcome* as this may vary, even where the same health issue is at stake. For example, an *outcome* may be experienced with different levels of severity, or may be likely to occur within different time periods.[5]

There are many kinds of risk agents that may give rise to health risks and a corresponding decision to enact SPS measures. The SPS Agreement – while it does not actually provide a definition of risk *per se*[6] – categorizes risk in the case of *human* life or health as that arising from 'additives, contaminants, toxins, disease-causing organisms in foods, beverages or feedstuffs'.[7] Examples include hormones and antibiotics in beef, pesticide residues in crops, aflatoxin in grains, Escherichia coli, Salmonella, Listeria, and botulism

---

[2] Helena Chmura Kraemer, Karen Kraemer-Lowe and David J. Kupfer, *To Your Health: How to Understand What Research Tells Us About Risk* (New York: Oxford University Press, 2005) at 5.

[3] *Ibid.*

[4] *Ibid.* Kraemer, Kraemer-Lowe and Kupfer note that the outcome may sometimes be desirable, for example, recovery or remission of a disease. However, for the purposes of the current study, the focus is on undesirable outcomes as it is such outcomes that prompt the necessity to protect health.

[5] *Ibid.* at 10.

[6] See however, Codex, which defines risk as 'a function of the probability of an adverse health effect and the severity of that effect, consequential to a hazard' (Codex Alimentarius Commission, *Procedural Manual*, Definitions of Risk Analysis Terms Related to Food Safety, 11th ed.); the OIE which refers to 'the likelihood of the occurrence and the likely magnitude of the consequences of an adverse event to animal or human health in the importing country during a specified time period' (OIE International Animal Health Code, 2001); and the IPPC which defines it as 'evaluation of the probability of the introduction and spread of a pest and of the associated potential economic consequences' (IPPC International Standard for Phytosanitary Measures No. 5, Glossary of Phytosanitary Terms).

[7] Annex A, Article 1(b).

toxin. It also covers risk arising from 'diseases carried by animals, plants or products thereof, or from the entry, establishment or spread of pests'.[8] Such diseases may include zoonoses such as BSE.

In the case of *animal or plant* life or health, reference is made in the SPS Agreement to risks arising from 'the entry, establishment or spread of pests, diseases, disease-carrying organisms or disease-causing organisms'.[9] Examples include insect pests that are introduced in produce such as the Mediterranean fruit fly, infectious agents such as the prions that cause mad cow disease, and weed pests like purple loose strife. It also covers risks to *animal* life or health arising from 'additives, contaminants, toxins, or disease-causing organisms in foods, beverages, or feedstuffs'.[10]

An important challenge facing domestic regulators in complying with, and adjudicators in ruling on, international trade rules is the constant evolution of understanding of risks to health as scientific advances provide the technical capacity to identify previously unsuspected risks.[11] The discovery of the health effects of lead provides a vivid example. Evidence concerning the toxicity of lead and its effect on human health has been refined many times since it was first identified as a risk in the 1950s.[12] Undoubtedly, continuing scientific developments in areas such as molecular and genetic epidemiology will continue to identify new health risks, and raise questions as to whether or not a new development presents a risk. WTO rules must be flexible enough to allow countries to deal appropriately with these new risks and uncertainties.[13] Where the existence or otherwise of a risk is uncertain, disputes will likely arise over the appropriate level of precaution that countries should be able to exercise in risk identification. Indeed, much of the current trade debate over genetically modified (GM) foods is due to a concern held by some that as scientific knowledge advances, risks may become evident that are not currently suspected or identifiable.

[8] Annex A, Article 1(c).

[9] Annex A, Article 1(a).

[10] Annex A, Article 1(b).

[11] McLaughlin Centre for Population Health Risk Assessment, McLaughlin Centre for Population Health Risk Assessment online: http://www.mclaughlincentre.ca/welcome/index.shtml (Date of Access: 12 July 2005).

[12] See generally, Christian Warren, *Brush With Death: A Social History of Lead Poisoning* (Baltimore, MD: John Hopkins University Press, 2000).

[13] McLaughlin Centre, *supra* note 11.

6.3 THE 'DIFFICULT' (OR 'AMBER') CASES

Situations embodying conflict between health and trade can be broadly identified and categorized as in Table 6.1. The table organizes cases according to a traffic light typology. The 'green' and 'red' categories are those where conflict between health and trade objectives is avoided through a predetermined setting of priorities by the SPS Agreement, and the 'amber' cases are those where conflict persists.

The red and green categories are clear-cut in the sense that the provisions of the SPS Agreement avoid any conflict or tension that would otherwise exist in respect of the health and trade objectives in question. In the 'green' case, there is a clear risk to health, which, if manifested, will reduce domestic welfare. In this situation, the SPS Agreement recognizes the importance of protecting health and allows nations to restrict trade in order to provide that protection. The negative welfare impacts on consumers negate the gains from trade. The 'green' light therefore signifies that the government ought to be able to go ahead and maintain the regulation.

Conversely, in the red case, the motivation for the trade-restrictive measure is to protect local industry, not to protect health. Such a measure would diminish consumer welfare by reducing trade but it would do nothing to protect health. Here again, conflict or tension is avoided between the objectives of domestic health regulation and trade rules. The 'red' light signifies that the regulation ought to be disallowed.

The task for WTO panels becomes more difficult in the 'amber' cases. Before examining these cases, it must be emphasized that there will often be overlap between the reasons for the measures as described in the table. For example, in any one case, both public risk perceptions and scientific uncertainty, as well as protectionism, may play a role, while risks may be any combination of low or high probability and minor or serious consequences. The examples given are greatly simplified and in the real world, it will often be extremely difficult to determine whether a risk is low or high probability, and what the severity of the consequences will be.

The objectives of trade and health conflict or come into tension with each other in the amber categories because it is arguable that the welfare gains from the health regulations in question are not sufficient to outweigh the welfare losses due to lost trade opportunities. It is instructive to comment here on the meaning of the term 'welfare'. The *Oxford Dictionary of Economics* defines it as 'enjoyment of the necessary resources for a worth-while life'.[14] It must be

---

[14] John Black, *Oxford Dictionary of Economics* (New York: Oxford University Press, 2002) at 502.

Table 6.1 Situations of conflict between health and trade objectives

	Reason for restrictive measure	Example of regulatory response	Existence of and reason for tension or conflict
Green	Clear case of health risk, i.e., high probability of risk eventuating and risk has serious consequences	Import ban on food contaminated with E-coli bacteria	Conflict or tension avoided: the negative welfare impact of E-coli poisoning negates the welfare gains from trade
Amber, e.g., #1	Risk that is high probability but consequences are not serious[1]	Ban on GM nuts because they cause minor allergic reactions in some people	Conflict or tension between objectives persists: it is not clear that the risk is serious enough to warrant trade restrictions
Amber e.g., # 2	Risk that is low probability but consequences are not serious	Ban on GM milk where there is a small chance that some people will be intolerant to it due to changes in the enzyme structure	Conflict or tension between objectives persists; it is not clear that the risk is serious or likely enough to warrant trade restrictions
Amber e.g., # 3	Risk that is low probability but has serious consequences	Ban on beef due to outbreak of 'mad cow disease' in country of origin, where the risk, albeit of low probability, is of contracting a severe brain-wasting illness	Conflict or tension between objectives persists: it is not clear that the risk is likely enough to eventuate to warrant trade restrictions
Amber, e.g., #4	Public risk perceptions that differ from expert perceptions of the same risk	Import ban on irradiated mangoes in response to public perceptions that a risk to health exists	Conflict or tension between objectives persists: public risk perceptions differ from those of experts and there is no scientific evidence of a risk which would warrant trade restrictions

Amber, e.g., #5	Scientific uncertainty	Import ban on mayonnaise treated using nanotechnology to improve emulsification in response to scientific uncertainty regarding the risk associated with the use of nanotechnology *or* Ban on 'energy' drinks with high levels of caffeine, taurine, and glucuronolactone where the effects of these ingredients reacting together in the body is uncertain	Conflict or tension between objectives persists: it is not clear whether there is actually a risk to health which warrants trade restrictions
Amber, e.g., #6	Mixed motives	Some elements of risk to health as described in #1–5, combined with protectionist intent	Conflict or tension between objectives persists: the measure both seeks to protect health and to shield local producers from competition
Red	Clear case of protectionism	Rule that forbids margarine from being coloured yellow, whether domestically produced or imported (where margarine is not domestically produced but butter is)	Conflict or tension avoided: there is no health risk, therefore no welfare reduction by allowing imports of yellow margarine

Note: [1] This example assumes expert risk assessment, as opposed to public perception of risk. Likewise for examples #2 and #3.

noted that a number of economists do not include non-measurable variables when they use the term 'welfare'. However, it is argued that health, as defined by the WHO, cannot always be measured in a quantitative manner. As such, the term welfare as used here is intended to refer to both those variables that can be measured, such as income, and those that cannot.

In amber example #1, there is conflict or tension with respect to GM nuts because the consequences of the allergenicity are not very serious. In such cases, is a protective measure that restricts trade necessary to protect health? Should a risk that would only have a minor adverse health impact justify a trade-restrictive measure that negates the potential welfare gains from trade? A key question here is how the 'seriousness' of a threat should be judged; that is, what constitutes 'minor'? Is avoidance of such consequences necessary to protect health?

In amber example #2, there is conflict or tension because the risks to health are neither very serious nor very likely to eventuate. Again, the question arises whether a protective measure that restricts trade is necessary. Should a risk that is unlikely to eventuate and would only have a minor adverse health impact justify a trade-restrictive measure that negates the potential welfare gains from trade? Here again, key questions are, how the 'seriousness' of a threat should be judged, but also what constitutes 'low probability'?

In amber example #3, there is conflict or tension because the risk, while potentially having a serious negative impact on health, is unlikely to eventuate. Perhaps only very few cattle have been infected, and further, transmission of the disease is unlikely to occur even where an animal is infected. How should a panel determine when the probability of a risk occurring is high enough to justify a trade-restrictive measure? This will be one of the more common amber situations as there is likely to be greater demand for regulation where potential risks are serious. From a more cynical point of view, these cases provide better cover for protectionist measures.

In amber example #4, it may be arguable that there is no literal risk to health at all and that public sentiment is unnecessarily demanding regulation when experts would not make such a recommendation. However, consumers' subjective sense of well-being may be diminished as a result of the negative weight they attach to irradiated food due to their particular risk perceptions.[15] It is arguable on the one hand that governments should not be able to justify import bans on the grounds of public sentiment where it reflects risk perceptions that are based on misconceptions, mistakes, or irrational judgments that

[15] For example, consumers may have adopted any one or more of various heuristics for assessing risks, or may have been misled by the media or other information sources. See discussion in Chapter 5.3.2.3.

lead to over-estimation of a risk.[16] The argument is strongest where there is no observable risk to health or if the risk is of such low probability, or the expected harm so minor that the welfare gains from regulation are not clear. On the other hand, it might be argued that public risk perceptions are no less valid than those of experts and should be respected. This last point is the subject of discussion in Chapter 9.

Amber example #5 involves cases where there is scientific uncertainty as to the extent and/or nature of the risk to health claimed by the regulating country. Such uncertainty will pervade many cases in the amber categories and will often also be an element in the categories listed in amber examples #2 to #4. The existence of scientific uncertainty presents conflict or tension between health and trade objectives because if the nature and extent of the risk are not known, how can it be determined whether the risk justifies a trade-restrictive measure that is necessary to protect health and that negates the potential welfare gains from trade? Questions thus arise as to the proper approach to take, including to what extent countries are justified in taking a precautionary stance.

Finally, in amber example #6 – *mixed motives for regulation* – there is conflict or tension because while health is being protected, so is domestic industry to the detriment of trade liberalization. Many regulations serve more than one purpose and as Chang notes, protectionist intent may be a subtle matter of degree.[17] This makes it difficult for panels to disaggregate the regulatory purpose so as to determine whether or not the regulation ought to be allowed to stand. Where it can be determined that both health and domestic industry are being protected, panels will essentially have to decide whether one or other of these objectives ought to be prioritized. This book argues that where both genuine health protection and protectionism are apparent, the regulation should be allowed so long as it is the least trade-restrictive measure available and is not arbitrary or discriminatory in its application. This would appear to be in accordance with the SPS Agreement's approach to the question of dealing with conflict or tension between health and trade objectives where the risks to health are deemed to be necessary.

---

[16] For example, consumers may have adopted any one or more of various heuristics for assessing risks, or may have been misled by the media or other information sources.

[17] Howard F. Chang, 'Risk Regulation, Endogenous Public Concerns, and the Hormones Dispute: Nothing to Fear But Fear Itself?' (2004) 77 S. Cal. L. Rev. 743 at 29. See also Michael Trebilcock and Julie Soloway, 'International Policy and Domestic Food Safety Regulation: The Case for Substantial Deference by the WTO Dispute Settlement Body under the SPS Agreement' in Daniel Kennedy and James Southwick, eds, *The Political Economy of International Trade* (Cambridge: Cambridge University Press, 2002) at 542. Trebilcock and Soloway make reference to 'political log-rolling, posturing, and dissembling, and the potential for regulatory capture'.

7. Resolving the tension: balancing trade and health objectives in the WTO

7.1 INTRODUCTION

This chapter summarizes the normative basis that will guide the analysis in the following chapters. It does so by setting the parameters of what would be a desirable balance between health protection and trade liberalization objectives in 'amber' cases under the SPS Agreement. The perspective taken acknowledges both the value of international trade liberalization and the value that should be accorded to human, animal, and plant health.

The value of international trade liberalization was highlighted in Chapter 5's discussion of the theoretical foundations underlying trade theory. The analysis found that the objections to international trade do not seriously undermine either the validity of trade theory, nor the case for gains to be had from trade. However, it did find a valid argument that the gains may in some cases be exaggerated and noted the perspectives put forward by Dunkley, Schumacher, and others who argue for the importance of recognizing non-economic values such as health in addition to the economic gains from wealth maximization.

7.2 THE WTO AND RULES OF INTERNATIONAL LAW

Chapter 5 examined rationales for regulation and introduced the non-economic imperatives of regulations designed to protect health. In the case of human health, such imperatives arise both from within the state based on notions of a social contract and government obligations, and from outside – from norms and instruments of international law. While such imperatives arguably impose obligations on governments, the question arises as to how they fit within the WTO's legal framework. Does WTO law recognize and accord a role to concepts of international law such as the right to health? This is a crucial issue as we move forward to consider how WTO panels and the Appellate Body ought to approach 'amber' cases. Thus, this section briefly explores the place of rules of international law in the WTO.

The international legal system is a decentralized system without a clear hierarchical relationship between different legal norms.[1] The question thus arises of how to resolve conflict between different legal norms that have equal status as valid rules of international law. In a 2006 study on fragmentation of international law, the International Legal Commission (ILC) expressed a presumption against normative conflict, stating that when 'several norms bear on a single issue they should, to the extent possible, be interpreted so as to give rise to a single set of compatible obligations'.[2] As noted, the SPS Agreement seeks to advance seemingly conflicting norms: the advancement of trade liberalization on the one hand, and the right of Members to protect health on the other. While conflict is avoided in obvious cases of measures that are necessary to protect against risks to health, it persists in the 'amber' cases where the necessity of regulatory measures is contentious. The question then arises, can norms contained in treaties such as the ICESCR and ICCPR be used to help interpret WTO rules in cases of conflict between the SPS Agreement's objectives?

In the first instance, reference should be had to the Vienna Convention on the Law of Treaties which provides in Article 31(3)(c) that in interpreting treaties, reference must be had to 'any relevant rules of international law applicable to the parties'. In *EC – Biotech*, the Panel took a very narrow approach to Article 31(3)(c), finding that it was only applicable where the rules of international law in question were applicable to *all* WTO Members. This approach has been criticized and it is not clear to what extent, if at all, it will be followed in future cases.[3] In other cases, the Appellate Body has suggested that non-WTO legal norms have legitimate uses in WTO dispute settlement. For example, it has used international environmental law to interpret the scope of the Article XX(g) exception to the GATT (*Shrimp–Turtle*) and to establish whether a species is endangered (*Shrimp–Turtle*) as required by Article XX(g). Also in *Shrimp–Turtle*, the Appellate Body referred to international law (the Rio Declaration) as being reflective of broader agreement in the international community, even where not all parties were bound by that law.

---

[1] International Law Commission, *Conclusions of the Work of the Study Group on the Fragmentation of International Law: Difficulties Arising from the Diversification and Expansion of International Law* (New York: United Nations, 2006) at para 1(4).

[2] *Ibid.*

[3] See for example, Caroline Henckels, 'GMOs in the WTO: A Critique of the Panel's Legal Reasoning in EC-Biotech' (2006) Melbourne Journal of International Law 7:2 278. Nathalie Bernasconi also critiqued the decision during a presentation to the 7th Annual WTO Conference, British Institution of International and Comparative Law, Grey's Inn, London, 21–2 May 2007. Bernasconi argued that the Panel's interpretation has the capacity to lead to increased fragmentation of international law.

In a recent survey by Howse and Teitel of the relationship between the WTO and the ICESCR, the authors find a place for recognition of the right to health within the WTO's framework. As they argue, the 'WTO should not be lightly assumed to have undertaken in the first place obligations that are inconsistent with the right to health, when those obligations can be read and applied otherwise, consistent with the Vienna Convention'. Further, they suggest that panels and the Appellate Body can, through interpretation and implementation, advance consistency of the WTO with the ICESCR. [4]

Thus, it may be concluded that WTO rules should rightfully be read in a manner which is consistent with the various incarnations of the right to health found in international law (as discussed in Chapter 5). Not only is this consistent with the lack of hierarchy within the international legal system, but it is normatively desirable in order to advance the realization of intrinsic human rights.

7.3 A NORMATIVE FRAMEWORK FOR RESOLUTION

Having considered the theory underlying trade and the regulatory state, and noting the validity and importance of the objectives of both trade liberalization and regulation to protect health, Chapter 6 sought to articulate when these objectives may conflict. This is a critical first step and needs to be considered in the light of the SPS Agreement's provisions. The SPS Agreement acknowledges the importance of both health protection and trade liberalization through its recognition of the right of Members to regulate to protect health. However, there are a number of situations in which the Agreement does not clarify whether Members have the right to impose trade-restrictive measures. These have been referred to as the 'amber' cases and involve:

1. Risks that are of *high probability but* have *consequences that are not serious;*
2. Risks that are of *low probability and* have *consequences that are not serious;*
3. Risks that are of *low probability but* have *serious consequences;*
4. *Public risk perceptions that differ from expert risk perceptions*;
5. *Scientific uncertainty* concerning the existence of a risk; and/or
6. *Mixed motives* for a trade-restrictive regulation.

---

[4] Robert Howse and Ruti G. Teitel, *Beyond the Divide: The Covenant on Economic, Social and Cultural Rights and the World Trade Organization* (Geneva: Friedrich Ebert Stiftung, 2007) at 21.

Each of these situations involves conflict or tension between the objectives of health protection and trade liberalization because it is arguable that the welfare gains from the health measures in question are not sufficient to outweigh the welfare losses due to foregone gains from trade. These situations can be contrasted with those envisaged by the SPS Agreement where the welfare gains from health regulations in cases of clear and serious risk are deemed sufficient to outweigh the welfare losses from lost trade opportunities (the 'green' cases).[5] Where these 'amber' situations arise, it is suggested that the following framework will assist in striking an appropriate balance between the conflicting objectives.

Within the category of 'amber' cases, there are two situations: (1) where governments have an obligation arising from international law and human rights to protect health, that is, where human health or that of wild flora and fauna is at risk; and (2) the remaining cases where governments have no such obligation. In both situations, panels must ask whether the regulating country has met the SPS Agreement's scientific requirements (*risk assessment*). They must also ask whether, in determining its appropriate level of protection and choosing a measure (*risk management*), the country has complied with the provisions requiring Members to take into account the objective of minimizing negative trade effects (Article 5.4), to avoid discrimination or disguised restrictions on trade (Articles 2.3 and 5.5), and to ensure that the measure taken is not more trade restrictive than necessary (Article 5.6). The question of balancing health and trade objectives arises most acutely when assessing a Member's compliance with these constraints on its risk management activities.

It is suggested that for panels to strike an appropriate balance between health and trade in the first type of 'amber' cases, that is, where governments have an obligation to protect health, they ought to follow an orientation for WTO law that is based on widely agreed norms of human rights and international law[6] and adopt a bias towards health protection objectives. This would have two consequences as set out below.

First, it would impact decisions concerning whether a country has met its obligations regarding its risk management activities by minimizing the negative impacts on trade and ensuring that the measure is not more trade restrictive than necessary. It would do so by allowing countries to manage risks in a way that avoids Type I Error Costs, but accepts Type II Error Costs. Type I

---

[5] Shapiro argues that some products are so harmful to health that trade in them should not be subject to WTO rules at all. She writes in particular that banning trade in cigarettes would have significant public health benefits which ought to override the trade losses incurred. See Ira S. Shapiro, 'Treating Cigarettes as an Exception to the Trade Rules' (2002) XXII:1 SAIS Review 87.

[6] Howse and Teitel, *supra* note 4 at 28.

Error Costs result from allowing the importation of products that pose a real health risk. They arise from policies made in response to *false negatives* – risks thought to be minor or unlikely to manifest themselves that do arise and turn out to be serious. Type II Error Costs, on the other hand, involve banning benign products. They arise from policies made in response to *false positives* – risks thought to be serious and/or highly likely that turn out to be minor or non-existent. Both types of costs may cause harm to society: Type I Error Costs may result in harm to health from the imported products; while Type II Error Costs include high costs to consumers, restricted consumer choice, lost innovations from new products, and public cynicism about exaggerated risk.[7] However, it is argued here that in 'amber' cases involving risks to human health or wild fauna and flora – the consequences of which are often irreversible – a bias towards health objectives supports avoidance of Type I Error Costs and acceptance of Type II Error Costs. Where the use of such a bias might assist panels is, for example, in the case of risk that is low probability but has potentially serious consequences. In such a case, all other factors being equal, adopting a bias towards health would result in health being privileged over trade.

This is not to suggest that a panel should never find a health measure in violation of Articles 2.3, 5.5, or 5.6 or to impose an impossibly high standard for complaining countries. Rather, what it does suggest is that in 'amber' cases where it is disputed whether or not the welfare gains from the SPS measure in question are sufficient to outweigh the welfare losses due to lost trade opportunities and there is no clear indication one way or the other, then the inherent value of health and people's concerns about health ought to justify a decision that leans towards favouring health protection by avoiding Type I Error Costs and accepting Type II Error Costs. There can be no black and white answers in such cases, however, and each case must be considered on its facts. The greater the probability of risk eventuating and/or the more severe the potential consequences, the stronger the argument for biasing health and allowing countries to avoid Type I Error Costs and/or accept Type II Error Costs.

Second, a bias towards health objectives would impact panel decisions as to whether there is discrimination or a disguised trade restriction (Articles 2.3 and 5.5). It would do so in the following way. In looking to see if there is discrimination or a disguised trade restriction, panels ought to identify what Josling, Roberts, and Orden refer to as a country's 'socially optimal level of protection'. The authors define this as a situation where the gains from trade (as conventionally measured) just balance the expected loss from the conse-

---

[7] Jonathon B. Wiener, 'Whose Precaution After All? A Comment on the Comparison and Evolution of Risk Regulatory Systems' (2003) 13:207 Duke J. Comp. & Int'l L. 207 at 223.

quences of the importation of pest, disease, or other risk factor.[8] Where the level of protection is socially optimal, there should be neither 'over-protection' (where a relaxation of the regulation would be socially beneficial) nor 'under-protection' (where more restrictive import protocols would increase expected net social benefits).[9] Identifying the socially optimal level of protection is a useful concept in sifting out disguised protectionism from genuine health protection; over-protection is a likely indicator of the presence of protectionism, while under-protection is a likely indicator that there would be a negative impact on health.[10]

Where the health bias would have an impact is that in identifying the socially optimal level of protection, panels should include consideration not only of quantifiable gains and losses, but also of those that cannot be measured, such as human rights and public values. In particular, they should take notice of a country's health objectives, along with the fact that the government is obliged to protect health, and the inherent value and importance of doing so. Thus, all other things being equal, a measure to protect against a low probability risk might not be found to constitute a disguised restriction on trade when the value of health protection has been taken into account.

In the second situation, that is, where governments are under no obligation to protect health, there is no normative argument to be made for the value of health that would justify adopting a bias towards health protection. Rather, the focus in enquiries concerning a country's determination of its appropriate level of protection should be solely on the factors outlined in Article 5.3, namely, the potential damage in terms of loss of production or sales in the event of the entry, establishment, or spread of a pest or disease; the costs of control or eradication in the territory of the importing Member; and the relative cost-effectiveness of alternative approaches to limiting risks.

It must be noted that the notion of a bias towards health objectives as discussed above presents a concern from a trade liberalization perspective because such an approach could conceivably be abused by a country wishing to protect its producers. The positive analysis of regulation discussed above demonstrates significant opportunities for capture of science and the regulatory process by vested interests. Therefore, it is important that panels adopting

---

8 Tim Josling, Donna Roberts and David Orden, *Food Regulation and Trade* (Washington DC: Institute for International Economics, 2004) at 30.

9 *Ibid.* at 29–31.

10 As the authors note, identifying the socially optimal level of protection is an empirically difficult task, even for domestic regulators. It will be even more difficult for WTO disputes panels which must rely on evidence submitted by the parties. However, the amber cases are difficult cases, and panels must have some form of principled guidance in navigating through the facts. Josling, Roberts and Orden, *ibid.*

such a bias are confident that the regulating country has taken appropriate steps to assess the risk in question. The importance of this is one of the key factors supporting the maintenance of the SPS Agreement's science-based framework.

The suggested approach is justified by the public interest imperatives for health protection regulation and human rights and international law approaches to health as identified in Chapter 5. International trade theory can arguably accommodate such a bias towards health objectives. The alternative perspectives of theorists such as Schumacher discussed in Chapter 5 emphasize the importance of maintaining the integrity of non-quantifiable values such as health and suggest that they may need to be pursued at the expense of production and income maximization if necessary. It is maintained here that this perspective is compatible with orthodox economic theory.

Finally, note should be made of cases that see conflict between developed and developing countries. Developed countries can afford to be far more concerned with less serious and more remote risks to health than can developing countries that must deal with basic health issues arising from factors such as food shortages, lack of clean water, and a high rate of tropical diseases. This can cause conflicts in a number of circumstances. Developing countries are often unable to comply with the regulatory requirements of developed countries or will have difficulty in doing so. Not only does this have potential to harm their export sales, but where a developing country is unable to export its crops or other products, the income loss for those who would have otherwise been able to sell their products will impact negatively on their health through an inability to buy goods such as food, medicine, and shelter.

This poses a dilemma. How do we balance the rights of citizens in countries at different levels of development to be protected from health risks that are pertinent to them? On the one hand, introducing a bias towards health objectives into the balancing exercise may result in a developed country claiming an entitlement to introduce stringent restrictions to protect health, even while the population of affected developing countries are denied the most basic necessities essential for health. Yet on the other hand, denying a rich country the right to protect the health of its population cuts against the notion protected by the SPS Agreement that no country should be prevented from protecting the health of its citizens and against human rights as relevant to that country's population.[11] It may also restrict democracy if public sentiment concerning risk to health is ignored.

If we accept that human health is a fundamental human right, it could be

---

[11] In this regard, some have warned against a 'race to the bottom' where low standards in developing countries trigger a weakening of regulatory standards across

argued – from the perspective of global health – that in cases where countries have conflicting health interests, effort ought to be made to reach a resolution which results in the greatest health gain. Yet this is an enormously difficult issue to deal with in practice. From a political perspective it would be inappropriate to require WTO panels to deal with issues of weighing lives across countries. The SPS Agreement has at least the potential (as yet unfulfilled) to provide some assistance in this regard. It recognizes the issues facing developing countries in Article 10 which provides, *inter alia*, that in preparing and applying SPS measures, Members shall take account of the special needs of developing country Members, and in particular of least-developed country Members. This Article, if fully implemented, would seem to provide some leeway in the situation described for a developing country to argue that an overly strict measure was not applied with its special needs in mind. While this is an issue of enormous importance, it will not be considered further here as it requires considerable empirical research as to how the objective of Article 10 can be most effectively promoted.

all countries. See Sungjoon Cho, 'Linkage of Free Trade and Social Regulation: Moving Beyond the Entropic Dilemma' (2005) 5 Chicago J. Int'l L. 625 at 644. Evidence suggests, however, that such fears are mostly unfounded. Michael J. Trebilcock, 'Critiquing the Critics of Economic Globalization' (Paper presented to the Law and Economics Workshop Series, Law and Economics Programme, Faculty of Law, University of Toronto, 2004). Citing Alan O. Sykes, 'International Trade and Human Rights: An Economic Perspective', University of Chicago, John M. Olin Law & Economics Working Paper No. 188 (2D series), pp. 15–18; Jagdish Bhagwati and Robert Hudec, eds, *Unfair Harmonization: Prerequisite for Free Trade?* Vols I & II (Cambridge, MA: MIT, 1997). See generally David Vogel, *Trading Up: Consumer and Environmental Regulation in a Global Economy* (Cambridge, MA: Harvard University Press, 1995).

PART III

Regulating to protect health: where and how?

8. Setting the standards: home or away?

8.1 INTRODUCTION

The Uruguay Round Agreements extended the reach of international trade law into areas that had previously been seen as the exclusive domain of domestic governments, including regulatory decision-making in matters of health protection. As discussed, the SPS Agreement asks countries to regulate in a way that reduces the incidence and potentially distortive effects of non-tariff trade barriers. Inevitably therefore, its application impinges to a certain extent on state sovereignty. Questions of interpretation of the SPS Agreement raise questions about the appropriate extent of this infringement.

In and of itself, the concept of sovereignty – being a reference to political control over a given territory – provides little guidance to interpretation of the SPS Agreement beyond the observation that health protection is at the heart of a nation's sovereign powers. We might accept that becoming part of the international trading regime necessitates a certain level of infringement, but how do we judge when that infringement becomes too great? It is in this regard that the principle of subsidiarity may prove helpful as a way of thinking about what is the most appropriate institutional setting in which to tackle the policy- and rule-making challenges arising from the growing commercial interdependence of nations.[1]

This chapter will introduce the principle of subsidiarity as a conceptual tool to help us think not only about where decisions about health risks are most appropriately made but also what standard of review might be appropriate when WTO panels and the Appellate Body are called upon to examine a country's compliance with the SPS Agreement. It is, as Carozza argues, a helpful conceptual alternative to the idea of state sovereignty.[2]

[1] Pierre Sauve and Americo Beviglia Zampetti, 'Subsidiarity Perspectives on the New Trade Agenda' (2000) 3:1 Journal of International Economic Law 83 at 84.

[2] Paolo G. Carozza, 'Subsidiarity as a Structural Principle of International Human Rights Law' (2003) 97 American Journal of International Law 38 at 40.

8.2 THE PRINCIPLE OF SUBSIDIARITY

The principle of subsidiarity holds that policy decisions should be made at the lowest level of government capable of effectively addressing the problem at hand.[3] It has a long intellectual history,[4] dating back as far as classical Greece and resurfacing among medieval scholars including Thomas Aquinas. Johannes Althusius developed the principle in connection with theories of the secular federal state in the seventeenth century, and it appeared in the writings of a variety of political actors and theorists such as Montesquieu, Locke, Tocqueville, Lincoln, and Proudhon. In the latter part of the nineteenth century, it was used by Catholic social theorists looking for a middle ground between the perceived excesses of both *laissez-faire* liberal capitalist society and Marxian socialist alternatives. In 1931, Pius XI endorsed the idea of subsidiarity, writing that it is a 'grave evil and disturbance of right order to assign to a greater and higher association what lesser and subordinate organizations can do . . . the more perfectly a graduated order is kept among the various associations, in observance of the principle of "subsidiary function," the stronger social authority and effectiveness will be [and] the happier and more prosperous the condition of the State'.[5]

Subsidiarity thus originated as a principle applicable in the domestic context, where, as Carozza explains, each level of society is responsible for helping the 'lower' one freely to accomplish its aims, and it would contravene that relationship if the 'higher' association arrogated to itself those tasks which can be effectively undertaken by a group that is closer to the individual.[6] This is known as 'negative subsidiarity'. 'Positive subsidiarity', on the other hand, is where the state is justified in intervening in situations where 'lower' forms of organization cannot achieve their desired ends by themselves.[7]

Today, the principle of subsidiarity is most commonly discussed in the context of regional politics in the EU. Here the debate is about how powers should be shared between Member States and the centralized Community institutions. It is enshrined in Article 5(2) of the EC Treaty which provides that

---

3 Sauve and Zampetti, *supra* note 1 at 85.

4 The brief historical outline here is taken from Carozza, *supra* note 2. For a comprehensive history of the concept, see Chantal Millon-Delsol, 'L'Etat Subsidaire: Ingérence et non-ingérence de l'etat: Le Principle de subsidarité aux fondements de l'histoire européenne' (1992) 15–27. Also, Ken Endo, 'The Principle of Subsidiarity: From Johannes Althusius to Jacques Delors' (1994) 44 Hokkaido Law Review 2064.

5 Pius XI, *Quadragesimo Anno: Encyclical Letter on Reconstruction of Social Order* (15 May 1931), in 3 *The Papal Encyclicals 1903–1939*. Cited in Carozza, *supra* note 2 at 41.

6 *Ibid.* at 44.

7 *Ibid.*

'the Community shall take action, in accordance with the principle of subsidiarity, only if and in so far as the objectives of the proposed action cannot be sufficiently achieved by the Member States and can therefore, by reason of the scale or effects of the proposed action, be better achieved by the Community'. It thus contains elements of both negative subsidiarity – discouraging Community-level action where objectives can better be achieved at the state level, but encouraging such action where it would be more effective.

Subsidiarity takes on a different flavour when we move to the broader international context.[8] Different considerations apply, for example, in the international economic arena where the WTO is an inter-governmental organization whose main actors are sovereign states.[9] However, the fundamental concept remains constant, that is, whether in a given case it is both more desirable and/or effective to conduct activity at the multilateral level. In particular, is it desirable and/or effective to shift responsibility for governance of health regulation (through standard-setting) to the multilateral level?[10] The next section takes up this task in a consideration of the merits or otherwise of greater international harmonization of health standards. The subsidiarity principle should be seen as a foundation for this discussion; it is, as Carozza argues, a 'conceptual and rhetorical mediator between supranational harmonization and unity, on the one hand, and local pluralism and difference, on the other'.[11]

8.3 REGULATORY HARMONIZATION

8.3.1 What is Regulatory Harmonization?

Consideration of the appropriate level for regulatory decision-making regarding health leads us to a discussion of the merits of regulatory harmonization. Chapter 6 identified a number of situations (the so-called 'amber cases') where trade and health objectives conflict. Each of these situations involves, to a greater or lesser degree, regulatory divergences between countries in their response to risk situations. Clearly, such divergence occurs when decisions are made at the national level. This section questions whether one way in which such conflict might be diffused and regulatory differences reconciled, is by negotiating an end to regulatory differences through international harmonization. It briefly outlines arguments for and against harmonization, and examines provision for harmonization under the SPS Agreement.

8 Sauve and Zampetti, *supra* note 1 at 86.
9 *Ibid.* at 87.
10 *Ibid.* at 93.
11 Carozza, *supra* note 2 at 40.

The term regulatory harmonization has been given various interpretations, all of which contrast with the notion of deference to national sovereignty. A narrow definition, as suggested by Sykes, says that regulatory harmonization involves countries negotiating agreements to follow the same substantive regulations.[12] A broader definition is offered by Leebron who sees it as 'making the regulatory requirements or governmental policies of different jurisdictions identical, or at least more similar'.[13] Essentially, it is about reducing the differences in the laws and policies of two jurisdictions.[14]

Leebron identifies four broad types of harmonization: (i) rules that regulate the outcome, characteristics, or performance of, *inter alia*, economic goods; (ii) governmental policy objectives such as maintaining a high standard of public health; (iii) principles intended to influence or constrain the factors that are taken into account in making policies or rules (for example, the 'polluter pays' principle adopted by the EC addresses the issue of who should bear the cost of pollution); and (iv) harmonization of institutional structures and procedures such as provisions in international agreements that require certain procedures for enforcement of domestic laws.[15]

A broad view of regulatory harmonization as described by Leebron is adopted here. It may not be necessary for countries to follow the same substantive regulations in order to assist in diffusing conflict between trade and health; lesser forms of harmonization also have potential to achieve such a goal.

8.3.2 Arguments for and against Regulatory Harmonization

A key factor in many trade disputes concerning health regulations is regulatory divergence between countries, or what has sometimes been termed 'regulatory regionalism'.[16] As noted, a policy of regulatory harmonization implies the goal of reducing regulatory divergences. Bhagwati and Hudec divide arguments for regulatory harmonization into those of a philosophical nature and those of an economic nature.[17] On the philosophical side, harmonization plays

---

[12] Alan O. Sykes, 'The (Limited) Role of Regulatory Harmonization in International Goods and Services Markets' (1999) 2 J. Int'l Econ. L. 49 at 50.

[13] David W. Leebron, 'Lying Down with Procrustes: An Analysis of Harmonization Claims' in Jagdish Bhagwati and Robert E. Hudec, eds, *Fair Trade and Harmonization: Prerequisites for Free Trade? Vol. 1: Economic Analysis* (Cambridge, MA: MIT Press, 1996) at 43.

[14] *Ibid.*

[15] *Ibid.* at 44.

[16] See Chapter 4, section 4.5.

[17] Bhagwati and Hudec, *supra* note 13.

a role in achieving a threshold of fairness in international trade.[18] Regulatory diversity is seen as undesirable because those producers whose countries have lesser regulatory burdens have an unfair advantage in international trade. Bhagwati and Hudec argue for example that the use of fairness is a central feature of US demands for harmonization, noting that 'since the United States is a major player in trade negotiations whether multilateral or bilateral, it follows that the American demands also tend to seek to remake the world in its own image'.[19] The notion of fairness is of particular relevance in the case of health regulations where exporters and governments may see harmonization as critical to decreasing barriers to trade. However, there are compelling arguments to be sceptical of its merits. Trebilcock and Howse address the argument in the context of environmental standards where it is sometimes claimed that a country can gain an advantage in trade from lower environmental standards.[20] Their arguments are equally relevant to the health context. First, they argue that such claims implicitly assume that the importing state's environmental standards are optimal from both a domestic and a global perspective. This assumption, they suggest, merely reflects the bias of the importing country towards its own regulatory approach. Second, higher environmental standards may actually confer a competitive advantage in some sectors, where they may create incentives for the development of environmental technologies that can then be exported to other countries as their demand for environmental protection increases. Third, environmental standards and costs must be distinguished as even if standards were harmonized, different countries would face different costs in meeting them. As such, they find that the argument for harmonization is incoherent. They conclude that for unfairness to have a normatively defensible meaning, it must entail the violation of some neutral, objective baseline for the balance of benefits and burdens that governments create for industries. These arguments are particularly compelling when we look at them through the lens of the subsidiarity principle. They form a case for keeping regulatory decision-making powers at the national level due to the validity of different approaches to regulation, the potential competitive advantage for individual

---

18 *Ibid.* at 16. Other philosophical arguments referred to by Bhagwati and Hudec are first, the notion of obligations beyond borders (largely of relevance to environmental and labour standards) where people feel that they owe obligations to others or to humanity itself; and second, the objective of distributive justice where harmonization is seen as necessary to ensure that unskilled labour in developed countries is not adversely affected by liberalized trade with developing countries which maintain lower standards.

19 *Ibid.* at 18.

20 Michael J. Trebilcock and Robert Howse, *The Regulation of International Trade*, 3rd ed. (London: Routledge, 2005) at 508.

countries from adopting unique standards, and the fact that different countries face different costs in meeting standards.

On the economic side, Bhagwati and Hudec argue first that structural changes in the world economy, including economic globalization, support a move towards regulatory harmonization. They write that globalization has diminished the comparative advantage held by many industries, as technological know-how has converged among industrialized countries. As a result, the comparative advantage held by many industries is fragile and can move across countries almost randomly. Those investing in and managing these industries tend to be sensitive to any possibility of an 'unfair' advantage gained by their foreign rivals and demand harmonization of those elements of domestic policies and institutions that might give their rivals an advantage.[21]

Second, Bhagwati and Hudec suggest that harmonization might be justified as a means of dealing with externalities where rules adopted by one country result in costs for other countries, such as where pollution crosses national borders. Weak SPS regulations in one country might expose citizens of another country to the risk of disease and in the case of animal and plant diseases, might put economic interests at jeopardy. For example, weak regulations controlling the spread of pathogenic avian influenza in Country A will impact citizens of Country B if birds from Country A migrate and cause illness among birds and/or humans in Country B. A related argument is that made by Leebron who talks about 'political economies of scale', and suggests that because there is an optimal set of standards for all jurisdictions, resources can be better utilized and coordinated by shifting regulatory decision-making to a centralized forum.[22] From a subsidiarity perspective, the argument here is that one country cannot achieve the goal of health protection from either a domestic or global perspective if it maintains responsibility for standard-setting at the national level, hence a shift to a multilateral forum would be both desirable and more effective.

A third economic argument is based on economies of scale. Leebron argues that if a manufacturer faces significantly different requirements in each jurisdiction for which it manufactures, it will not be able to achieve economies of scale beyond its market share in one jurisdiction.[23] This is largely because legal information costs arise for each jurisdiction and must be recovered through sales. Harmonization is able to assist in this regard by reducing information costs, enabling market entry even for relatively small sales, and

---

21 Bhagwati and Hudec, *supra* note 13 at 22.

22 Leebron, *supra* note 13 at 63. Leebron suggests that this is not really a claim for harmonization so much as it is a claim for a certain process when it is thought that separate processes would, or should, produce similar standards or results.

23 *Ibid.* at 62.

thereby justifying a shift of responsibility from the domestic to the international level.

While these arguments for regulatory harmonization arguably all have some merit and at first glance suggest that moving regulatory decision-making to a multilateral level might be both desirable and effective, there are also compelling reasons to take a conservative view of harmonization of health regulations. Trebilcock and Howse take such a view, arguing that the focus of attention on domestic policies that might constitute non-tariff barriers to trade should principally be related to two objectives: (i) elaborating on the principles of negative integration that have historically characterized the GATT approach, that is, national treatment combined with the Article XX exceptions; and (ii) structuring ground rules pursuant to which mutually beneficial agreements between Member States can be reached over policy harmonization or convergence that are non-coercive and non-discriminatory *vis-à-vis* other trading partners.[24]

Trebilcock and Howse base their argument on four key premises.[25] The first is related to Breton's notion of 'competitive governments' where government actions cannot always be explained in public choice terms and are instead seen as being intensely competitive in a number of areas including policy priorities and design.[26] Breton argues that harmonization reduces competition between governments, and that in a system of extreme harmonization (he uses the EU as an example), supranational institutions exhibit a 'democratic deficit' and are liable to monopolize or cartelize government policy-making. As such, he finds them to be highly imperfect mechanisms for preference revelation. On this basis, Trebilcock and Howse argue that to the extent comparative advantages are functions of government policies, regulatory heterogeneity can be considered one of the conditions for the existence of comparative advantage. Exploiting differences in government policies is no less legitimate than exploiting differences in natural endowments.[27] From a subsidiarity perspective, it is therefore not desirable to shift responsibility for regulations away from the national level.

---

[24] Michael Trebilcock and Robert Howse, 'A Cautious View of International Harmonization: Implications from Breton's Theory of Competitive Governments' in Gianluigi Galeotti, Pierre Salmon and Ronald Wintrobe, eds, *Competition and Structure* (Cambridge, UK: Cambridge University Press, 2000) 386 at 392. The SPS Agreement, with its 'post-discriminatory' approach, arguably already goes beyond the principles of non-discrimination with its requirement that countries base their SPS measures on scientific evidence.

[25] Trebilcock and House, *ibid.* at 388.

[26] Albert Breton, *Competitive Governments* (Cambridge, UK: Cambridge University Press, 1996).

[27] Michael Trebilcock and Robert Howse, 'Trade Liberalization and Regulatory Diversity: Reconciling Competitive Markets with Competitive Politics' (1998) 6 Eur. J. L. & Econ. 5 at 14.

Second, Trebilcock and Howse argue that international trade rules should minimize the extent to which harmonization can be induced by either judicial fiat or threats of unilateral sanctions and instead encourage mutually beneficial agreements on policy convergence. Third, they argue that from a practical viewpoint, there are significant barriers to international harmonization. Noting the emphasis that tends to be placed on the European experience by advocates of greater regulatory harmonization, they argue that the same experience is simply not replicable in most other institutional contexts.[28] They argue that deep integration on a wider scale will be difficult to achieve because it would require either the existence of a hegemonic power with ability to impress its will on other smaller and weaker states, or willingness among members to cede substantial aspects of their domestic political sovereignty to supranational political institutions which beyond the particular circumstances of the EU is unlikely. Clearly, to the extent that this scenario is likely to eventuate,[29] harmonization would not be desirable for those countries; it runs directly counter to the principle of subsidiarity. Also from a practical perspective, problems with harmonization suggest that it would not be overly effective. For example, there is strong potential for ambiguities as rules require interpretation and application, as well as incoherence and vagueness as a result of the necessity for compromise in reaching the harmonized standard in the first instance.[30] Another problem is that of distortion. For example, if harmonization is required for bacteria levels and other contaminants in poultry products but not meat or fish, this may result in an increase in the price of poultry. As a result, consumption may be skewed away from chicken, possibly resulting in increased rather than decreased risk levels.[31]

Fourth, Trebilcock and Howse argue that the analysis of regulatory barriers to trade is often complicated by both theoretical and empirical uncertainty about their effects on social welfare. This is due to the fact that regulations are designed to serve a wide array of values and concerns. For example, SPS measures that service protectionist interests may also serve public health objectives and thus be welfare-enhancing from a domestic perspective. This contrasts with traditional impediments to trade such as tariffs and quotas

---

[28] Trebilcock and Howse, *supra* note 24 at 391.

[29] Some would argue that it has already occurred to a certain degree in the context of Codex which has been criticized for a lack of inclusion of developing countries and for an over-representation of industry interests. See Natalie Avery, Martine Drake and Tim Lang, 'Codex Alimentarius: Who Is Allowed In? Who Is Left Out?' (1993) 23:3 The Ecologist 110; and Natalie Avery, *Cracking the Codex: An Analysis of Who Sets World Standards* (London: National Food Alliance, 1993).

[30] Leebron, *supra* note at 13 at 65.

[31] *Ibid.* at 49.

which can be shown theoretically and empirically to be welfare-reducing from both a global and domestic perspective.

As noted, national policies for managing health risks diverge widely, even when they are supposedly based on the same scientific research.[32] A key factor in tempering the arguments for regulatory harmonization is that many of these differences between nations have inherent value, including those that result from factors such as cultural, religious, ethical, and moral values. Harmonization can only be achieved at the cost of eliminating or reducing these differences. If differences have inherent value, then that value must be taken into account in determining whether it is appropriate to take responsibility away from states in order to pursue harmonization.

Leebron argues that differences between national regulatory policies can be regarded as either substantively legitimate or procedurally legitimate. They are *substantively* legitimate if the differences in policy are justified by differences in the substantive concerns and values that inform policy, for example, the rejection by Islamic nations of non-Halal meat.[33] They are *procedurally* legitimate if we regard the process by which they were adopted as establishing their legitimacy, whether or not the difference could be justified by reference to differing values.[34] For example, a nation may defend its policy choice on the basis of a procedurally sound risk assessment carried out in accordance with the requirements of the SPS Agreement. While a claim for harmonization is not defeated by the mere fact that a difference is 'legitimate', Leebron suggests that this fact alerts us to some competing considerations and invites a more critical analysis of the claim.[35] It is argued here that where regulatory divergence results from differing public sentiment, it is critical to consider the potential welfare losses that would be incurred in the event of harmonization if citizens feel that something important to their value system has been surrendered.

A claim for harmonization will arguably be strongest where regulatory differences are the result of less 'legitimate' factors, such as protectionist capture, bureaucratic indifference, and information failures. In these cases, regulatory differences are the source, rather than a reflection, of comparative advantage. In this sense the nation's comparative advantage might be seen as

---

[32] Sheila Jasanoff, 'Technological Risk and Cultures of Rationality' in National Research Council, ed., *Incorporating Science, Economics, and Sociology in Developing Sanitary and Phytosanitary Standards in International Trade: Proceedings of a Conference* (Washington DC: National Academy Press, 2000) 65 at 66.

[33] Leebron suggests looking at differences in endowments, technologies, preferences, institutions, and coalition formation. *Supra* note 13 at 68.

[34] *Ibid.* at 67.

[35] *Ibid.* at 75.

artificial and unfair, a distortion created by illegitimate governmental policy.[36] Sykes argues that in an ideal world there would be no regulatory heterogeneity driven by such illegitimate factors.[37] However, in reality, these factors must be taken into account. He also argues that there may be cases where even legitimate regulatory heterogeneity is not worth its cost because its benefits are small and its trade effects large.[38]

Even in cases where regulatory diversity is arguably illegitimate, Sykes argues that harmonization is not required other than in exceptional cases, for example, matters of technical compatibility or cases where the substance is so harmful to the environment that it is optimal to ban it everywhere, suggesting chlorofluorocarbons as an example.[39] Instead, he suggests that regulatory diversity can be respected through principles of what he calls 'policed decentralization'.[40] Under a system of 'policed decentralization', national authorities are largely free to pursue their own policy objectives but must do so subject to legal constraints such as the non-discrimination principle, the 'sham' principle, transparency requirements, and benefit/burden balancing tests. Such rules may be accompanied by mutual recognition arrangements whereby the importing country recognizes the equivalence of the regulations in the exporting country so that exporters need only comply with the regulations in their home country. Sykes considers that if policed decentralization and mutual recognition are gainfully employed, there ought to be very little role for regulatory harmonization. This is because the principles of policed decentralization would have ensured that regulatory differences did not arise for illegitimate reasons.[41] As regards legitimate differences in national regulatory policy, he suggests that harmonization would seem inferior to mutual recognition.[42]

Sykes also considers that a 'policed decentralization' system may properly include deference to negotiated international standards with specified procedures for deviation. This is the approach taken in the SPS Agreement where

---

[36] *Ibid.* at 77.

[37] Sykes, *supra* note 12 at 52.

[38] *Ibid.*

[39] Alan O. Sykes, 'Regulatory Competition or Regulatory Harmonization? A Silly Question?' (2000) 3:2 J. Int'l Econ. L. 257 at 263.

[40] *Ibid.*

[41] Sykes, *supra* note 12 at 68. See also Roessler, who argues that the range of policies for which worldwide harmonization is desirable and attainable is likely to be very small. Frieder Roessler, 'The Distinction between Trade and Domestic Policies under the GATT' in Frieder Roessler, ed., *The Legal Structure, Functions and Limits of the World Trade Order, A Collection of Essays* (London: Cameron May, 2000) 119 at 142.

[42] Sykes, *ibid.* at 69.

deference is paid to international standards developed by the Codex, the OIE, and IPPC. An obligation on countries to employ international standards when they suffice to achieve the domestic objective will, in his opinion, avoid the unnecessary costs of regulatory differences. It must be remembered, however, that Sykes places a narrow meaning on the term 'regulatory harmonization'. It is arguable, therefore (and indeed, will be argued throughout the remainder of this book) that the approach of deferring to international standards with provision for deviation is a form of harmonization.

In summary, while there are arguments to both support and militate against claims for regulatory harmonization of SPS measures, looking at them through the lens of the subsidiarity principle leads to the conclusion that there are only very limited cases where it would be both desirable and effective to move to a full-blown system of regulatory harmonization. It is arguable that harmonization might help diffuse trade friction over trade-restrictive health measures by shifting the forum for standard-setting to a higher (that is, international) and more conciliatory environment, that is, from a contentious dispute settlement proceedings to negotiations to agree on common standards before a dispute arises. However, it also seems that this would only be true in a narrow range of cases where legitimate regulatory differences do not exist and multilateral institutions could therefore do a better job than domestic governments. Meanwhile, the system of 'policed decentralization' is sufficient to enable the SPS Agreement to play a role in correcting or constraining domestic policy choices that result from capture by domestic industries or private interests.

8.3.3 Harmonization under the SPS Agreement

The SPS Agreement was negotiated on the understanding that WTO dispute settlement panels are not the best venue to resolve complex scientific issues and it therefore encourages international cooperation in the design of SPS measures. In furtherance of this objective, the Agreement' s Preamble refers to the Members' desire to 'further the use of harmonized sanitary and phytosanitary measures between Members, on the basis of international standards, guidelines and recommendations developed by the relevant international organizations, including the Codex Alimentarius Commission (Codex), the International Office of Epizootics (OIE), and the relevant international and regional organizations operating within the framework of the International Plant Protection Convention (IPPC), without requiring Members to change their appropriate level of protection of human, animal or plant life or health'.

In Annex A(2), the SPS Agreement defines harmonization as 'the establishment, recognition and application of common sanitary and phytosanitary measures by different Members'.

Article 3.1 of the Agreement encourages Members to 'base' their SPS measures on international standards, guidelines and recommendations, where they exist. The Appellate Body has interpreted the term 'based on' as meaning something less than 'conform to'.[43] The consequence of this interpretation is that harmonization under the SPS Agreement is not complete harmonization, but rather there is room for deviation from the relevant standard. It is, in Sykes's words, a system of 'policed decentralization'. However, the extent of room for deviation has not been clarified. As the Appellate Body noted in the *EC – Hormones* case, the Preamble, among other articles, makes it clear that harmonization of SPS measures on the basis of international standards is a goal, to be realized in the future.[44] However, pursuant to Article 3.2, where countries are willing to ensure that their SPS measures do 'conform' to international standards, guidelines, or recommendations, then they will be deemed to be necessary to protect human, animal, or plant life or health, and presumed to be consistent with the relevant provisions of the Agreement, as well as of GATT 1994.

The term 'international standards, guidelines or recommendations' is defined in Annex A as being those established by, for food safety, the Codex; for animal health, the OIE; and for plant health, the IPPC. These organizations have the common objective of developing standards, guidelines, and other recommendations through a consensus-based approach. Regarding the setting of such standards, guidelines, or recommendations, Article 3.4 requires Members to participate 'within the limits of their resources' in the relevant international bodies.

Where Members do not wish to base their measures on international standards, Article 3.3 gives them the alternative option of setting a level of protection that is higher than that implicit in the international standard. This higher level of protection must be justified by scientific evidence and based on a risk assessment as required by Article 5.1. Codex, OIE, and IPPC have a role with regard to the procedure followed by countries when carrying out risk assessments. Article 5.1 specifically states that Members must take into consideration the risk assessment procedures developed by the 'relevant international organizations', which implicitly includes Codex, IPPC, and OIE.

The SPS Committee plays a role in harmonization efforts, being mandated by Article 12.2 to encourage Members to base their measures upon international standards, guidelines, or recommendations. Article 12.3 requires the Committee to maintain close contact with relevant international organizations in the field of SPS protection in order to secure the 'best available scientific evidence and technical advice for the administration of this Agreement and in

---

43 *European Communities – Measures Affecting Meat and Meat Products* (1998), WTO Doc. WT/DS26/AB/R, WT/DS48/AB/R (Appellate Body Report). [*EC/Hormones* AB Report]

44 *EC – Hormones* AB Report, *ibid.* at para 165.

order to ensure that unnecessary duplication of effort is avoided'. To this end, its meetings are attended by representatives of the Codex, OIE, and IPPC, as well as accredited bodies such as WHO.

Article 12.4 requires the SPS Committee to develop a procedure to monitor the process of international harmonization and the use of international standards, guidelines, or recommendations. In October 1997, it adopted a provisional monitoring procedure to achieve this end.[45] The procedure required Members to submit, 30 days in advance of SPS Committee meetings, concrete examples of what they consider to be problems with a significant trade impact that they believe are related to the use or non-use of relevant international standards, guidelines, or recommendations. Members were asked to describe the nature of each of these trade problems and state whether they were the result of the non-use of an appropriate existing international standard, guideline, or recommendation; or the non-existence or inappropriateness of an existing international standard, guideline, or recommendation, that is, it is outdated, technically flawed, etc.[46] In doing so, the procedure seeks not only to identify where there has been a major trade impact, but also to help identify, for the benefit of the relevant international organizations, where a new standard, guideline, or recommendation is needed or where a current one is not appropriate for its purpose and use.

Prior to July 1999, the SPS Committee considered nine standards-related issues raised by Members.[47] However, after July 1999, Members have used the procedure even less frequently. At several meetings, the Chairman has noted that by not identifying new concerns, Members are missing an opportunity to inform the standard-setting bodies of their needs.[48] In 2004, the Committee amended the procedure to decrease the time for submission of examples to 10 days prior to Committee meetings. However, the procedure has remained underused.[49]

---

[45] WTO Committee on Sanitary and Phytosanitary Measures, Summary of the Meeting held on 15–16 October 1997, WTO Doc. G/SPS/R/9/Rev.1 at para 21.

[46] WTO Committee on Sanitary and Phytosanitary Measures, Procedure to Monitor the Process of International Harmonization (1997), WTO Doc. G/SPS/11.

[47] WTO Committee on Sanitary and Phytosanitary Measures, Note by the Secretariat, Revised Draft Annual Report, Procedure to Monitor the Process of International Harmonization (1999), WTO Doc. G/SPS/W/94/Rev.2.

[48] See for example, WTO Committee on Sanitary and Phytosanitary Measures, Summary of the Meeting held on 21–2 June 2000, WTO Doc. G/SPS/R/19 at para 70. WTO Committee on Sanitary and Phytosanitary Measures, Summary of the Meeting held on 8–9 November 2000, WTO Doc. G/SPS/R/20 at para 68. WTO Committee on Sanitary and Phytosanitary Measures, Summary of the Meeting Held on 19–21 March 2002, WTO Doc. G/SPS/R/26 at para 119.

[49] The Chairman has suggested that Members should make better use of the

The SPS Agreement arguably encompasses each of the four types of harmonization identified by Leebron. His first category is typified by the establishment of international standards that regulate the characteristics of food and other products which may pose a risk to human, animal, or plant health as a result of pests, diseases, disease-carrying (or -causing) organisms, contaminants, toxins, or additives. For example, pursuant to the Codex Alimentarius, avocados must be, *inter alia*, clean, practically free of any visible foreign matter, and must not be affected by rotting, or deterioration such as to make them unfit for consumption.[50]

The second category – government policy objectives – may be seen in the objectives of the Codex, OIE, and IPPC. Through these organizations, member governments have agreed to the overarching policy objectives of protecting health and facilitating trade. For example, the OIE includes among its objectives: to ensure transparency in the global animal disease and zoonosis situation, to safeguard world trade, and to provide a better guarantee of the safety of food of animal origin.

Leebron's third category – principles that influence the factors taken into account in making rules or policies – is used in the SPS Agreement which aims to constrain the factors taken into account in making SPS measures by requiring (in the absence of adherence to international standards) that measures be based on scientific evidence. Within the international standard-setting organizations themselves, this type of harmonization is also evident. For example, prior to the implementation of the SPS Agreement, the US had sought international recognition of the principle that food standards be based only on scientific evidence.[51]

The fourth type of harmonization identified by Leebron involves harmonization of institutional structures and procedures. In this regard, the SPS Agreement requires Member States to comply with certain procedural guidelines in the operation of control, inspection and approval procedures, including national systems for approving the use of additives or for establishing tolerances for contaminants in food.[52]

---

monitoring procedure. See for example WTO Committee on Sanitary and Phytosanitary Measures, Summary of the Meeting held on 9–10 March 2005, WTO Doc. G/SPS/R/36 at 168.

[50] Codex Standard for Avocados, Codex Stan. 197-1995, AMD. 1-2005.

[51] Leebron notes that this proposal was defeated in the Codex by a 28 to 13 vote (with nine abstentions). Leebron, *supra* note 13 at 44.

[52] Article 8.

8.4 MUTUAL RECOGNITION

The SPS Agreement makes provision for mutual recognition of standards in Article 4(1) which states that 'Members *shall* accept the sanitary and phytosanitary measures of other Members, even if these measures differ from their own or from other Members trading in the same product, if the exporting Member objectively demonstrates to the importing Member that its measures achieve the importing Member's appropriate level of sanitary or phytosanitary protection'.

Essentially, Article 4 provides for the negotiation of a formal agreement to recognize, on a case-by-case basis, the SPS measures or testing and conformity requirements of a trading partner as equivalent to that existing domestically. Article 4(2) seems to envisage a series of preferential agreements, eliminating trade barriers on a bilateral rather than on a most-favoured-nation basis.[53] Members are encouraged to provide information on their experiences with equivalence to the SPS Committee. However, to date, only two Members have done so. In June 2005, Brazil reported that since 1996, Brazil, Argentina, Uruguay, and Chile had established a committee on hygiene and health of fisheries products. The committee's role had included establishing a single health certificate for fisheries products that were traded amongst the four countries.[54] In March 2006, Egypt reported that it had contacted some of its trading partners to propose the establishment of quarantine offices to test and inspect products in a more timely and cost-efficient manner.[55] Clearly, not all agreements are notified to the Committee. For example, the EC and Switzerland have signed an agreement which addresses equivalence of legislation with regard to various agricultural products including seeds, animal feed, and plants.[56]

Swinbank suggests that practice to date has shown that developing countries may not have the technical capacity to demonstrate that their SPS measures meet the importing Member's standards. He thus notes the suspicion that Article 4 has more validity between equal (developed) countries than for the developing world.[57]

---

[53] Alan Swinbank, 'The Role of the WTO and the International Agencies in SPS Standard Setting' (1999) 15:3 Agribusiness 323 at 330.

[54] WTO Committee on Sanitary and Phytosanitary Measures, Summary of the Meeting Held on 29–30 June 2005, WTO Doc. G/SPS/R/37 at para 94.

[55] WTO Committee on Sanitary and Phytosanitary Measures, Summary of the Meeting Held on 29–30 March 2006, WTO Doc. G/SPS/R/40, at para 70.

[56] Agreement between the European Community and the Swiss Confederation on trade in agricultural products, Official Journal of the European Communities, L114/132, 30.4.2002.

[57] Swinbank, *supra* note 53 at 330.

8.5 CONCLUSION

This chapter has examined the case for shifting the focus of standard-setting in relation to matters of sanitary and phytosanitary health from the domestic level to the multilateral level. Looking at the arguments raised through the lens of the principle of subsidiarity (which provides that action be taken at lowest level of government capable of achieving the desired objective), it can be concluded that in the area of health protection, there is only a limited case to be made for international regulatory harmonization and a correspondingly strong case for an emphasis on local decision-making. The benefits of regulatory heterogeneity – recognition of domestic preferences – not only support the 'policed decentralization' approach taken by the SPS Agreement but also militate against further efforts to increase harmonization of standards, and support arguments for significant deference to domestic regulatory decision-making agencies in the exercise of review by WTO panels and the Appellate Body (a matter which will be examined in Chapter 13).

9. Perception of risks: the role of public perceptions

9.1 INTRODUCTION

Regulations to protect health reflect a government's regulatory response to risk and it is this response which WTO panels and the Appellate Body are called upon to judge under the SPS Agreement's science-based framework. An examination of domestic regulatory decision-making in response to risk will allow a closer assessment of the appropriateness of the science-based framework and the charge made by some that it interferes unduly with countries' regulatory sovereignty.

It has been observed that in many cases, trade disputes over health regulations are attributable to the existence of regulatory divergence between countries. A number of factors account for such divergence, including how risks are perceived in different societies. This chapter draws on literature in the growing field of risk perception to examine attitudes to risk in Western society, as well as how it is perceived by lay people as compared to experts, and how differences between public and expert perceptions may be dealt with by regulators.

9.2 RISK IN WESTERN SOCIETY

The rise in Western society of health-related regulation over the past 30 years is primarily attributed to what Sunstein coins '1970s environmentalism'.[1] This movement was motivated in large part by the 1962 publication of Rachel Carson's classic book *Silent Spring*, which documented, among other things, the risks associated with pesticides and insecticides. A key aspect of the book was the notion that a new technology which seems harmless and beneficial might have serious long-term effects on the environment, wildlife, and human health.

1 Cass R. Sunstein, *Risk and Reason: Safety, Law, and the Environment* (Cambridge, UK: Cambridge University Press, 2002) at 11.

In the wake of the 1970s environmentalism movement, a number of commentators have identified a growing culture of risk aversion.[2] In particular, public concern has manifested itself in response to new and innovative man-made products and industrial practices including modern agricultural practices. Beck writes that we are living in a 'risk society' in which the risks of scientific and technological innovations to individuals, society, and the environment are perceived to have escaped the control of society, its institutions, and national boundaries.[3] From a risk-society perspective, the industrial and technological developments that define modernity are no longer equated with progress, as used to be the case. Instead, science, resulting technology, industry, and even government, are seen as the main 'producers' of risk and the 'bads' that we perceive to be threatening.[4] Beck argues that this risk society questions the continuing appropriateness of traditional concepts of authority based on hierarchy and neutral expertise. The science that created the new dangers, he suggests, cannot be relied upon to protect us from those dangers.

Also writing about the phenomenon of risk in Western society, Lowrance argues that there are many hazards that while we know enough, scientifically, to 'worry', we do not have enough knowledge about those hazards to know how much to worry. For example scientists can detect tiny traces of pesticides in food, but in many cases cannot determine whether the chemicals will have any adverse health effect. Lowrance finds that 'in our knowing so much more, though imperfectly, and aspiring to so much more, we have passed beyond the sheltering blissfulness of ignorance and risk-enduring resignation. This has given rise to considerable social apprehensiveness, which is affecting both the outlook of individuals and the functioning of institutions.'[5] The problem identified by Lowrance is central to the 'amber' cases where science is uncertain and/or public fears demand regulation leading to a regulatory response by one country that leads to trade-restrictive measures and in turn to trade friction.

Neal puts a slightly different spin on Beck's position, suggesting that we live in a 'non-risk society'.[6] He finds that people are obsessed with, and intolerant of, any risks, great or small, and that the culture of risk aversion in Western societies is so strong and pervasive that heavy-handed and/or irra-

---

[2] See generally Dick Taverne, *The March of Unreason: Science, Democracy, and the New Fundamentalism* (Oxford: Oxford University Press, 2005).

[3] Ulrich Beck, *World Risk Society* (Cambridge, UK: Polity Press, 1999).

[4] Ulrich Beck, *Risk Society* (London: Sage, 1992).

[5] William W. Lowrance, *Modern Science and Human Values* (New York: Oxford University Press, 1986) at 20.

[6] Mark Neal, 'Risk Aversion: The Rise of an Ideology' in Laura Jones, ed., *Safe Enough? Managing Risk and Regulation* (Vancouver: The Fraser Institute, 2000) 13 at 14.

tional responses to risk are not seen as a cause for concern.[7] He speaks of a 'death-denying, anti-risk culture' where questions of who to blame arise whenever unexpected death, illness, or injury occurs.[8] In this culture, 'blame for commission' is directed at those who manufactured or caused the harmful process or product, while 'blame for omission' is directed at regulators who have failed to regulate to prevent occurrence of the event in question.[9] The spectre of the second type of blame arguably provides an incentive for regulators to 'play it safe'.

The prevailing risk mentality in Western nations is difficult to understand because it is at odds with empirical evidence that shows people are living longer than at any time in history. The following quote from Douglas and Wildavsky is a perfect summation of the situation and raises questions that will be considered in the following section on risk perception:[10]

> How extraordinary! The richest, longest lived, best protected, most resourceful civilization, with the highest degree of insight into its own technology, is on its way to becoming the most frightened.
>
> Is it our environment or ourselves that have changed? Would people like us have had this sort of concern in the past? . . . Today there are risks from numerous small dams far exceeding those from nuclear reactors. Why is the one feared and not the other? Is it just that we are used to the old or are some of us looking differently at essentially the same sorts of experience?

9.3 PUBLIC VS. EXPERT RISK PERCEPTIONS

The trend noted in Chapter 4 to introduce public participation as a formal aspect of the regulatory process increases potential for public perceptions to influence decision-making.[11] Concern has been expressed that the public tends to overestimate the probability of some risks and underestimate others.[12] For example, on the evidence currently available, cigarette smoking poses a

---

7 *Ibid.* at 18.

8 *Ibid.*

9 *Ibid.* at 21.

10 M. Douglas and A. Wildavsky, *Risk and Culture* (London: University of California Press, 1982). Cited in P. Slovic, 'Perception of Risk' (1987) 236:4799 Science 280 at 280.

11 Sheila Jasanoff, 'Technological Risk and Cultures of Rationality' in National Research Council, ed., *Incorporating Science, Economics, and Sociology in Developing Sanitary and Phytosanitary Standards in International Trade: Proceedings of a Conference* (Washington DC: National Academy Press, 2000) 65 at 69.

12 Jeremy D. Fraiberg and Michael J. Trebilcock, 'Risk Regulation: Technocratic and Democratic Tools for Regulatory Reform' (1998) 43 McGill L.J. 835 at 841.

greater scientific risk to health than eating hormone-treated meat, yet in many European countries, the first is tolerated and the second is banned.[13] This lack of correlation with science-based risk assessments has sometimes led decision-makers and scholars to conclude that consumers are ignorant, sometimes hysterical, and misinterpret statistical data.[14]

Yet the picture is not black and white; many factors play a role in public risk perceptions. Researchers across a number of disciplines have sought to understand what influences people's risk perceptions, arguing that those who promote and regulate health and safety need to understand the ways in which people think about and respond to risk.[15] Such an understanding is also important in the application of international trade rules when thinking about how much credence countries ought to be allowed to give to public sentiment in justifying trade-restrictive measures under the SPS Agreement. Three key drivers can be identified from the literature: psychological, social, and cultural. These are discussed below.

9.3.1 Psychological Mechanisms ('Heuristics')

A widely cited approach to understanding public risk perception is the psychometric approach developed by Slovic which examines psychological factors that play a part in people's risk perception. Using a psychometric paradigm, this approach uses psychological scaling and multivariate analysis techniques to produce quantitative representations or 'cognitive maps' of risk attitudes and perceptions. Chapter 5 briefly introduced the concept of heuristics which people use to help them understand and make sense of risks. These heuristics, the argument goes, are necessary in order for people to make the most efficient use of their limited cognitive abilities and time, and their inability to cope with large amounts of information.[16] Cognitive research reveals various situations

---

[13] Gary P. Sampson, 'Risk and the WTO' in David Robertson and Aynsley Kellow, eds, *Globalization and the Environment: Risk Assessment and the WTO* (Cheltenham, UK and Northampton, MA: Edward Elgar Publishing Limited, 2001) 15 at 15. Another example given by Lave is that of asbestos. He finds that public opinion typically considers asbestos to be a more serious concern – even when it is in good repair in buildings – than risks that pose a greater real threat, such as radon in some homes, or aflatoxin in corn. Lester B. Lave, 'Health and Safety Risk Analyses: Information for Better Decisions' (1987) 236:4799 Science 291 at 293.

[14] Susan B.T. Wilkinson, Gene Rowe and Nigel Lambert, 'The Risks of Eating and Drinking' (2004) 5 European Molecular Biology Organization Reports S27 at S29.

[15] P. Slovic, 'Perception of Risk' (1987) 236:4799 Science 280 at 280.

[16] A.R. Pratkanis and E. Aronson, *The Age of Propaganda: The Everyday Use and Abuse of Persuasion* (New York: W.H. Freeman, 1992). Cited in Wilkinson, Rowe and Lambert, *supra* note 14 at S30.

in which people tend to use heuristics, including when they are overloaded with information; when they have little knowledge or information on the topic; and when a specific shortcut comes to mind easily. Slovic pays particular attention to availability, anchoring, and case-based decisions.[17] He also identifies three related beliefs held by many people. First, that risk is an 'all or nothing' matter; something is either safe or it is dangerous and there is no middle ground. This leads to a lack of interest in the statistical probability of harm, which, combined with a tendency to focus on the worst-case scenario,[18] leads to what Sunstein refers to as *'probability neglect'*. That is, with respect to risks of harm, vivid images and concrete pictures of disaster can 'crowd out' other kinds of thoughts, including that the probability of disaster is tiny. He cites, for example, studies that show that when people discuss a low-probability risk, their concern rises even if the discussion consists mostly of apparently trustworthy assurances that the likelihood of harm is infinitesimal; that people show 'alarmist' bias: when presented with competing accounts of danger, people tend to believe the more alarming account; and that visualization or imagery greatly influences people's reactions to risks.[19]

The second belief identified by Slovic is that it is possible to eliminate risk altogether (a 'zero-risk' mentality). And third, that nature is benevolent, that is, man-made products and activities are more likely to be dangerous than natural processes (despite the many dangers that exist 'in nature').[20] Sjoberg considers this to be closely related to moral beliefs.[21] Technologies that are seen as disturbing and that interfere with nature and natural processes tend to be perceived as risky, and sometimes as immoral.[22] Sjoberg also identifies other factors that influence the public's perception of risks, including what he terms 'cynical suspiciousness', 'new age beliefs' about a spiritual element of existence, beliefs in the physical reality of the soul, denial of relevance of analytical thinking and scientific knowledge, and belief in various paranormal phenomena.[23]

Slovic argues that the use of heuristics is valid in some cases but in others can lead to large and persistent biases.[24] Along with the other beliefs identified, this

---

17 See discussion of heuristics in Chapter 5 at section 5.3.2.3.

18 Dan M. Kahan, *et al.*, 'Fear and Democracy: A Cultural Evaluation of Sunstein on Risk' (2006) 119 Harv. L.R. 1071 at 9.

19 Cass R. Sunstein, 'Book Review – The Laws of Fear' (2001) 115 Harv. L. Rev. 1119 at 1142.

20 Sunstein, *ibid.* at 1128.

21 Lennart Sjoberg, 'Principles of Risk Perception Applied to Gene Technology' (2004) 5: Special Issue EMBO Reports 547 at S48.

22 *Ibid.*

23 *Ibid.* at S50.

24 Slovic, *supra* note 15 at 281.

can have serious implications for risk assessment and regulation. As Slovic states: 'research on basic perceptions and cognitions has shown that difficulties in understanding probabilistic processes, biased media coverage, misleading personal experiences, and the anxieties generated by life's gambles cause uncertainty to be denied, risks to be misjudged (sometimes overestimated and sometimes underestimated), and judgments of fact to be held with unwarranted confidence'.[25] In this Sunstein concurs, arguing that law and policy are likely to be adversely affected by people's use of mental shortcuts. He finds this is particularly evident in democracies where officials are likely to respond to public alarm.[26] For example, Chapter 12 discusses public fears in Europe of health risks from the meat of hormone-injected cattle and more recently, from GM crops, which have led to regulatory responses that some argue are out of proportion to the true level of risk.[27]

Interestingly, Slovic finds that experts' judgments are prone to similar biases as the public, especially when they have to go beyond the limits of available data and rely on intuition.[28] For the most part however, he finds that when experts judge risk, their responses correlate highly with technical estimates of annual fatalities and are not affected by qualitative factors. On the other hand, lay people are more likely to make an estimate that is correlated with non-technical and quantifiable hazard characteristics such as its catastrophic potential. As such, their estimates will differ from those of experts.[29]

Slovic finds further that the public is influenced by emotion and affect,[30]

---

[25] *Ibid.*

[26] Sunstein, *supra* note 19 at 1127.

[27] Of course, the opposite phenomenon may also occur, that is, some risks may be neglected because the public do not see them as serious, perhaps because the risks in question are 'unspectacular events which claim one victim at a time and are common in non-fatal form' or because they have not manifested themselves in recent memory, leading to complacency. Paul Slovic, *The Perception of Risk* (London: Earthscan Publications Ltd, 2000) at 107.

[28] Slovic, *supra* note 15 at 281.

[29] *Ibid.* at 283. Slovic's assertions concerning the gap between lay and expert risk perceptions have been challenged by Sjoberg who argues that they are based mainly on data from the 1970s and have not been supported by later work. Instead, he argues that experts' perceptions of risk are actually not that different from public perceptions. Experts, he argues, make risk judgments on the basis of factors and thought structures that are similar to those of the public, but their level of perceived risk will be drastically lower than that of the public. Sjoberg, *supra* note 21 at S48.

[30] Paul Slovic, 'Trust, Emotion, Sex, Politics, and Science: Surveying the Risk-Assessment Battlefield' in Dennis J. Paustenbach, ed., *Human and Ecological Risk Assessment: Theory and Practice* (New York: John Wiley & Sons, Inc., 2002) 1377 at 1385.

worldviews, ideologies, and values.[31] He describes an 'affect' heuristic where people have an emotional, 'all-things-considered' reaction to certain processes and products. It operates as a mental shortcut for a more careful evaluation.[32] He finds this heuristic evidenced by data showing that when asked to assess the risks and benefits of a particular product, people tend to say that risky activities carry low benefits and that beneficial activities carry low risks. People hardly ever find that the same product or activity is both highly beneficial and quite dangerous, or lacking in both benefits and danger. 'Affect' is described as a subtle form of emotion defined as a positive (like) or negative (dislike) evaluative feeling toward an external stimulus (for example, a hazard such as smoking).[33]

Slovic identifies a number of risk characteristics that make them more or less acceptable to people. He sorts these characteristics into two categories, the first being the extent to which a hazard is *'dreaded'* (severe, likely or uncontrollable, judged by perception of controlling the risk) and the second being how *'known'* the hazard is (known to science, new, or has delayed effects).[34] People will, for example, tolerate a higher probability of injury or death from activities that they feel they can meaningfully control (for example, smoking, eating, driving automobiles) than from activities that make them feel powerless or that they distrust (for example, nuclear power, pesticide use, biotechnology).[35]

Slovic also discusses what he perceives as a lack of trust of authorities on the part of many lay people, which leads to reluctance to accept technical

---

[31] For a full discussion of these influencing factors, see Slovic, *ibid.*

[32] Slovic, *supra* note 27. See also Sunstein, *supra* note 19 at 1144. Sunstein questions whether the 'affect' heuristic is really a heuristic at all, that is, does it substitute an easy question for a harder one? He suggests that in most cases, affect is probably a result, at least in part, of a rough assessment of risks and benefits using the availability heuristic, intuitive toxicology, other mental shortcuts, and value judgments.

[33] Slovic, *supra,* note 27. See also Dan M. Kahan and Donald Braman, 'Cultural Cognition and Public Policy' (Yale Law School Public Law Working Paper No. 87, 2005) at 8. Kahan and Braman discuss 'cognitive-dissonance avoidance' whereby individuals find it comforting to believe that what is 'noble' is also benign, and what is 'base' is dangerous.

[34] Paul Slovic, B. Fischhoff and S. Lichtenstein, 'Facts and Fears: Understanding Perceived Risk' in R.C. Schwing and W.A. Albers, eds, *Societal Risk Assessment: How Safe is Safe Enough?* (New York: Plenum, 1980) 181. Cited in Wilkinson, Rowe and Lambert, *supra* note 14 at S30.

[35] See generally Baruch Fischhoff, *et al.*, *Acceptable Risk* (Cambridge, UK: Cambridge University Press, 1981). Douglas and Wildavsky, *supra* note 10. Branden B. Johnson and Vincent Covello, *The Social and Cultural Construction of Risk* (Dordrecht: Reidel, 1987). Deborah G. Mayo and Rachelle D. Hollander, eds, *Acceptable Evidence: Science and Values in Risk Management* (New York: Oxford University Press, 1991).

estimates of risk.[36] In this regard, Deane argues that perceived risks will be amplified in the eyes of the public where government or regulatory bodies have a history of secrecy or where there is a lack of public involvement in policy-making.[37] An example of this can be seen in the handling of the BSE outbreak in the UK where the government initially took the approach of reassuring people that there were no risks to human health from the outbreak, leading to a shattering of trust in government when it was later revealed that evidence pointed to a link between BSE and variant Creutzfeldt-Jakob disease (vCJD) in humans and that the government had covered up this evidence earlier on following the outbreak. This distrust has by all accounts persisted among the British population.[38]

9.3.2 Social Mechanisms

Sunstein discusses the social forces that drive people to fear (or not) certain hazards.[39] He premises his argument on the common-sense claim that most lay people lack knowledge of risks and must therefore rely on information received from others in making assessments.[40] He refers to both *informational cascades* and *reputational cascades*. Informational cascades arise where people with little personal information about a matter base their own beliefs on the apparent beliefs of others, a likely scenario in the context of risk perception.[41] Sunstein suggests that when informational cascades occur, people might become fearful of something that carries small risks. Regulatory divergence among countries may be explained by the fact that certain information is, for largely arbitrary reasons, able to spread quickly in one place but not another.[42]

Reputational cascades occur when people are influenced by others not because they think they are more knowledgeable, but because they want to earn social approval and avoid disapproval.[43] As Sunstein writes, 'if many

---

36 Slovic, *supra* note 30 at 1390.

37 Christine R. Deane, 'Public Perceptions, Risk Communication and Biotechnology' in David Robertson and Aynsley Kellow, eds, *Globalization and the Environment: Risk Assessment and the WTO* (Cheltenham, UK and Northampton, MA: Edward Elgar Publishing Limited, 2001) 106 at 110.

38 See generally Scott C. Ratzan, *The Mad Cow Crisis: Health and the Public Good* (London: UCL Press, 1998).

39 Sunstein, *supra* note 19 at 1125. Sunstein notes that Slovic recognizes that social forces can heighten the effect of the availability heuristic, but criticizes Slovic for paying too little attention to the social forces at work.

40 *Ibid.* at 1130.

41 Sunstein, *supra* note 1 at 86.

42 *Ibid.* at 87.

43 *Ibid.*

people are alarmed about some risk, you might not voice your doubts about whether the alarm is merited, simply in order not to seem obtuse, cruel, or indifferent. And if many people believe that a certain risk is trivial, you might not disagree through words or deeds, lest you appear cowardly or confused. Sometimes people take to speaking and acting as if they share, or at least do not reject, what they view as the dominant belief.'[44] Relevant in this regard are the remarks of a British sociologist during the BSE scare in the UK. After publicly raising questions about the health threats from mad cow disease, he suggested that if you raise those doubts publicly, '[y]ou get made to feel like a pedophile'.[45]

Informational and reputational cascades can produce public demand for regulation, even if the risks are minimal. Again, it is not only lay people who are subject to such pressures. As Sunstein notes, lawmakers are particularly vulnerable to reputational pressures and may support risk regulation that they privately believe is unnecessary.[46] Sunstein finds that these cascade effects are related to what he calls *group polarization*, a process where people engaged in discussion about a particular risk end up thinking a more extreme version of what they initially thought.[47]

Another important influencing factor on the public's risk perception which can arguably be classified under the heading of 'social mechanisms' is the news media which, in Slovic's words, 'rather thoroughly document mishaps and threats occurring throughout the world'.[48] It is beyond the scope of this book to fully explore the extent of the media's influence on public risk perception but the literature points to the conclusion that it is significant and in many cases results in a distorted public perception of risk and a corresponding demand for government regulation.[49] The media's coverage of an issue is often driven more by commercial needs than by a desire to accurately

---

44 *Ibid.*

45 Sunstein, *supra* note 19 at 1113. Citing Andrew Higgins, 'It's a Mad, Mad, Mad-Cow World' *Wall Street Journal* (12 March 2001) A13.

46 Sunstein, *supra* note 1 at 87.

47 *Ibid.* at 88.

48 Slovic, *supra* note 15 at 280.

49 See for example Miljan who found several concerns regarding the Canadian media's coverage of risks in a study of media coverage of cancer; namely: that news stories place a disproportionate emphasis on environmental factors such as pesticides that cause cancer and very little attention to causes that are within the control of individuals such as smoking and diet; that journalists tend to (over)represent the degree of scientific acceptance of a given risk; that journalists have inappropriately tended to acquiesce with activists; and that they tend to downplay or ignore new evidence that challenges the original story. Lydia Miljan, 'Unknown Causes, Unknown Risks' in Laura Jones, ed., *Safe Enough? Managing Risk and Regulation* (Vancouver: The Fraser Institute, 2000) 31 at 34.

inform.[50] Anti-risk activists and NGOs also have a significant impact on public risk perception by exploiting public anxieties; as well as on the response of regulators by exerting pressure for regulation.[51]

9.3.3 Cultural Cognition or 'Normative Bias'

While psychological and social factors explain much of what goes on when individuals make decisions about risk, a further set of factors can also be identified, namely, those that arise from people's cultural worldviews. Kahan and Braman argue for a theory of cultural cognition that helps explain why disagreement about public policies in and across societies occur not on a random basis but across distinct social groups – racial, sexual, religious, regional, and ideological.[52] Kahan *et al.* then apply this theory more specifically to the question of risk.[53] According to their 'cultural cognition' theory, 'culture is prior to facts' in societal disputes over risk.[54] In a *normative* sense, this means that cultural values determine what significance individuals attach to the consequences of risk regulation. In a *cognitive* sense, it means that cultural values shape what individuals believe the consequences of risk regulation to be. In other words, individuals shape their factual beliefs to support their preferred vision of 'the good society'.[55]

Discussions of public risk perception that draw on psychological and social mechanisms to explain people's mistakes tend to assume that if people were made cognizant of scientifically sound information, then they could be expected to make sound risk policy decisions upon which there would be no disagreement. The cultural cognition theory disputes this basic assumption, suggesting that it will be futile to provide people with scientific information because in determining whether such information is sound, 'individuals will

---

[50] Peter Cook, 'Science and Society' (Paper presented to the Conference on Public Science in Liberal Democracy: The Challenge to Science and Democracy, University of Saskatchewan, Saskatoon, 14-16 October 2004).

[51] See Neal who identifies various strategies used by activist groups in this regard, including initial exaggeration of a hazard; equating coincidence with causality; using tactics to shock and scare; selective citation of scientists to give the impression that a (non-existent) consensus of opinion exists; portraying nature as benign and desirable, while the 'artificial' as harmful and bad, even when the opposite is true; and portraying consumers as innocent and naïve and companies as conspirators. Neal, *supra* note 6 at 23.

[52] Kahan and Braman, *supra* note 33.

[53] Kahan *et al.*, *supra* note 18. See also Dan M. Kahan and Paul Slovic, 'Cultural Evaluations of Risk: "Values" or "Blunders"?' (2006) 119 Harv. L. Rev. 166.

[54] Kahan *et al.*, *ibid.* at 17.

[55] *Ibid.*

inevitably be guided by their predetermined cultural evaluations of these activities'.[56]

Kahan *et al.* link disputes over risk to three groups of values that form competing worldviews: egalitarian, individualistic and hierarchic.[57] The *egalitarian* worldview is sensitive to risk and finds, for example, that abatement of environmental hazards justifies regulatory action. The *individualistic* worldview, in contrast, dismisses claims of risk as specious, consistent with their commitment to the autonomy of markets and other private orderings. The *hierarchic* worldview is also sceptical of risk, because warnings of imminent catastrophe are seen as threatening the competence of social and governmental elites.[58] In a recent survey of 1800 Americans, Kahan *et al.* provide empirical evidence for the theory of cultural cognition using these worldviews. They find, for example, that the more egalitarian and solidaristic people become, the more concerned they are about global warming, nuclear power, and pollution generally, whereas the more hierarchical and individualistic they become, the less concerned they are.[59]

Sunstein criticizes the cultural cognition theory and queries the use of the term 'cultural cognition'. He suggests that it is more plausible to say that people have different normative positions which bias their judgments on questions of fact.[60] He also argues that in any event, cultural conflicts are only likely to arise regarding what he refers to as 'hot risks' such as abortion or gun control (and arguably GM foods).[61] He argues further that even when such disagreements matter, bounded rationality is also important. Unlike Kahan *et al.* who find that cultural cognition is an alternative to bounded rationality, Sunstein sees it as a reflection of bounded rationality and a part of the general framework that it offers.[62] He sees three means by which 'culture' can

---

[56] Kahan and Braman, *supra* note 33 at 3. Kahan *et al.* do suggest, however, that regulators should look at devising risk communication strategies that make it possible for citizens to accept new information, and ultimately change their minds, without experiencing a threat to their cultural identities. *Ibid.* See also Kahan and Slovic, *supra* note 53.

[57] Kahan *et al.*, *ibid.* In doing so, they rely extensively on the work of Douglas and Wildavsky, *supra* note 10.

[58] Kahan *et al.*, *supra* note 18 at 17. Citing Douglas and Wildavsky, *ibid.*

[59] Kahan *et al.*, *ibid.* at 20. See also *National Risk and Culture Survey*, http://research. yale.edu/culturalcognition/content/view/45/89.

[60] Cass R. Sunstein, 'Misfearing: A Reply' (2006) 119 Harv. L. Rev. 1110 at 1112. Sunstein also argues that it is not clear that Kahan *et al.* actually establish that culture produces different factual judgments about the magnitude of social risk. Kahan and Slovic refute this argument in a subsequent article. Kahan and Slovic, *supra* note 53.

[61] Sunstein, *ibid.*, at 1115. Sunstein compares for example these types of 'hot' risks to the risks from power saws, bridges, earthquakes, fire, and trains.

[62] *Ibid.* at 1117.

contribute to judgments about risk: (i) social influence where cascade effects and group polarization help to constitute relevant 'cultures'; (ii) 'normative bias' by which people tend to seek out and believe evidence that supports their own antecedent views; and (iii) status competition (where people think that their status is on the line and press their views on a given issue as part of that competition).[63]

Kahan *et al.* argue that where cultural worldviews pervade popular risk assessments, a genuine commitment to democracy forbids simply dismissing such perceptions as products of 'bounded rationality'.[64] Sunstein, on the other hand, argues that citizens can only be seen to be making a reasonable request if they are not blundering on the facts, but are responding to 'coherent visions of the good society and the virtuous life'.[65] But as Chapter 10 will reveal, the difficulty lies in determining when a citizen has 'blundered'. Often, there will be too many scientific uncertainties to allow an easy determination as to whether such a 'blunder' has occurred. Sunstein does not address this point, seeming to suggest instead that science will provide us with the answers. But as we shall see, science is largely about interpreting facts; who then is to say in marginal cases whether someone has blundered?

Whether one sympathizes with Kahan *et al.* regarding the role of cultural cognition, or, like Sunstein, finds that bounded rationality lies behind cultural cognition, some conclusions can be drawn at this point that will lead the discussion forward:

- People's risk perceptions are impacted by a variety of psychological, social, and cultural factors. In many cases, it will be difficult (if not impossible) to fully unpack these influences.
- A purely 'quantitative' or technical approach to risk perception is likely to overlook important concerns held by lay people.
- In some cases, lay people might want governments to take precautions that experts would consider to be excessive. Whether or not governments ought to respond to lay perceptions in their regulatory decision-making is a difficult and contested question both domestically and in the context of trade disputes.

---

63 *Ibid.* at 1118 (arguing that Kahan *et al.* overstate the role of status competition but that in the event it does play a role, a form of bounded rationality is involved).

64 Kahan and Slovic, *supra* note 53.

65 Sunstein, *supra* note 60 at 1123.

9.4 REGULATING IN RESPONSE TO PUBLIC PERCEPTIONS

From a trade liberalization perspective, if one views public risk perceptions as flawed, any influential role played by those perceptions in the domestic regulatory decision-making process will be considered problematic and will raise important questions about the legitimacy of the resulting regulation. For example, Fraiberg and Trebilcock relate public risk perception to what they describe as the dismal record of regulatory performance in the area of health and safety. They cite a study by Robert Hahn which found 'reason to believe that most regulations implemented in the US since 1990 would not pass a cost-benefit test'.[66] Similar studies are cited by Slovic, who finds that great disparities in monetary expenditures designed to prolong life may be traced to public perceptions of risk.[67]

Despite this observation, Slovic argues that discrepancies between expert and public perceptions of risk should not simply be explained away by suggesting that the public is irrational or foolish. Rather, he finds room for respecting public risk perceptions. He refers to the public's 'rival rationality' *vis-à-vis* experts, and suggests that each side must 'respect the insights and intelligence of the other'.[68] He bases this conclusion on the fact that lay people focus on more factors (largely qualitative) than just the number of lives at stake.[69] He suggests that because people pay attention to such crucial factors as whether a risk is dreaded and/or known, they are in a way capable of thinking better, and more rationally, than experts.[70] That is, people's 'basic conceptualization of risk is much richer than that of experts and reflects legitimate concerns that are typically omitted from expert risk assessments'.[71] Slovic's arguments lead to the logical conclusion that a good system of risk regulation

---

[66] Robert Hahn, 'Regulatory Reform: What Do the Government's Numbers Tell Us?' in Robert Hahn, ed., *Risks, Costs and Lives Saved: Getting Better Results from Regulation* (New York: Oxford University Press, 1996) 225. Cited in Jeremy D. Fraiberg and Michael J. Trebilcock, 'Risk Regulation: Technocratic and Democratic Tools for Regulatory Reform' (1998) 43 McGill L.J. 835. The authors suggest that as the US has some of the most highly developed risk regulation institutions in the world, it would be surprising to find different results in other jurisdictions.

[67] Slovic, *supra* note 30 at 1377. See also Stephen Breyer, *Breaking the Vicious Circle: Toward Effective Risk Regulation* (Cambridge, MA: Harvard University Press, 1993) at 10.

[68] *Ibid.* See also Paul B. Thompson, 'Risk Objectivism and Risk Subjectivism' (1990) 1:3 Risk 22.

[69] Slovic, *supra* note 27 at 231.

[70] *Ibid.*

[71] *Ibid.*

should err on the side of 'populism' as opposed to be being more strongly 'technocratic'.[72] This is consistent with the cultural cognition model of Kahan *et al.* Given that this model sees risk judgments as being inextricably bound up with cultural values, it counsels a regulatory model that seeks to include public input through procedures that are 'genuinely deliberative, open, and democratic'.[73]

Slovic's approach has garnered support from commentators such as Wilkinson *et al.* who suggest that in many cases consumers' beliefs seem to be based on alternative, sensibly informed frameworks and that differences between lay risk perceptions and expert judgments may be a consequence of alternative framings of, or emphasis on, the issue.[74] They argue that while consumers do not understand 'risk' in the same way as do scientists; their non-statistical, more qualitative, and heuristic-based risk perceptions are often arguably quite sensible. Pointing to the case of GM foods, they suggest that when there are familiar and relatively risk-free alternatives to GM foods available, who is to say that consumer avoidance of GM foods is illogical?[75] Similarly, Jasanoff argues that doing justice in a democratic society requires courts to respect the public's normative positions even if they 'cut against the managerial preferences of the nation's scientific and technological elite'.[76]

Sunstein is one of the strongest critics of Slovic's view that the public displays a 'rival rationality'.[77] He questions why, if lay people really have a richer rationality, people's concerns differ so notably between countries?[78] For Sunstein, the conclusion that people deal poorly with the topic of risk is sufficient justification for a more 'technocratic' approach to regulatory decision-making where decision-makers should follow science and evidence rather than the public.[79] Thus, while he concedes that experts do indeed make mistakes, he considers that in their area of expertise they are more likely to be right more of the time than are lay people.

---

[72] *Ibid.*

[73] Kahan *et al.*, *supra* note 18 at 51.

[74] Wilkinson, Rowe, and Lambert, *supra* note 14 at S30.

[75] *Ibid.*

[76] S. Jasanoff, *Science at the Bar* (Cambridge, MA: Harvard University Press, 1995) at 13.

[77] Sunstein, *supra* note 19 at 1147. (In particular, Sunstein takes issue with Slovic's study methodology.) See also Howard Margolis, *Dealing with Risk: Why the Public and the Experts Disagree on Environmental Issues* (Chicago: University of Chicago Press, 1996) at 71. Margolis argues that Slovic's psychometric paradigm does not explain divisions between experts and lay people; the bottom line being that experts know the facts and ordinary people do not.

[78] Sunstein, *supra* note 19 at 1148.

[79] *Ibid.* at 1123.

9.5 PUBLIC RISK PERCEPTIONS AND THE WTO

This chapter has sought to bring together the major strands of theory that help explain how lay people perceive risk. This in turn tells us something about regulatory decision-making. Slovic's *psychological* model tells us that regulations made in response to public risk perceptions may reflect biases and thus be over-reactive to small risks (or conversely, under-reactive to larger risks). Such regulations are at their core unlikely to be intentionally protectionist of local producers, although they may well have that effect.

The Baptist–Bootlegger effect may arise where producers wanting to keep out foreign products join forces with concerned members of the public to lobby for strict regulations. The *social factors* emphasized by Sunstein reveal more potential for protectionist interests to sway public opinion and regulators in support of regulation that impacts foreign producers, and again, the Baptist–Bootlegger effect may act to strengthen such protection. The *cultural cognition* theory tells us that cultural worldviews may also play a role in determining what an individual sees as constituting a risk or not.

Each of the 'amber' categories identified in Chapter 6 may involve a government's response to public risk perceptions. In the first three categories,[80] people may place what appears to be excessive importance on risks that carry a low probability of eventuating and/or have consequences that are not serious. In the fourth category,[81] people may perceive there to be a risk and demand regulation, even where experts come to the opposite conclusion. In the fifth category,[82] the presence of scientific uncertainty may confuse the public who may be guided by cultural, psychological, and/or social factors in judging the risk in question. In the sixth category, regulations reflect a combination of the above factors, together with intended protectionism.

A country that enacts trade-restrictive health regulations must meet the tests set forth by the SPS Agreement which aims primarily to ensure that trade is not distorted by overly strict and/or protectionist regulations. The test is one of 'scientific evidence' and a requirement that the measure be based on a 'risk assessment'. If one accepts the proposition that public risk perceptions have validity and ought to be considered in regulatory decision-

---

[80] (1) High probability, consequences that are not serious; (2) low probability, consequences that are not serious; (3) low probability, serious consequences.

[81] Where public risk perceptions differ from expert risk evaluation.

[82] Scientific uncertainty as to whether or not there is a risk to health.

making, the question arises as to what place is left for these values in the domestic decision-making process by the science-based trade rules. Further consideration of this point requires an understanding of the nature of science and the discipline of risk assessment, topics to which the discussion turns in the next chapter.

10. Analysis of risks: the role of science

10.1 INTRODUCTION

This chapter contributes to an assessment of the appropriateness of the WTO's science-based framework by examining the role of science in regulatory decision-making about risk. After some general comments about the relationship between science and law in Section 10.2, it examines the rationale for the SPS Agreement's scientific evidence test, namely, the claim that science is independent and objective. As the next chapter notes, some critics of the SPS Agreement argue that science suffers from a lack of objectivity which undermines its role as a tool in regulatory decision-making and as a benchmark in international trade rules.

Section 10.4 explores the specifics of risk regulatory decision-making through an examination of the risk analysis process (which includes risk assessment, risk management, and risk communication). In particular, it asks whether risk assessment is truly an objective exercise that ought to be elevated to such a position of significance within the Agreement. The final section questions what role public opinion can (and should) play in regulatory decision-making on risk.

10.2 SCIENCE AND LAW

The debate over the role of science in the WTO echoes much of what has been said about science's relationship with the law in domestic legal systems. Debates over the role of science in domestic law concern struggles over the authority of knowledge.[1] For example, whose knowledge should count as valid science, according to what criteria, and as applied by whom? When should lay understandings of phenomena take precedence over expert claims to superior knowledge? Should expert views about risk and cost prevail or should they give way to countervailing non-expert values? These questions are all pertinent to resolution of disputes under the SPS Agreement.

[1] S. Jasanoff, *Science at the Bar* (Cambridge, MA: Harvard University Press, 1995) at 19.

In a piece entitled 'Law and Science: A Dialogue on Understanding', Markey sets out an imagined conversation between 'Law' and 'Science' where Law has attended a meeting of the American Association for the Advancement of Science's annual meeting, and Science has attended a meeting of the American Bar Association. The dialogue is worth reproducing at some length for its insights into some of the perceived problems with using scientific evidence in law. Commenting on the inability of each other's followers to understand them, Law and Science remark:[2]

> Science: . . . they could do better at their profession if they understood me. Not to understand me is to understand the world only in part. I am the servant of humanity. As the seeker of physical knowledge, I make discoveries that allow development of the technology to sustain humanity on a planet of finite resources . . . Your followers' ignorance of me, my role, and my methods is appalling. Now, I'll put to you the same question, why should my followers, who are busy doing so much for mankind, take the time to try to understand you and your role?
>
> Law: (sternly) Because you are not 'the' servant of humanity; you are but one; I am another. Your followers, who deal mainly with things seen, would contribute so much more if they understood that I deal with things no less real – like mercy, compassion, philosophy, freedom, and justice – although only the effects of those are seen. I am the only activity that can assure that you do produce a truly higher, not a cruel, standard of living. Your followers must understand me if perverted uses of that knowledge are to be avoided. If you were humanity's only servant, you would create a society ruled entirely by cold, hard, despotic, physical facts.

Commenting on public scepticism about both law and science, the conversation continues:

> Law: Our masters, thinking you infallible, objective, dedicated, able to remake the world, enthroned you with no felt need to understand you. Now, when the need for understanding is even greater, your technological juggernaut is moving too fast. Many of our masters are ambivalent about technology. Unthinking awe has become uniformed antipathy. Well-publicized concerns for health, the environment, and quality of living – remember Love Canal, Three Mile Island, Thalidomide – have caused many of our masters to turn from awestruck adulation to cynical skepticism.
>
> . . .
>
> Law: Our masters need us both, but they are not well served by philosophically illiterate technologists and technologically illiterate lawyers. Where do we start in trying to encourage mutual understanding?
>
> Science: Ah, yes, I think understanding begins with recognizing differences. My purpose is to increase physical knowledge and ways to use it. Yours is to resolve disputes, provide justice, and exercise social control. I rest on the material; you on the moral, ethical, and philosophical. I say what can be done; you say whether it should. I determine; you compare. I describe; you prescribe. I equip; you guide.

---

[2] William A. Thomas, ed., *Science and Law: An Essential Alliance* (Boulder, CO: Westview Press, 1983) at 3.

The comments of Law in the exchange echo arguments put forward by critics of SPS Agreement's science-based framework.[3] The argument that scientific knowledge ignores core human values is paralleled by arguments that privileging the role of science in the WTO excludes important values and social and economic factors. Just as the imaginary conversation recognizes the importance of mutual understanding in the domestic context, it is argued here that mutual understanding is critical to WTO law.

The conversation also alludes to a long-standing view of law and science as being in conflict. Jasanoff, for example, argues that there is a 'culture clash' between lawyers (or bureaucrats) and scientists, and that the culture of law should strive as far as possible to assimilate itself to the culture of science when dealing with scientific issues.[4] The considerable differences between scientific and legal thinking can be observed in their approaches to fact-finding. In particular, science favours *progress*, its focus is on resolving substantive issues and finding the 'truth' of the matter. Law, on the other hand, focuses on *process* and often views 'the fact of the matter' as uncertain. Law is designed to manage uncertainty and resolve conflicts for which there is no answer.[5] At the same time, law tends to view science as capable of offering clear answers, that is, the truth of the matter. However, in the case of risk regulation (as in other areas), scientific truths are rarely ascertainable and science must, like law, manage uncertainty and conflicts to which there are no clear answers.

At this point, something should be said about the meaning of the term 'uncertainty' in the context of risk. The term 'certainty' refers to a situation where there is the ability to fully characterize the different possible outcomes (such as types of harm) and to confidently state or determine their relative likelihood. 'Uncertainty', on the other hand, exists where various possible outcomes are clear but where – due to lack of information – it is difficult to attach objective, or even subjective, probabilities to them.[6] Both terms must be

---

3 See discussion in Chapter 11, section 11.3.

4 Jasanoff, *supra* note 1 at 7.

5 Jerry L. Mashaw, 'Law and Engineering: In Search of the Law-Science Problem' (2003) 66:4 Law & Contemp. Probs 135 at 138. Science also focuses on *process*; however, it will not view a matter as resolved in the face of uncertainty, therefore *progress* is necessary to continue searching for the answer.

6 Andy Stirling, Ortwin Renn and Patrick van Zwanenberg, 'A Framework for Precautionary Governance of Food Safety: Integrating Science and Participation in the Social Appraisal of Risk' in E. Fisher, J. Jones and R. von Schomberg, eds, *Implementing the Precautionary Principle* (Cheltenham, UK and Northampton, MA: Edward Elgar, 2006) at 287. See also F. Knight, *Risk, Uncertainty and Profit* (London: Houghton Mifflin Co., 1921). Cited in William A. Kerr, 'Science-based Rules of Trade – A Mantra for Some, An Anathema for Others' (2003) 4:2 The Estey Centre Journal of International Law and Trade Policy 86 at 91.

contrasted with 'ambiguity' where the problem is not with probabilities but in agreeing appropriate questions to ask, values to adopt, priorities to assign, or limits to impose in the definition of the possible outcomes themselves.[7]

Uncertainty, as the following discussion will detail, is ubiquitous in science, meaning that in international trade disputes over health regulations, the law must deal with scientific uncertainty and resolve conflicts to which there are no clearly ascertainable answers. This is a task which the law should be capable of dealing with; however, it must first recognize the uncertainty.

Jasanoff disputes the oft-cited proposition that the law's obligation in relation to science is a simple, two-step prescription where courts first seek out the findings of mainstream science and then incorporate them into their decisions.[8] To the contrary, she argues that scientific claims, especially those that are implicated in controversies, are 'highly contested, contingent on particular localized circumstances, and freighted with buried presumptions about the social world in which they are deployed'.[9]

Another relevant aspect of the science/law relationship referenced in the conversation cited above is what Posner refers to as the 'scientific illiteracy' of lawyers, many of whom he finds deliberately 'turn their back on science' when they decide to attend law school.[10] He argues that if lawyers and judges have no interest in and feel for science, they are unlikely to attend closely to either the dangers or the opportunities that modern science creates. Given this, he suggests that lawyers need to become comfortable with scientific questions; 'comfortable not in the sense of knowing the answers to difficult scientific questions or being able to engage in scientific reasoning, but in the sense in which most antitrust lawyers today, few of whom are also economists, are comfortable in dealing with the economic issues that arise in antitrust cases. They know some economics, they work with economists, they understand that economics drives many outcomes of antitrust legislation, and as a result they can administer – not perfectly but satisfactorily – an economically sophisticated system of antitrust law'.[11]

---

7 Stirling, Renn and van Zwanenberg, *ibid.*

8 Jasanoff, *supra* note 1 at xiv.

9 *Ibid.* at xv.

10 Richard A. Posner, *Catastrophe: Risk and Response* (New York: Oxford University Press, 2004) at 96.

11 *Ibid.* at 200. Posner suggests that unlike economics, science involves phenomena of which ordinary people have no intuitive sense and that law students should therefore undergo basic scientific education so that they are comfortable with scientific methods, attitudes, usages, and vocabulary.

10.3 WHY USE SCIENCE? THE NOTION OF OBJECTIVITY

This section examines in more detail the underpinnings of Jasanoff's claim that science is not simply a producer of objective truths about the world. The SPS Agreement places faith in the notion of science as an 'objective' discipline that is free of values and leads to definite knowledge.[12] The term 'objectivity' has been described as a notion 'used to connote emotional detachment of observer from object observed, to mean that measurements could be independently verified by other observers, to imply that facts were allowed to fall wherever they might'.[13] The characterization of science as objective has traditionally been accepted outside, as well as inside, the scientific establishment. As Atik writes: 'Science, more than religion, more than traditional ways, is unassailable within its domain . . . And science is more than mere reason; it is also experience built upon a continuing interaction with the physical world in a progressive, historical process'.[14]

Ziman seeks to explain what makes science so appealing:[15]

> Objectivity is what makes science so valuable in society. It is the public guarantee of reliable disinterested knowledge. Science plays a unique role in settling factual disputes. This is not because it is particularly rational or because it necessarily embodies the truth: it is because it has a well-deserved reputation for impartiality on material issues. The complex fabric of democratic society is held together by trust in this objectivity, exercised openly by scientific experts. Without science as an independent arbiter, many social conflicts could be resolved only by reference to political authority or by a direct appeal to force.

In 1942, Merton suggested that scientific practice is governed by a set of unwritten social norms which lead to the conclusion that science is objective and independent. *Communalism* requires that research findings should be regarded as 'public knowledge'. *Universality* requires that contributions to science should not be excluded because of nationality, religion, social status, or other irrelevant considerations. The idea that academic scientists have to be *disinterested* means that in presenting their work publicly they must discount any material interests that might prejudice their findings and adopt a humble,

---

[12] Gregory N. Derry, *What Science Is and How It Works* (Princeton, NJ: Princeton University Press, 1999) at 207.

[13] William W. Lowrance, *Modern Science and Human Values* (New York: Oxford University Press, 1986) at 48.

[14] Jeffery Atik, 'Science and International Regulatory Convergence' (1996–7) 17 Nw. J. Int'l L. & Bus. 736 at 738.

[15] John Ziman, 'Is Science Losing its Objectivity?' (1996) 382 Nature 751 at 754.

neutral, impersonal stance that hides their enthusiasm for their own ideas. *Originality* is the norm that keeps science progressive and open to intellectual novelty. *Scepticism* is the basis for academic practices such as critical controversy and peer review. It stresses the systematic testing of research claims in terms of rational qualities such as logical consistency and practical reliability.[16] Merton's ideas are useful in summarizing familiar characteristics of science. Nevertheless, they have been subject to much criticism and the perception of science as objective and neutral is far from being universally accepted. Instead, it is subject to challenge both in academia, and in the wider world. The academic challenge focuses on whether objective knowledge is possible and the extent to which science is socially constructed. In the wider world, the challenge focuses more on the actual use of science in different contexts, including in regulatory decision-making.

Turning to the academic challenge, Derry notes that while the opposing camps in the academic debate are not well defined, it is possible to paint a broad picture where the *traditionalists* favour Western culture, values that are absolute, and truth; while the *postmodernists*, on the other side, favour multiculturalism, relativism, and a worldview in which truth does not exist. Conflict between these camps raises some important questions, for example, is the achievement of scientific consensus a rational process, or a political process? Are scientific theories accurate reflections of a pre-existing reality, or merely social constructions based on negotiations between interest groups? Are the knowledge claims of science more valid than knowledge produced by other ways of knowing, or are all such claims equally valid?[17]

Postmodernism can be described as a rejection of 'scientism', which is the claim that science deals in certainties. The postmodernist critique denies the possibility of objectivity and of definite knowledge.[18] It contends that science wrongly claims to describe the physical world that surrounds us objectively and truthfully.[19] Postmodernists argue that science is a product of Western culture; if we deny the primacy of Western culture, then science is just another claimant (among many) for the validity of its results.[20] They argue further that scientists have not 'discovered' laws of nature, but have 'constructed' them; and that science is wrong in claiming that the results of research can be inde-

---

16 *Ibid.* at 751.

17 Derry, *supra* note 12 at 207–8.

18 *Ibid.* at 207.

19 Dick Taverne, *The March of Unreason: Science, Democracy, and the New Fundamentalism* (Oxford: Oxford University Press, 2005) at 193.

20 As American philosopher, Richard Rorty, has said, 'there is no truth, but only truths. Truths is what your contemporaries allow to you get away with'. Derry, *supra* note 12 at 207.

pendent of local cultural constraints or of moral and ideological motivations. Postmodernists reflect the views of relativists that all knowledge and values are relative to some particular standpoint, such as the individual, their culture and era, etc.[21] They find that all points of view are equally valid, and that scientific truth is only one of many truths – just 'one story among many'.

Most famous for having called into question the enterprise of science is Thomas Kuhn who wrote *The Structure of Scientific Revolutions* in 1962.[22] Kuhn's central premise was that science does not progress by a linear accumulation of new knowledge, but instead undergoes periodic revolutions which he calls 'paradigm shifts', in which the nature of scientific inquiry within a particular field is abruptly transformed. Kuhn also argued that rival paradigms are incommensurable, in other words, that it is not possible to understand one paradigm through the conceptual framework and terminology of another rival paradigm. While there has been much debate as to whether or not Kuhn's views had relativistic consequences, his work was embraced by postmodernists wishing to discredit or attack the authority of science.

Postmodernism has in turn been subject to much criticism, with some scientists arguing that postmodernists have little understanding of either the results or the methods of science.[23] Disagreement between the postmodernists and their critics gained particular notoriety in 1996 when a physicist from New York University, Alan Sokal, wrote a parody imitating a 'typical' postmodern paper and then published it in a leading postmodernist journal, *Social Text*. The paper, entitled 'Transgressing the Boundaries: Toward a Transformative Hermeneutics of Quantum Gravity', purported to argue against the 'Enlightenment idea' that there exists an external and knowable world. Physics, Sokal implied, was simply another field of cultural criticism.[24]

Sokal later explained the rationale for the parody, noting his belief '. . . that there exists an external world, that there exist objective truths about that world, and that my job is to discover some of them. (If science were merely a negotiation of social conventions about what is agreed to be "true", why would

---

[21] It is worth noting that postmodernism was also a reaction among anthropologists against colleagues who proclaimed the superiority of their own culture over the other cultures they studied. This led to the inference that facts are only true in relation to a particular culture, in other words, that there is no such thing as objective truth, because no one culture is superior to another. Taverne, *supra* note 19 at 196.

[22] Thomas S. Kuhn, *The Structure of Scientific Revolutions*, 1st ed. (Chicago: University of Chicago Press, 1962).

[23] Derry, *supra* note 12 at 208.

[24] Ruth Rosen, 'A Physics Prof Drops a Bomb on the Faux Left' *Los Angeles Times* (23 May 1996) A11.

I bother devoting a large fraction of my all-too-short life to it? I don't aspire to be the Emily Post of quantum field theory.)[25] And further:[26]

> My aim wasn't to defend science from the barbarian hordes of lit crit or sociology . . . rather, my goal is to defend what one might call a scientific worldview – defined broadly as a respect for evidence and logic, and for the incessant confrontation of theories with the real world; in short, for reasoned argument over wishful thinking, superstition and demagoguery.

Sokal argues that nothing is gained from denying the existence of objective scientific knowledge.[27] To illustrate, he distinguishes between, *inter alia*, sociological, ethical, and ontological questions regarding the application of quantum mechanics to modern technology. *Sociological* questions (for example, to what extent is our (true) knowledge of computer science, quantum electronics, solid-state physics and quantum mechanics – and our lack of knowledge about other scientific subjects, for example, the global climate – a result of public-policy choices favoring militarism?) and *ethical* questions (for example, ought society to forbid (or discourage) certain applications of computers?) are serious questions but, he argues, they '*have no effect whatsoever on the underlying scientific (ontological) questions:* whether atoms (and silicon crystals, transistors and computers) really do behave according to the laws of quantum mechanics (and solid-state physics, quantum electronics and computer science).[28]

Lowrance argues, on the other hand, that any imperious claims to objectivity in fact-finding should be suspect. He considers the fatal challenge to the notion of objectivity to be as follows: 'By what criteria could one possibly confirm this self-claim to correctness, other than by subjectively judging the persons, reputations, and procedures involved? Judging evidence on its internal logical merits is a separate issue, and does not hinge on anything that could be called objectivity. Performing analyses by committee is no guarantee of detachment, either. The term objectivity should be abandoned.' It is better, he argues, to aim for studies that can be defended as being 'precisely defined,

---

[25] Alan D. Sokal, 'Transgressing the Boundaries: An Afterword' (1996) 43:4 Dissent 93.

[26] Alan Sokal, 'A Plea for Reason, Evidence and Logic' (1997) 6:2 New Politics 126.

[27] *Ibid.*

[28] For example, if the worldwide community of solid-state physicists, following what they believe to be the conventional standards of scientific evidence, were to hastily accept an erroneous theory of semiconductor behaviour because of their enthusiasm for the breakthrough in military technology that this theory would make possible.

strictly and honestly reported, state-of-the-art, reliable, independently confirmable, judiciously balanced, broadly critiqued, authoritative'. These, he argues, are socially evaluated criteria for which there can be no non-social substitute.[29]

Like so many debates, there are no clear answers here and the situation is well summed up by Derry who argues that extreme views do not do justice to the subtle complexities involved in questions of the proper relationship between science and values. He argues that science and values are related to each other, but the relationship is not hierarchical.[30] Baber also has some useful insights in this regard; he argues that science is both constructed and natural. He describes it as being like a map; it represents a reality but it is also a political and social construct. Thus, he argues that there is no need to choose one over the other.[31]

This book will take the approach that there are indeed objective facts to be discovered, and that the scientific disciplines have much to be recommended in providing a baseline of legitimacy for health regulations. Yet, it also recognizes that pure objectivity is often neither possible nor plausible in the health sciences, particularly in the regulatory arena where uncertainties abound. While scientific facts may exist on a neutral basis, their interpretation will often be subject to values. To this end, much of their legitimacy derives from the scientific procedures and protocols followed. It is worth noting that some scientific disciplines are arguably more resistant than others to challenges. Atik writes that the 'hardness' of a hard science refers in part to the degree of systemization it enjoys and in part to the universality of its core claims. Biology is generally conceded to be 'softer' than mathematics, physics or chemistry, and so may be more open to the operation of relativity. Human biology is likely to be 'softer' again. The human body, animals, plants, and the interplay of social factors are all complex systems, about which strong scientific assertion often breaks down, introducing the possibility of multiple outcomes.[32] This proposition highlights the problems faced in international trade disputes given that the science which underlies health regulation is for the most part so-called soft science. As will be argued in Chapter 11, it also shows the importance of proper procedures and protocol in scientific studies.

---

[29] Lowrance, *supra* note 13 at 48.

[30] Derry, *supra* note 12 at 156.

[31] Zaheer Baber, 'Science in an Illiberal Democracy: The Genesis and Implications of Singapore's Biomedical Hub' (Paper presented to the Conference on Public Science in Liberal Democracy: The Challenge to Science and Democracy, University of Saskatchewan, Saskatoon, October 14–16, 2004).

[32] Atik, *supra* note 14 at 747.

Gaps in scientific knowledge have been referred to as 'trans-science', that is, 'questions which can be asked of science yet cannot be answered by science'.[33] Walker describes trans-science as uncertainty where scientists cannot even perform experiments to test the hypothesis.[34] Various constraints account for this, including those that are technological, informational, and ethical.[35] Wagner suggests that trans-science can be placed on the end of a spectrum, where the other end holds mechanistic science, being science characterized by almost universal agreement among scientists on certain theories such as the rate of acceleration of falling objects on earth.[36] In the middle of Wagner's spectrum is scientific judgment where two scientists may dispute the proper methodology or the proper interpretation of the same data with their dispute based on differences in scientific judgment.[37] Trans-science, on the other hand, involves one of two conditions, either scientists would ultimately agree that selection of the most appropriate hypothesis among a range of possible alternatives is based not on data or scientific experimentation, but on non-scientific factors; or the magnitude of the difference between the two positions is substantial. Where trans-science issues arise, regulatory decision-makers must use policy considerations to fill in the gaps left by science.[38]

Regarding the wider public challenge to science, Skogstad and Hartley find that societies divide within and across one another in their views of the authority of science and their beliefs regarding what constitutes an appropriate role for science in public policy processes.[39] Overall, they find that in liberal democracies, the traditional authority that scientific knowledge has enjoyed as a basis of legitimate policy-making is eroding. Science has come under attack for being elitist and non-responsive to public concerns.

Recent surveys show a range of attitudes towards science. In a recent Eurobarometer survey, for example, it was found that on average only 17 per cent of respondents in the EU had faith in science to resolve any problem (37

---

33 A. Weinberg, 'Science and Trans-science' (1972) 10:2 Minerva 209 at 209.

34 See discussion in Weinberg, *ibid.*

35 Wendy Wagner, 'The Science Charade in Toxic Risk Regulation' (1995) 95:7 Colum. L. Rev. 1613 at 1621. She notes, for example, that it is not ethically acceptable to test potentially toxic chemicals directly on humans.

36 *Ibid.* at 1620. Wagner notes that such theories are characterized by well-established hypotheses, tests that have been run thousands of times, and non-controversial data interpretation.

37 *Ibid.*

38 See discussion of science policy in section 10.4.2.

39 Grace Skogstad and Sarah Hartley, 'Science and Policy-Making: The Legitimate Conundrum' (Paper presented to the Conference on Public Science in Liberal Democracy: The Challenge to Science and Democracy, University of Saskatchewan, Saskatoon, 15–16 October 2004).

per cent in Greece, 25 per cent in Spain, 20 per cent in Portugal, 8 per cent in Finland and the Netherlands, 7 per cent in Denmark, and 3 per cent in Sweden).[40] On the other hand, a 2005 survey in the UK found that most people (86 per cent) think science makes a good contribution to society. Further, it was found that a clear majority of people (around two-thirds) trust scientists, although it was noted that certain newspaper readers seemed to have lost trust in scientists. This survey also showed significant variation of trust in scientists working for different organizations, with those working for industry or government much less trusted than those working in universities or for charities.[41] In the US, a 2007 survey of 1500 New York residents found that they lean heavily on scientists as they form opinions on agricultural biotechnology.[42]

Key reasons Skogstad and Hartley identify for the erosion of trust in science which they perceive are a political cultural shift which has seen a decline in deference to authority and the elevation in importance of democratic norms of public participation, public deliberation, and government accountability; along with new understandings of science which stress its limitations as a basis of decision-making.[43] One particular aspect of science that may be seen to reduce its legitimacy in the public sphere is its vulnerability to capture by interest groups. Michaels and Monforton argue that opponents of public health regulations often try to 'manufacture uncertainty' by questioning the validity of scientific evidence on which the regulations are based. They also argue that these opponents use the label 'junk science' to discredit research that threatens powerful interests.[44] Michaels and Monforton argue that the label 'junk science' was invented and publicized in the US to denigrate science supporting environmental regulation and was spawned by industrial interests. They argue that these interests are well-funded and use public forums effectively to attack the scientific basis of

---

[40] Eurobarometer 55.2, December 2001. See also The Royal Society, National Forum for Science – 'Do We Trust Today's Scientists?' A Report by Wendy Barnaby, 6 March 2002.

[41] Ipsos MORI, 'Science in Society: Findings from Qualitative and Quantitative Research' (Conducted for the Office of Science and Technology, Department of Trade and Industry, UK, 2005).

[42] 'Survey Examines Americans' Trust in Science, Approach to Scientific Issues' (News Release, 1 May 2007, University of Wisconsin, online at: www.news.wisc.edu/releases/13733, date accessed: 22 October 2007).

[43] Skogstad and Hartley, *supra* note 39.

[44] The term 'junk science' was coined by Huber who claimed that plaintiffs' claims in toxic tort actions are often based on 'junk science', which he described as 'the mirror image of real science, with much of the same form but none of the substance'. P.W. Huber, *Galileo's Revenge* (New York: Basic Books, 1991).

public health standards.[45] However, it is conceivable that in some situations, those using the term 'junk science' to oppose regulations may be justified in doing so. The truth is likely to be found somewhere in the middle, with science vulnerable to capture by both those who advocate and those who oppose regulation.

Capture by interest groups is an important consideration in any discussion about science in the WTO. Kerr argues that underlying the notion of a science-based system is the assumption that scientists are independent. In the context of the SPS Agreement, the reliance on science assumes that those making decisions with trade policy implications are independent of political and other vested interests.[46] Arguing that this assumption is flawed, Kerr cites the example of biotechnology where much research is undertaken within private industry. He suggests that as scientists in private industry have a direct economic stake in the research they undertake, either as entrepreneurs or as employees of the firms engaged in research, regulatory compliance, and commercialization, their independence is open to question.[47] Posner makes a similar argument, suggesting that a common mistake is to suppose that scientists are 'such admirable people that they can be safely entrusted with the ultimate responsibility for guiding scientific research'. This is a mistake, he argues, because scientists are no more admirable than any other type of worker, and their judgment is as likely to be hampered by career, financial, and ideological considerations as those of non-scientists.[48]

Kerr argues that if a science-based system is to function properly, then there must be a sufficient public regulatory scientific establishment to check the claims of private-sector scientists.[49] He argues further that even if governments appoint scientists on the basis of their ability and do not interfere in their deliberations, there is no way to ensure that they cannot be bought by vested interests. He does concede, however, that the same is true of any alternative group that might be charged with making decisions regarding science policy.[50]

---

[45] David Michaels and Celeste Monforton, 'Manufacturing Uncertainty: Contested Science and the Protection of the Public's Health and Environment' (2005) 95:S1 American Journal of Public Health S39 at S43.

[46] William A. Kerr, 'Science-based Rules of Trade – A Mantra for Some, An Anathema for Others' (2003) 4:2 The Estey Centre Journal of International Law and Trade Policy 86 at 89.

[47] *Ibid.* at 90.

[48] Posner, *supra* note 10. Posner is referring specifically to scientists' policy analysis, yet his argument would seem to be a valid one wherever scientists – knowing the potential policy implications – must exercise judgment in interpreting data.

[49] Kerr, *supra* note 46 at 90.

[50] *Ibid.* at 91.

To conclude on the question of the objectivity of science, the claim put forward by some postmodernists that all knowledge is relative is difficult to justify in the light of history and scientific knowledge. It is accepted here that there is in nature a core of objectivity, although in many cases it is also noted that such objectivity may not be known to us and indeed, given the state of knowledge at any given point in time, may not be knowable, particularly in the softer sciences. It is also accepted that values will play a role in scientific studies, particularly where data gaps or uncertainties exist.[51] Further, even where there is an objective truth that current scientific knowledge is capable of revealing, it will likely not be knowable to lay people in the absence of explanation from scientists. When this factor is combined with gaps and uncertainties in knowledge, it is not surprising that science is vulnerable to capture by both those opposed to and those supportive of regulation in any given situation. These characteristics of science do not, it is argued, necessarily diminish its value as a benchmark for determining the validity of trade-restrictive regulations. As one commentator has argued, it remains the best system we have of coordination and negotiation.[52] They do, however, require that WTO panels be aware of and realistic about the way in which science operates. This includes recognition of the complexity and uncertainty of science and the role that values play in reaching decisions. More on this will be said in the following examination of the use of science in the risk analysis process.

10.4 RISK ANALYSIS: THE ROLE OF SCIENCE AND VALUES

Regulatory decision-making processes regarding risk are generally looked upon as containing three elements: risk assessment, risk management, and risk communication. Taken together, these elements are commonly referred to as risk analysis. *Risk assessment* is the process of identifying and estimating the risk associated with an option, including evaluation of the likelihood of an event and of the consequences if that event were to occur. *Risk management* is the process of identifying, documenting, and implementing measures to reduce risk; in other words, reducing the likelihood and/or consequences of something going wrong. *Risk communication* is the process of interactive

[51] See discussion of scientific uncertainty below in section 10.4.1.

[52] Ian C. Jarvie, 'The Democratic Deficit of Science and its Possible Remedies' (Paper presented to the Conference on Public Science in Liberal Democracy: The Challenge to Science and Democracy, University of Saskatchewan, Saskatoon, 14–16 October 2004).

exchange of information and opinions concerning risk between risk analysts and stakeholders.[53]

While commonalities in the risk analysis process can be identified, there will always be variations within and across countries. For example, there are differences in the kinds of evidence that governments and the public consider suitable as a basis for decisions. Standards of proof and persuasion differ across countries, along with preferences for particular methods of technical analysis.[54] Where experts are employed to give evidence, differences exist between the manner in which they are selected, for example, whether the main criterion is their technical qualifications or their institutional affiliations and experience.[55]

Traditionally, commentators and reports have emphasized the importance of separating risk assessment from risk management.[56] They considered that scientific questions can be isolated and addressed in an objective manner through risk assessment methodologies at the beginning of the regulatory

---

[53] Michael J. Nunn, 'Allowing for Risk in Setting Standards' in David Robertson and Aynsley Kellow, eds, *Globalization and the Environment: Risk Assessment and the WTO* (Cheltenham, UK and Northampton, MA: Edward Elgar Publishing Limited, 2001) 95 at 95.

[54] Sheila Jasanoff, 'Technological Risk and Cultures of Rationality' in National Research Council, ed., *Incorporating Science, Economics, and Sociology in Developing Sanitary and Phytosanitary Standards in International Trade: Proceedings of a Conference* (Washington DC: National Academy Press, 2000) 65 at 71. Comparative studies have noted for example that the US tends to prefer formal and quantitative analytic methods. This approach reflects the notion of 'scientism', the idea that facts and values are distinct entities, and that facts, unlike values, are beyond dispute. On the other hand, European countries have given greater credence to more qualitative appraisals based on the weight of the evidence, as well as to non-scientific criteria. Daniel Lee Kleinman and Abby J. Kinchy, 'Boundaries in Science Policy Making: Bovine Growth Hormone in the European Union' (2003) 44:4 The Sociological Quarterly 577 at 586.

[55] Jasanoff suggests that on the whole, the US policy process places emphasis on experts' technical competency more than their institutional or political background. She argues that, by contrast, expert advisory bodies in other industrial nations are often more explicitly representative of particular interest groups and professional organizations. Jasanoff, *supra* note 54 at 73.

[56] Jessica Glicken Turnley, 'Risk Assessment in Its Social Context' in Dennis J. Paustenbach, ed., *Human and Ecological Risk Assessment: Theory and Practice* (New York: John Wiley & Sons, Inc, 2002) 1359 at 1359. Turnley cites various sources in this regard, including: National Research Council, *Risk Assessment in the Federal Government: Managing the Process* (Washington DC: National Research Council, 1983); M. Douglas and A. Wildavsky, *Risk and Culture* (London: University of California Press, 1982); P.C. Stern and H.V. Fineberg, *Understanding Risk: Informing Decisions in a Democratic Society* (Washington DC: Committee on Risk Characterization, National Academy Press, 1996).

process, while pure policy choices are confined to the second stage.[57] However, as the discussion here will highlight, the distinction between risk assessment and risk management, while intuitively appealing, is far from simple and is increasingly being discredited.

A few words should be said here about the SPS Agreement's treatment of risk assessment and risk management. There is no explicit division in the Agreement between risk assessment and risk management. However, the Agreement's rules do cover the normal activities that one would expect to see under both of these categories. Thus, broad distinctions may be drawn in regards to which rules are applicable to what activities. As has been noted in Part I, the Agreement directs countries to perform a risk assessment. It provides rules around the performance of risk assessment through its definition of what actually constitutes a risk assessment (in Annex A(4)) and specifies various factors which must be taken into account in conducting a risk assessment, including available scientific evidence (Article 5.2). In terms of the risk management phase of risk analysis, the SPS Agreement provides that once a risk assessment has been conducted and there is evidence of a risk, countries must ensure, *inter alia*, that negative effects on trade are minimized (Article 5.4), that the least-trade-restrictive option is taken (Article 5.6), and that there is no discrimination or disguised restriction on trade (Articles 2.3 and 5.5). While the Agreement does not state so specifically, these rules all apply to what would be considered the risk management phase of risk analysis.

The following sections focus on the risk assessment and risk management phases of the risk analysis process in the domestic context. While recognizing the difficulties in separating the phases into distinct elements in practice, they are discussed separately here in order to provide clarity to the discussion. The elements of a typical risk assessment and risk management process are examined; with a view to assessing whether and how risk assessment correlates to the notion of objective and independent scientific evidence. Given the SPS Agreement's requirement for measures to be based on scientific evidence and for performance of a risk assessment, one might expect that the process of risk assessment is one that would provide the 'objective' scientific evidence necessary to meet that requirement. However, this is not the case.

10.4.1 Risk Assessment

Risk assessments aim to estimate the likelihood of an adverse effect on

---

[57] Lowrance, *supra* note 13 at 121.

humans, animals, or plants (as well as ecological systems) posed by a specific level of exposure to a chemical or physical hazard.[58]

Risk assessment gained popularity in the US in the context of health regulation in the 1980s in order to help regulators make better decisions in the face of large quantities of scientific and medical data concerning potential health hazards posed by physical and chemical agents in the environment.[59] While the concept had been around in the context of health since at least the 1930s, its methodology evolved significantly in the 1980s and has continued to do so, such that risk assessments are now widely accepted practice in the US and most other developed countries.[60] The main purpose of a risk assessment is to provide relevant information to policy-makers and decision-makers. The information obtained from a risk assessment will not be the only information used in the final policy (risk management) decision, with other pertinent factors including social, economic, and political factors. This is because the information obtained from a risk assessment cannot answer all questions, for example, whether or not the risk is acceptable.[61] This is well illustrated by a simple example. A scientific risk assessment may help in setting a standard designed to limit the probability that an individual will develop cancer after a lifetime of exposure to a particular chemical substance to no more than one chance in a million. But the choice of the one-in-a-million goal is a policy choice that cannot be determined by science.

Risk assessment may be either quantitative or qualitative in nature. Where a *qualitative* approach is taken, probability is expressed in words, for example, the probability is high, low, negligible, etc. In *semi-quantitative* approaches, numerical values (for example, the prevalence of the disease of concern) are applied at each point for which data are available. In fully *quantitative* approaches, numerical data are applied at all points of the pathway of entry and establishment.[62] Many risk assessments are qualitative because data on relevant hazards and on exposure pathways are incomplete.[63] Complete and

---

[58] Dennis J. Paustenbach, 'Primer on Human and Environmental Risk Assessment' in *supra* note 56 at 6. Paustenbach explains that risk assessments of environment hazards are – in contrast to the analysis of the likelihood of the occurrence of any undesirable event such as an earthquake – extremely dependent on the degree of exposure. Thus, assessments of health hazards posed by sources such as food contaminants are dependent on both the potency of the agent and the level of exposure.

[59] Paustenbach, *ibid.* at 3.

[60] *Ibid.* at 4.

[61] Marion Wooldridge, 'Risk Assessment and Risk Management in Policy-making' in Robertson and Kellow, *supra* note 53 at 93.

[62] Nunn, *supra* note 53 at 98.

[63] David Wilson and Digby Gascoine, 'National Risk Management and the SPS Agreement' in Robertson and Kellow, *supra* note 53 at 166.

reliable quantitative data are usually only available for all steps in a risk assessment (that is, for all points in the potential pathway or pathways of entry and establishment of a potential disease of concern) in relatively simple cases. As well, in complicated situations with multiple possible scenarios that each have only an extremely small probability of occurrence (as is often the case in import risk assessments), the mathematics of fully quantitative assessment is problematic and not well defined. These situations can only be assessed using a qualitative or semi-quantitative approach. Even when they are possible, more quantitative approaches tend to be very resource-intensive, requiring skilled staff, large amounts of data, sophisticated computing resources, and a large time investment. Yet, despite this, a trend is arguably observable to use more quantitative approaches on the rationale that such approaches are 'better' or 'more scientific' than less quantitative approaches. Some argue that this is misguided. For example, Nunn writes that 'a poor quantitative risk assessment (e.g., one using poor data or using inappropriate quantitative techniques) can be quite misleading and far less scientific than a good semi-quantitative or qualitative assessment'.[64]

The specifics of risk assessment vary from agency to agency, and depend on the type of hazard (for example, chemical or microbiological or other imported risk) and the subject of the risk (human, animal, or plant).[65] As well, risk assessment techniques tend to be more sophisticated in dealing with some hazards than with others. The discipline is generally well developed with respect to hazards posed by chemicals in food and the environment, but less so for biological hazards.[66] Nevertheless, four basic elements of risk assessment can be identified, namely: hazard identification; dose exposure; exposure assessment; and risk characterization.

Hazard identification is 'the process of determining whether human exposure to an agent could cause an increase in the incidence of a health condition (cancer, birth defect, etc.) or whether exposure by a nonhuman receptor, for example, fish, birds, or other wildlife, might adversely be affected'. A key aspect of the determination is an examination of the nature and strength of the evidence of causation, in other words, did the exposure cause the adverse effects?[67] If the suspected hazardous substance in question is a chemical, then

---

64 Nunn, *supra* note 53 at 98 at 100.

65 Van Schothorst notes, for example, that risk assessment of foods has been developed for chemical hazards rather than for microbiological ones. M. van Schothorst, 'Microbiological Risk Assessment of Foods in International Trade' (2002) 40 Safety Science 359 at 359.

66 John D. Stark, 'An Overview of Risk Assessment' in National Research Council, *supra* note 54 at 60.

67 National Research Council (NRC), Committee on the Institutional Means for

basic information about toxicology is needed. If it is an organism, basic information about its biology and life history are required.[68]

With regard to human health, a key problem with hazard identification is a lack of definitive human data showing causation of adverse health effects. Where possible, epidemiological[69] studies will be used to show a positive association between an agent and a disease. However, it is rare that convincing causal relationships can be identified with a single epidemiological study and as such, epidemiologists usually weigh the results from several studies to determine whether there is a consistent pattern or response among them. Epidemiologists attempt to reach consensus regarding causality by weighing the evidence; however, different experts will weight the data differently and consensus often is not achieved.[70]

There are a number of difficulties in obtaining definitive epidemiological evidence, and as a result, reliance is often placed on studies of laboratory animals or other test systems.[71] A fundamental assumption of toxicology research is that results from animal studies are applicable to humans. A key disadvantage of animal studies, however, is that humans are not animals and are hardly ever exposed to the chemical at the same dose levels as those administered in animal studies, raising questions about the relevance of the results of studies conducted near the maximum tolerance dose (MTD).[72] The MTD is usually just below that which causes frank (overt) toxicity and humans will never be chronically exposed to such large doses. As such, observations of toxicity in laboratory animals usually require two acts of extrapola-

---

Assessment of Risks to Public Health Commission on Life Sciences, 'Risk Assessment in the Federal Government: Managing the Process' (National Academy Press: Washington DC, 1983). Cited in Paustenbach, *supra* note 58 at 6.

[68] Stark, *supra* note 66 at 53.

[69] Epidemiological studies involve data from groups of people who have been exposed to an agent.

[70] Dennis J. Paustenbach, 'Hazard Identification' in Paustenbach, *supra* note 56 at 88.

[71] Problems associated with epidemiological studies include the existence of too few subjects for confident conclusions; failure to control for important confounding factors; no or little data on exposure; exposure levels many times greater than the standards being considered; inadequate diagnoses; subjects lost to follow up; subjects who are qualitatively different from the population to be protected; a long latency period between exposure and disease. Lester B. Lave, 'Health and Safety Risk Analyses: Information for Better Decisions' (1987) 236:4799 Science 291 at 292.

[72] Animal data will in many cases be of uncertain relevance to humans either due to mechanism of action, pharmacokinetic differences, metabolic differences, or of concordance of a target organ. It should be noted that the maximum tolerance dose (MTD) is a concept that is relevant to hazards posed by carcinogens, however; where non-cancer hazards are in question, the concept of MTD is not used. Paustenbach, *supra* note 70 at 89.

tion to predict the human health hazards; interspecies extrapolation and extrapolation from high test doses to lower environmental doses. In most cases, there is limited information on a chemical's metabolic profile in humans; this makes it extremely difficult to identify the animal species and toxic response most likely to predict the human response accurately.[73] As a result, it is customary to assume that in the absence of clear evidence that a particular toxic response is not relevant to humans, any observation of toxicity in animal species is potentially predictive of response in at least some humans.[74]

The second element is *dose-response assessment* which is the process of using scientific data to characterize 'the relation between the dose of an agent administered or received and the incidence of an adverse health effect in exposed populations and [to] estimat[e] the incidence of the effect as a function of exposure to the agent'.[75] This process considers factors such as intensity of exposure, age pattern of exposure, and other variables that might affect response, such as sex, lifestyle, and other modifying factors. A dose-response assessment usually requires extrapolation from high to low doses, and from animals to humans, or a laboratory animal species to a species of wildlife. The results of such extrapolations can often be contentious, yet they are critical for setting standards for exposure to potentially toxic substances.[76]

The third element is *exposure assessment*, which is the process of measuring or estimating the intensity, frequency, and duration of human, animal, or plant exposure to an agent present in the environment or of estimating hypothetical exposures that might arise from the release of new chemicals into the environment. Exposure assessment is used in identifying options for control and predicting the effects of available technologies for controlling or limiting exposure. Complete exposure assessments include consideration of the magnitude, duration, schedule, and route of exposure; the size, nature, and classes of the human or animal populations exposed; and uncertainties in the estimates. All routes of exposure need to be considered, for example, the likelihood of contact with the chemical through exposure to contaminated soil, water, air, and/or food.[77]

---

73 Differences among animal species, or even strains of the same species, with respect to the metabolic handling of the chemical, can account for toxicity differences. Paustenbach, *ibid.*

74 *Ibid.* at 90.

75 Dennis J. Paustenbach, 'Dose-Response Modelling for Cancer Risk Assessment' in Paustenbach, *supra* note 56 at 151.

76 *Ibid.*

77 Stark, *supra* note 66 at 55. One area that has caused difficulty in the past is quantitatively accounting for indirect pathways of exposure, for example, the ingestion

The variations that may occur in exposure assessment can be illustrated by experience in the US where the practice of exposure assessment has changed over time. From the late 1970s, US environmental regulatory policy encouraged or mandated the use of conservative approaches when conducting exposure assessments; however, around 1985, concern arose that the use of conservative factor assumptions was producing unrealistically high estimates of exposure and that the cost of achieving the recommended clean-up levels was becoming unreasonable.[78] While exposure assessment is likely to contain less uncertainty than other steps in a risk assessment,[79] assumptions or inferences will still be required – even when actual exposure-related measurements exist.[80] Often, data will not be available for all aspects of the exposure assessment and these data may be of questionable or unknown quality. In these situations, the assessor will have to rely on judgment, inferences based on analogy with similar chemicals and conditions, estimation techniques, etc. As a result, the exposure assessment will be based on a number of assumptions with varying degrees of uncertainty.[81] For example, the Canada Pest Management Regulatory Agency keeps risk estimates regarding cancer conservative by generally overestimating exposure and risk and by using 'worst case' assumptions, for example, assuming that 100 per cent of the crop would be treated at

of particulate emissions that are deposited onto plants and soil and are subsequently eaten by grazing animals. Future research in this area will likely change views about the hazards posed by some chemicals. As well, different exposure parameters can have different impacts on final risk estimates, for example, consumption of methyl tert-butyl ether (MTBE) through drinking water has been found a more significant route of exposure than inhalation through vapours during showering. Dennis J. Paustenbach, 'Exposure Assessment' in Paustenbach, *supra* note 56 at 270 and 306.

[78] Paustenbach, *ibid.* at 192 and 269.

[79] Paustenbach notes that since about 1995, the ability to perform exposure assessments has matured to a degree that they will usually possess less uncertainty than other steps in the risk assessment; that is, scientists are generally able to quantify chemical concentrations in various media, and the resulting uptake by exposed persons, animals, or plants, if they account for all factors that should be considered. Paustenbach, *ibid.* at 191.

[80] *Ibid.* at 268.

[81] Uncertainty must be distinguished from variability. As noted, uncertainty represents a lack of knowledge about probabilities, while variability arises from heterogeneity across people, places, or time. Uncertainty can lead to inaccurate or biased estimates, whereas variability can affect the precision of the estimates and the degree to which they can be generalized. Uncertainty and variability have different ramifications for science and judgment. For example, science may require decision-makers to judge the probability of exposures being over- or underestimated for members of the exposed population, while variability requires them to deal with the fact that different individuals are subject to exposures both above and below any of the exposure levels chosen as a reference point. Paustenbach, *ibid.* at 261.

the maximum application rate, or that 100 per cent of the pesticide deposited on the skin would penetrate through the skin.[82]

The final element is *risk characterization*, which attempts to make sense of the available data and describe what it means to a broader audience.[83] It is the integration of the first three elements and the synthesis of an overall conclusion about risk that is useful for decision-makers.[84] In other words, it serves as the intermediary between risk assessment and risk management.[85] Technically, it is 'the process of estimating the incidence of a health effect under the various conditions of human exposure described in the exposure assessment'.[86]

Risk characterizations include both quantitative estimates and qualitative descriptors of risk,[87] as well as discussions about key model assumptions and data uncertainties.[88] Williams and Paustenbach (2002) argue that despite the importance of accurate and defensible risk characterizations, this step is often given insufficient attention in health risk evaluations and that many risk characterizations fall short of providing decision-makers with all of the relevant information needed to make an informed decision. The main reason for this is that unlike the first three elements, it is not possible to rely exclusively on guidance documents and formulas to properly capture the importance of the analysis.[89] Instead, risk assessors have to weigh and integrate many factors including the biology of the chemical, interspecies susceptibility, conflicting animal and human studies, statistics, regulatory history, and acceptable risk criteria.[90] Williams and Paustenbach identify the key shortcoming of risk characterizations as the failure to provide detailed information regarding uncertainties in the analysis, noting that this leads to the potential for misinterpretation of findings or false impressions of the level of confidence in reported risk estimates.[91]

The preceding discussion highlights the ubiquity of data gaps and uncertainties in risk assessment. Uncertainty may arise with respect to each element

---

[82] Pest Management Risk Agency, *Technical Paper: A Decision Framework for Risk Assessment and Risk Management in the Pest Management Risk Agency* (Ottawa: Pest Management Risk Agency, Health Canada, 2000) at 8.

[83] Pamela R.D. Williams and Dennis J. Paustenbach, 'Risk Characterization' in Paustenbach, *supra* note 56 at 293.

[84] *Ibid.* at 298.

[85] US National Academy of Sciences/National Research Council 'Red Book', 1983. Cited in Williams and Paustenbach, *ibid.* at 295.

[86] Paustenbach, *supra* note 58 at 9.

[87] Qualitative elements include the exercise of judgment in the aggregation of population groups with varied sensitivity and different exposure.

[88] Williams and Paustenbach, *supra* note 83 at 293.

[89] *Ibid.*

[90] *Ibid.*

[91] *Ibid.* at 294.

of a risk assessment, including in relation to data findings, extrapolation, and the parameters to be used in models of health risk.[92] As one commentator notes with respect to estimating environmental risks to human health: '[it]is as if you had no idea whether your wallet contained enough money to pay for coffee or to pay the national debt, and no way of finding out'.[93] The National Research Council (NRC) in the US has noted that 'because our [scientific] knowledge is limited, conclusive direct evidence of a threat to human health is rare'.[94] Uncertainty is not limited to risks to human health. In the case of plant pests, for example, Campbell suggests that given current levels of scientific knowledge (and funding) in the US, it is impossible to evaluate the likelihood or probability that any of the multitude of insets, fungi, or plant progagules – any of which could be inadvertently imported with foreign goods – might invade any of the country's agricultural or natural ecosystems.[95]

Some authors suggest that given these many difficulties, risk assessment is only in fact suitable in very limited situations, namely, where risks are not serious, uncertain, complex, large in scale, or socio-politically ambiguous.[96] In reality, however, problems tend to be addressed through the use of assumptions. Indeed, Powell writes that 'gaps in the data and holes in our understanding of the science routinely cause risk assessors to make assumptions'.[97] The NRC has estimated that a typical risk assessment consists of about 50 separate assumptions and extrapolations meaning that many risk assessments can be neither proven nor disproved scientifically.[98] Assumptions may be risk-averse, risk-tolerant, or risk-neutral. Powell notes that different assumptions can alter the outcomes significantly. He refers by way of example to different rules developed in US agencies to interpret the pathological evidence on dioxins that have resulted in the FDA's cancer potency estimate for dioxin being almost an order of magnitude smaller than the EPA's estimate and the CDC

---

[92] For a discussion of when uncertainty may arise, see Merrie G. Klapp, *Bargaining with Uncertainty* (Westport, CT: Auburn House, 1992).

[93] R. Cothern, 'Estimating Risk to Human Health' (1986) 20:2 Environmental Science and Technology 111–16. Cited in Mark R. Powell, *Science at EPA: Information in the Regulatory Process* (Washington DC: Resources for the Future, 1999) at 127.

[94] National Research Council, *Risk Assessment in the Federal Government: Managing the Process* (Washington DC: National Research Council, 1983) at 11.

[95] Faith Thomson Campbell, 'The Science of Risk Assessment for Phytosanitary Regulation and the Impact of Changing Trade Regulations' (2001) 51:2 BioScience 148 at 151.

[96] Stirling, Renn and van Zwanenberg, *supra* note 6 at 296.

[97] Powell, *supra* note 93 at 8.

[98] National Research Council, *supra* note 94.

falling somewhere in between.[99] Jasanoff relates the comments of a Dow Chemical executive that illustrates the same point. He said that 'if one wishes to eat fish caught in the Great Lakes, one had better do it in Canada. The fish is safe across the border, even though the dioxin residues have led US regulators to label it unfit for human consumption'.[100]

In some cases, assumptions are formalized through so-called *science policies* whereby the explicit use of inferences, choices, and assumptions are incorporated into risk assessments. Science policies reflect the broader goals of risk regulation, such as protecting human health.[101] An example is the US Environmental Protection Agency (EPA) which encourages risk assessors to follow guidance from the Agency's 'science policy' in conducting risk assessments. This policy contains 'determinations about how risk assessors should proceed when they encounter uncertainties involving multiple plausible accounts'. Because science policies usually specify which assumptions to use to bridge gaps in scientific knowledge, they are sometimes called 'inference guidelines' or 'default assumptions'. For example, in hazard identification, the EPA recognizes various default assumptions in its inferences from animal data to conclusions about the expected carcinogenicity of an agent in humans, including assumptions that positive effects in animal cancer studies indicate carcinogenic potential in humans, that effects seen in animals at the highest dose tested are appropriate to use in carcinogenicity assessment, and that there is a similarity in the basic pathways of metabolism that are relevant to species-to-species extrapolation of a cancer hazard.[102] An illustration of the diverse standards that can result from the use of different science policies can be seen in proposed US dioxin regulations. The use of strict default assumptions led

---

[99] Powell, *supra* note 93 at 128. The inconsistent estimates resulted from the agencies' applying the same linear cancer model but making a variety of different scientific assumptions and data treatments. The agencies' estimation procedures differed in how to extrapolate from rat to human (body weight or surface area); which pathology results were used; whether early mortality was taken into account; the assumed average human body weight (80 kg or 70 kg); and how the dose was measured (concentration in the tissue or administered dose). Using surface area to scale the administered dose between animals and humans leads to a higher potency estimate than does using body weight as a scaling factor.

[100] Sheila Jasanoff, 'American Exceptionalism and the Political Acknowledgement of Risk' in Edward J. Burger, ed., *Risk* (Ann Arbor, MI: University of Michigan Press, 1990) at 60.

[101] Vern Walker, 'Keeping the WTO from Becoming the "World" Trans-science Organization: Scientific Uncertainty, Science Policy, and Fact-finding in the "Growth Hormones Dispute"' (1998) Cornell Int'l L.J. 251 at 262.

[102] *Ibid*. at 262.

the EPA to propose a very restrictive limit that was two or three orders of magnitude higher than Canada's limit (which was based on the same data).[103]

In other cases, use of assumptions will be more of an *ad hoc* nature and risk assessment will be influenced by subjective values as risk assessors decide how the information they do have ought to be interpreted. For example, someone who believes that industry is overregulated might emphasize the point that uncertainties allow the possibility that a substance is not very dangerous. Someone who is more concerned about public health might instead emphasize the opposite point, that the uncertainty allows the possibility that the substance is more dangerous than can be documented.[104] Wildavsky and Douglas argue in this regard that it is impossible to expect scientists or experts to provide an objective assessment of risk. They state that: 'Everyone, expert and layman alike is biased. No one has a social theory above the battle . . . judgments of risk and safety must be selected as much on the basis of what is valued as on the basis of what is known'.[105]

Regardless of how assumptions are arrived at, it is instructive to note Powell's statement that risk assessment is really 'a negotiated consensus among experts' and that in many cases what is perceived as a 'scientific fact' remains an 'unvalidated, sometimes untestable, negotiated scientific judgment'.[106] It has also been said that in the face of data gaps and uncertainties, the process is not so much an exercise in finding scientific truths as it is a complex set of judgments about whether we believe the evidence presented.[107]

An important question for WTO law arises from the uncertainties and complexities inherent in risk assessment, and the use of science policies. Namely, how – in the context of a risk assessment – one can determine whether the decision made is a rational one? Crawford-Brown, Pauwelyn, and Smith suggest that rationality usually involves three aspects of science: (i) balancing categories of evidential reasoning; (ii) judging data and theories; and (iii) considering *desiderata* of rationality.[108] The first aspect, balancing categories of evidential reasoning, involves deciding how to weigh different risk estimates based on the persuasiveness of a particular mode of reasoning. Modes of reasoning range from direct empirical evidence, where evidence of effects or lack thereof is available, to existential insight which is rooted in an

---

103 Powell, *supra* note 93 at 329.

104 Derry, *supra* note 12 at 155.

105 Taverne, *supra* note 19 at 200.

106 Powell, *supra* note 93 at 127.

107 Telephone conversation with William Leiss, 27 February 2006.

108 Douglas Crawford-Brown, Joost Pauwelyn and Kelly Smith, 'Environmental Risk, Precaution, and Scientific Rationality in the Context of WTO/NAFTA Trade Rules' (2004) 24:2 Risk Analysis 461 at 465.

entirely subjective judgment of risk.[109] Second, risk assessors must determine the quality of data and theories used. Data quality will depend on a number of factors, including its statistical properties, methodology, reliability, relevance, and the level of scrutiny by the scientific community. The epistemic quality of a theory will depend on various criteria including empirical success, precision of fits to data, conceptual success in explaining previously poorly understood phenomena, research velocity, and pragmatic success. Finally, risk assessors have to weight lines of evidential reasoning based on broad guidelines for rationality. Bunge describes these as the seven *desiderata* of rationality and they include: (i) conceptual clarity for all terms used in discourse; (ii) logical deduction where possible; (iii) methodological rigour when applying tools such as statistical analysis; (iv) practicality to the extent that results can be reached with a reasonable allocation of resources; (v) ontological realism, or the incorporation of all important determinants of risk in the process of assessment; (vi) epistemological reflection; and (vii) valuation, or the assurance that the assessment determines the risk of loss to the most central values of society.[110]

The task of determining scientific rationality is complicated by a lack of codified rules in science and as a result, two scientists may arrive at different conclusions as to the rational support for a particular line of reasoning.[111] Despite the lack of rules, Crawford-Brown, Pauwelyn, and Kelly place some faith in the concept of an *epistemic threshold* whereby it can be asserted that no risk estimate should be considered as a candidate for scientific belief if it has not passed at least some minimum requirement of quality, in other words, it must have minimal epistemic status. They note that this threshold is one that is aspired to by the scientific community through the peer review process.

In conclusion, on the subject of risk assessment, the above examination has found that – contrary to the notion of science as a purely technocratic process which yields 'correct' and 'objective' answers based on scientific truths – the process is in fact a highly subjective process. As Powell observes, it is a 'messy business'.[112] The irony is that risk assessment, with its attendant uncertainties and inevitable value judgments, can be manipulated either to support or to oppose regulation. It can produce risk estimates which err on the side of

---

[109] *Ibid.* The authors identify five modes of reasoning: direct empirical evidence; semi-empirical evidence; empirical correlation; theory-based inference; and existential insight.

[110] *Ibid.* Citing M. Bunge, 'Seven Desiderata for Rationality' in J. Agassi and I. Jarvie, eds, *Rationality: The Critical View* (Dordrecht: Martinus Nijhoff Publishers, 1987) at 5–16.

[111] Crawford-Brown, Pauwelyn and Kelly, *ibid.* at 645.

[112] Powell, *supra* note 93 at 8.

risk acceptance, while at the other end of the scale, it can produce risk estimates that are extremely conservative and unwarranted by the scientific evidence available. If this aspect of science is not recognized by WTO adjudicators, there is a risk that almost any trade-restrictive SPS measure could be justified in the name of science if a domestic government plays its cards strategically.

10.4.2 Risk Management

Science cannot answer all of society's questions concerning risks to health. As a result, no regulatory process is wholly scientific. To the contrary, regulatory outcomes are usually designed to respond to a particular social, economic, or political context. It is widely recognized that while scientific analysis (through risk assessment) plays a key role in determining how to attain a given health objective, the choice of that objective reflects societal values as to which science may provide little, if any, guidance. In other words, science may inform the regulatory process but cannot, by itself, determine the result with particularity; this is the job of risk managers.[113]

Risk management is the process by which regulatory decision-makers decide what action to take in the face of estimates of risk arising from a risk assessment. It typically incorporates non-scientific factors that do not officially form part of the technical aspect of risk assessment, such as values, culture, politics, social factors (for example, the goals of public health protection), economic interests,[114] and public opinion.[115] It was defined more specifically as follows in the NRC's 1983 publication – *Risk Assessment in the Federal Government: Managing the Process* – which is better known simply as the 'Red Book':[116]

---

[113] David A. Wirth, 'The Role of Science in the Uruguay Round and NAFTA Trade Disciplines' (1994) 27:4 Cornell Int'l L.J. 817 at 833.

[114] For example, cost-benefit analysis which involves enumerating all tangible and intangible societal costs and benefits associated with a particular decision or option. These are generally valued in a common (typically monetary) unit. Alternatively, cost-effectiveness analysis may be undertaken in which the benefits of a programme are expressed as a unit of output or outcome, such as 'number of cases detected' or 'number of lives saved'. Williams and Paustenbach, *supra* note 83 at 346.

[115] See Wirth, *supra* note 113 at 834. See also Williams and Paustenbach, *ibid.*

[116] National Research Council (NRC), Committee on the Institutional Means for Assessment of Risks to Public Health Commission on Life Sciences, 'Risk Assessment in the Federal Government: Managing the Process' (Washington DC., National Academy Press (1983). Cited in Paustenbach, *supra* note 58 at 6.

> [T]he process of evaluating alternative regulatory actions and selecting among them. Risk management, which is carried out by regulatory agencies under various legislative mandates, is an agency decision-making process that entails consideration of political, social, economic, and engineering information with risk-related information to develop, analyze, and compare regulatory options and to select the appropriate regulatory response to a potential chronic health hazard. The selection process necessarily requires the use of value judgements on such issues as the acceptability of risk and the reasonableness of the costs of control.

The task of risk management is complicated where risk assessments lack certainty and contain value judgments. Wagner argues that scientific risk assessments tease policy-makers with the prospect of providing definitive guidance for regulatory management decision-making. However, in reality, the information that most scientific research provides to health regulators is incomplete and inconclusive, in both identifying and quantifying risks.[117] She identifies three specific problems in incorporating science (through risk assessment) into regulation (through risk management). First, the scientific studies relied upon might be of poor quality ('bad science'), or the agency might not include the best science in its analysis.[118] The second problem is one of transparency, where an agency fails to explicitly identify the separate roles scientific research and value choices play in reaching a regulatory decision. The third problem relates to the production of scientific research, which is important in anchoring regulation by reducing uncertainties.[119]

The more severe and the greater the number of uncertainties, the more the usefulness of science as a regulatory decision-making tool is diminished. As discussed, the existence of uncertainty is particularly pertinent in trade-related discussions. Decision-making in the face of scientific uncertainty has been described as the major battleground in the debate over science-based rules of trade.[120]

10.4.3 Precaution in Risk Analysis

Many commentators have argued that the so-called precautionary principle should be used to bridge the gap between scientific uncertainty and risk

117 Wendy E. Wagner, 'The "Bad Science" Fiction: Reclaiming the Debate over the Role of Science in Public Health and Environmental Regulation' (2003) 66:4 Law & Contemp. Probs 63 at 64.

118 Bad science can result from a variety of imperfections in research such as falsification of data, to researcher bias, or incompetence. Wagner, *ibid.*

119 *Ibid.* at 67.

120 Kerr, *supra* note 46 at 91.

regulation.[121] The term 'precautionary principle' comes from the German word *Vorsorgeprinzip* which is one of five fundamental principles recognized in German law as constituting the basis for environmental policy. Interpretation and implementation of the precautionary principle takes a number of forms. One of the best-known definitions is that set out in Article 15 of the Rio Declaration signed at the UN Conference on Environment and Development in Rio de Janeiro in 1992,[122] namely:

> In order to protect the environment, the precautionary approach should be widely applied by States according to their capabilities. Where there are threats of serious or irreversible damage, lack of full scientific certainty shall not be used as a reason for postponing cost-effective measures to prevent environmental degradation.

Another widely cited definition is the so-called 'Wingspread' definition which was proposed in 1997 by a group of US social scientists. Unlike the Rio Declaration, this version affirmatively states that action should be taken in the absence of scientific certainty.[123] It states that:[124]

> When an activity raises threat of harm to human health or the environment, precautionary measures should be taken even if some cause and effect relationships are not fully established scientifically.

Numerous other definitions of the precautionary principle exist, making it impossible to identify any one uniform description. This has led to criticisms that the principle is overused without a clear understanding of its meaning and consideration of its implementation.[125]

Sunstein argues that because the precautionary principle does not take into account the full set of risk trade-offs involved (given that risks are on all

---

[121] United Nations University Institute of Advanced Studies, *Trading Precaution: The Precautionary Principle and the WTO* (Yokohama: UNU IAS, 2005) at 3.

[122] United Nations Conference on Environment and Development (1992, Rio De Janeiro) *Final Declaration, Principle, 15.*

[123] Bernard Goldstein and Russellyn S. Carruth, 'The Precautionary Principle and/or Risk Assessment in World Trade Organization Decisions: A Possible Role for Risk Perception' (2004) 24:2 Risk Analysis 491 at 492.

[124] C.A. Raffensberger and J. Tichner, eds, *Protecting Public Health and the Environment: Implementing the Precautionary Principle* (Washington DC: Island Press, 1999). Cited in Goldstein and Carruth, *ibid.*

[125] United Nations University Institute of Advanced Studies, *supra* note 121 at 3. This issue has been discussed in the WTO Committee on Trade and Environment where it has been noted that the difficulty of further integrating precaution in the WTO lies in the lack of an internationally agreed definition of the precautionary principle. Meeting of the WTO Committee on Trade and Environment, 5–6 July 2000, WTO Doc. WT/CTE/M/24.

sides), it is literally incoherent and that it can only become operational because of identifiable features of human cognition. The principle can, he argues, become paralysing where it requires regulation to protect against risk and in doing so deprives society of significant benefits and hence produces a subsidiary set of harms. In other words, the precautionary principle does not help regulators decide which risks to regulate.[126] Taking this argument further, he suggests that the only way for the precautionary principle to work would be as an 'Anti-Catastrophe Principle' designed for special circumstances where it is not possible to assign probabilities to potentially catastrophic risks.[127] That is, where there are uncertainties, regulators might identify worst-case scenarios and choose the approach that eliminates the worst of these (an approach known as 'following maximin').[128] Sunstein's discussion of catastrophic risks would seem to exclude many of the health risks associated with SPS measures, focusing instead on potentially catastrophic risks such as global warming or terrorism.

If one accepts Sunstein's argument that there are risks on all sides, and short of an anti-catastrophe principle, is it possible to find any useful meaning in the so-called precautionary principle? Clearly it would be absurd to suggest that risk regulators should never act with caution. Even Sunstein concedes that both individuals and sensible governments will embrace some form of precaution in the face of many risks.[129] Hrudey and Leiss argue that 'done well, risk management is inherently precautionary in the sense that it should make use of effective risk assessment to predict, anticipate, and prevent harm, rather than merely reacting when harm arises'.[130]

Cosbey suggests that there are certain identifiable characteristics which can be labelled 'precautionary'. He identifies the following: (i) *preventative anticipation* (the notion of taking action in advance of full scientific proof of its necessity; (ii) *room for error* (because we are dealing with complex systems, a deliberate margin for error should be left open); (iii) *proportionality of*

---

[126] Cass R. Sunstein, *Laws of Fear – Beyond the Precautionary Principle* (Cambridge, UK: Cambridge University Press, 2005) at 29. Sunstein gives the example of a 'drug lag', produced when a government takes a highly precautionary approach to the introduction of new medicines onto the market. If a government insists on such an approach, it will protect people against harms from inadequately tested drugs, but it will also prevent people from receiving potential benefits if they were able to take the drug.

[127] *Ibid.* at 5.

[128] *Ibid.* at 109.

[129] *Ibid.* at 22.

[130] Steve E. Hrudey and William Leiss, 'Risk Management and Precaution: Insights on the Cautious Use of Evidence' (2003) 111:13 Environmental Health Perspectives 1577 at 1577.

response (the cost of proposed measures must not be out of proportion with the expected benefits, including costs avoided); (iv) *onus of proof* (should be on the proponent of any new product to provide an adequate level of evidence of its safety; (v) *a search for greater certainty* (any precautionary measures taken must be accompanied by a search for scientific uncertainty, and periodic re-evaluation of the measures in light of new evidence; (vi) *openness of process* (there must be transparency in the process of decision-making, timely distribution of information, and mechanisms for input from all those affected); (vii) *emphasis on finding alternatives* (if alternative products or technology exist that have the same valuable qualities without the same risks of negative effects, the alternatives are to be preferred).[131]

Hrudey and Leiss argue that the precautionary principle may be most relevant in cases where the problem is so poorly understood that there is no prospect for a risk assessment that has any degree of confidence attached.[132] The key question in such cases, they suggest, is not whether to be precautionary, but how precautionary we ought to be in specific cases.[133] On this account, the characteristics Cosbey identifies may be more or less visible depending on the circumstances of the particular case.

The application of these kinds of precautionary actions in *risk assessment* would recognize uncertainties and would entail that decisions regarding what evidence to believe reflect conservative or cautious approaches to risk. This would mean, for example, that in cases of uncertainty and doubt, greater weight would be given to evidence suggesting risk, and that the vulnerability of very sensitive members of the population be used to establish safety factors.[134] In *risk management*, precaution would entail allowing Type II errors where benign products are banned or strictly regulated. From an *international trade perspective*, recognition of the precautionary principle would entail giving the benefit of doubt to importing countries, rather than to the producer where there is any doubt about safety, even if the risks cannot be proved with certainty to exist.[135]

---

[131] Aaron Cosbey, *A Forced Evolution? The Codex Alimentarius Commission, Scientific Uncertainty and the Precautionary Principle* (Winnipeg: International Institute for Sustainable Development, 2002) at 11.

[132] Hrudey and Leiss, *supra* note 130 at 1580.

[133] *Ibid.* at 1577.

[134] Denise Prévost, 'What Role for the Precautionary Principle in WTO Law After *Japan – Apples*?' (2005) 2:4 Journal of Trade and Environment Studies 1 at 12. See also Stirling, Renn and Van Zwanenberg, *supra* note 6 at 297 who find also that a more elaborate appraisal is required in the case of risk assessments that deal with serious, complex, and large-scale risks.

[135] *EC – Measures Concerning Meat and Meat Products (Hormones) Complaint by the United States* (1997), WTO Doc. WT/DS26/R/USA (Panel Report), at para

Regardless of whether or not the 'precautionary principle' is accepted as part of international trade law to be applied in interpreting the SPS Agreement, the reality is that precaution in some form is a feature of regulatory decision-making in the face of risk. As such, the question of precaution is likely to continue to cause conflict.[136] It will cause conflict because of its potential to disguise trade barriers in the name of health protection. As such, there is likely to be contestation as to when precautionary action may be accepted such that the burden of proof shifts towards ensuring protection of health. This contestation will be complicated by the fact that what is deemed to be an appropriate precautionary approach will vary depending on the circumstances, including factors such as the legal and socio-political culture, the nature of the health problems, and the availability of precautionary measures.

Arguably, a generalization might be made that the threshold for precautionary action ought to be higher when the potential risks involve serious or irreversible harm (such as cancer in humans or biological pest invasions that would destroy plants[137]) and lower when the potential harm is only minor.[138]

10.4.4 Risk Assessment and Risk Management: Can they be Separated?

It was suggested in the introduction to this section that risk assessment and risk management cannot be as easily separated as has traditionally been thought. The NRC first made the formal distinction between risk assessment and risk management in its 1983 publication, the 'Red Book'. This distinction has played a key role in the organization of regulatory agencies both in the US and elsewhere. For example, the EU's recently developed food risk analysis model is designed to ensure a clear-cut separation between risk assessors, who discuss facts, and managers, who discuss values. The rationale behind the separation is to insulate scientific activity from political pressure.[139] More specifically, the European Commission set up the European Food Safety Authority in 2002 which was intended to be an independent scientific body lacking any actual decision-making power. Such a body was, it would hoped, guarantee the independence and objectivity of the scientific process and

---

4.202. *EC – Measures Concerning Meat and Meat Products (Hormones) Complaint by Canada* (1997), WTO Doc. WT/DS48/R/CAN (Panel Report), at para 4.212.

[136] See for example, Prévost, *supra* note 134.

[137] Thomson Campbell, *supra* note 95 at 149.

[138] United Nations University Institute of Advanced Studies, *supra* note 121 at 3.

[139] For a comprehensive discussion of the EU's food risk analysis model, see generally Alberto Alemanno, *Science and EU Risk Regulation: The Role of Experts in Decision-Making and Judicial Review* (Milan: Università Commerciale L. Bocconi, 2007).

remove value judgments from the risk assessment process, as well as enhance the democratic legitimacy of the actual risk management decision-making process.[140]

The EU experience to date has highlighted the enormous difficulties in trying to separate risk assessment and risk management, or in other words, science and values. As Alemanno writes, the supposed distinction between risk assessment and risk management lacks credibility as it is cut off from the reality of scientific and political work processes, and fails to recognize the many value judgments that are implicit in risk assessment.[141] Chalmers argues that the European Food Safety Authority is never simply conveying information, but is inevitably endorsing a particular ideological model of politics.[142]

The danger in maintaining a strict institutional divide between risk assessment and risk management is, as Alemanno points out, that judgments, uncertainties, and biases in the scientific assessments will go unrecognized.[143] From a trade law perspective, this creates difficulties for panels whose role is to sort out genuine health measures from those that have a protectionist intent.

Similar issues have arisen in the US where the NRC has revised its position on separation as set out in the Red Book. It has now accepted that risk analysis is an iterative process. Risk assessment is not followed in lock step by risk management and then risk communication. Rather, risk managers move back and forth between the parts of the risk analysis using different sciences, the cultural frameworks of the respective countries, and domestic and international political and economic considerations in making their decisions.[144] As Busch *et al.* state:

> . . . it is neither feasible nor appropriate to separate risk analysis into a purely technical phase (assessment) and a subsequent political phase (management). A strict division between the technical and the political impedes public deliberation on important dimensions of expert judgment and interferes with the recognition and resolution of both scientific and regulatory issues. Indeed, as one commentator has put it, 'both science and policy could be better served by recognizing the scientific

---

[140] Alemanno, *ibid.* at 7.

[141] *Ibid.* at 10.

[142] D. Chalmers, 'Food for Thought: European Risks and Traditional Ways of Life' (2003) 66 The Modern Law Review 532 at 543.

[143] Alemanno, *supra* note 139 at 11.

[144] National Research Council, ed., *Incorporating Science, Economics, and Sociology in Developing Sanitary and Phytosanitary Standards in International Trade: Proceedings of a Conference* (Washington DC: National Academy Press, 2000) at 21. See also P.C. Stern and H.V. Fineberg, *Understanding Risk: Informing Decisions in a Democratic Society* (Washington DC: Committee on Risk Characterization, National Academy Press, 1996).

> limits of risk-assessment methods and allowing scientific and policy judgments to interact to resolve unavoidable uncertainties in the decision-making process."[145]

This essentially means two things for panels and the Appellate Body in interpreting the SPS Agreement. First, while the Agreement requires countries to perform a risk assessment, that exercise cannot always be expected to reveal objective and politically untainted determinations. Thus, consideration must be given to how risk assessments can best be reviewed to ensure at least that they are not merely being used as a front for protectionist measures. This will be discussed in Chapter 11. Second, the risk management phase of risk analysis must always be expected to involve non-scientific factors and these must be reviewed within the bounds of the constraints outlined in Articles 2.3, 5.4, 5.5, and 5.6.

10.5 A ROLE FOR THE PUBLIC IN RISK ANALYSIS?

The discussion so far has challenged the traditional view that risk management is the only phase of risk analysis where value judgments enter the fray.[146] Even quantitative risk assessments may reflect not only scientific knowledge but also value judgments.[147] Such judgments enter virtually every stage of the risk assessment process. From the beginning, the initiation of a risk assessment is motivated by a perceived problem where something of social value (for example, public health) is under stress.[148] As the NRC argues, culture influences scientists' perceptions of what questions to ask and which risks to evaluate.[149] Priorities have to be set, for example, in determining which

---

145 Lawrence Busch, *et al.*, *Amicus Curiae Brief Submitted to the Dispute Settlement Panel of the World Trade Organization in the Case of EC: Measures Affecting the Approval and Marketing of Biotech Products* (WT/DS291, 292 and 293, 30 April 2004).

146 *Ibid.* at 4.

147 Nunn, *supra* note 53 at 100. Busch *et al.* refer to the role of policy values and culture in their *Amicus Curiae* brief submitted in the *EC – Biotech Products* case. In particular, they argue that risk assessments cannot be considered objective given the wider background assumptions and value commitments that are unavoidably embedded within scientific knowledge generated for policy applications. Busch *et al., ibid.* at 5.

148 Turnley, *supra* note 56 at 1371.

149 National Research Council, *supra* note 144 at 11. See also Busch *et al.* who argue that scientific risk assessments in particular national contexts are necessarily shaped by contingencies – scientific and cultural – which help determine the selection of particular analytic foci and strategies as relevant or valid. Busch *et al.*, *supra* note 146 at 5.

micro- organisms/chemicals/foods are of most concern and should be subject to risk assessment.[150] Beyond this, political and cultural contexts influence both 'the initial identification of hazard and subsequent attempts to assess the magnitude, seriousness, and distribution of potential harms'.[151]

Values will, as noted, come from scientists themselves; however; there is also the possibility of public sentiment playing a role throughout the risk analysis process. As Slovic argues, lay people have a rich conceptualization of risk and accordingly both public and experts have something to offer and must respect the other.[152] A number of critics have argued that risk assessment in the domestic context is 'undemocratic, morally bankrupt, and ethically impoverished'[153] and that it is a means to rationalize government decision-making and its alleged dominance by industry.[154] It is argued here that risk assessment should not be written off so quickly. First, where all the facts are known (a rare event indeed), risk assessment only characterizes risk, people make political and moral decisions about how much risk is acceptable. Risk assessment is used to inform decision-making; but risk decisions are, ultimately, public policy choices. Second, the more common case where a risk assessment is full of uncertainties speaks to the possibility for domestic governments to involve the public in shaping the approach taken by scientists in making assumptions and interpreting data. This is something that is increasingly being recognized around the world by both governments and scholars. For example, Stirling, Renn, and Van Zwanenberg argue that the greater the ambiguities, the greater the need for public participation.[155]

---

150 Van Schothorst, *supra* note 65 at 365.

151 Busch *et al.*, *supra* note 146 at 15.

152 P. Slovic, 'Perception of Risk' (1987) 236:4799 Science 280 at 285. The European Commission, in a recent report concerning 'quality of life', takes this view in arguing that risk analysis should take into account perception, costs, and benefits in determining quality of life. For example, noting that 'dread' is a stress factor that might impair health, it argues that qualitative risk characteristics and other quality of life indicators should be incorporated in risk assessments. European Commission – Health and Consumer Protection Directorate-General, *Final Report on Setting the Scientific Frame for the Inclusion of New Quality of Life Concerns in the Risk Assessment Process* (Brussels, 2003) at 12. See however, Sokal, who questions the desirability of 'local knowledges', asking '. . . when local knowledges conflict, *which* local knowledges should we believe? In many parts of the Midwest, the "local knowledges" say that you should spray more herbicides to get bigger crops. Its old-fashioned objective science that can tell us which herbicides are poisonous to farm workers and to people downstream . . . Another word for "local knowledges" is *prejudice*.' Alan Sokal, 'A Plea for Reason, Evidence and Logic' (1997) 6:2 New Politics 126.

153 Gail Charnley and E. Donald Elliott, 'Democratization of Risk Analysis' in Paustenbach, *supra* note 56 at 1406.

154 *Ibid.* at 1405.

155 Stirling, Renn and van Zwanenberg, *supra* note 6 at 287.

In a 2003 report prepared for Health Canada, Jardine *et al.* emphasize that stakeholder involvement and communication are key elements that should be included in a framework for human health and ecological risk assessment and management.[156] This must involve clearly communicating data sources and assumptions to the public (and other interested stakeholders).[157] Jardine *et al.* note that the importance of stakeholder involvement has been accepted in international fora, citing the outcome of a 2000 international workshop on risk-based decision-making which concluded that 'each step in risk assessment is heavily dependent on its specific cultural and regulatory context. To make results widely acceptable, the process needs to be contextualized in the sociocultural environment. It must also be framed within a participation process in which all stakeholders are involved early on in the process of characterizing and assessing risks rather than presenting a finalized solution to them'.[158]

Jardine *et al.* refer to a large number of instances where recognition has been given to the importance of public participation, including in the risk assessment process.[159] For example, the US Presidential/Congressional Commission on Risk Assessment and Risk Management, 1997, found that valuable information or perspectives may emerge during any stage of the process and that it is therefore important that the process be both iterative and flexible enough to accommodate changing or new information. The NRC has made similar arguments, suggesting in a 1996 report that risk management requires an 'analytic-deliberative' process whereby analysis and deliberation are complementary and integrated throughout risk characterization. Deliberation frames analysis, analysis informs deliberation, and the process benefits from feedback between the two. Each step of the process must, the report argued, have an appropriately diverse participation or representation of

---

156 Cindy G. Jardine, *et al.*, 'Risk Management Frameworks for Human Health and Environmental Issues' (2003) Part B:6 Journal of Toxicology and Environmental Health 569 at 570.

157 Diahanna L. Post, 'The Precautionary Principle and Risk Assessment in International Food Safety: How the World Trade Organization Influences Standards' (2006) 26:5 Risk Analysis 1259 at 1270.

158 Jardine *et al.*, *supra* note 156 at 612. Jardine *et al.* refer to the International Workshop on Promotion of Technical Harmonization on Risk-based Decision Making, held in Stresa and Ispra, Italy, May 2000 by the European Commission and its Directorate General Joint Research Centre. There is a growing body of literature on the issue of public participation in risk decisions. The report by Jardine *et al.* provides an excellent introduction to the topic. See also Gene Rowe and Lynne J. Frewer, 'Public Participation Methods: A Framework for Evaluation' (2000) 25:1 Science, Technology and Human Values 3.

159 See their report generally for a discussion of approaches in different countries. Jardine *et al.*, *ibid.*

the spectrum of interested and affected parties, of decision-makers, and of specialists in risk analysis.[160]

In practice, countries often do a poor job of involving the public, even during the risk management phase, and there are no easy answers as to how best to do so, with different situations requiring different responses.

---

[160] National Research Council, *Understanding Risk – Informing Decisions in a Democratic Society* (Washington DC: National Academy Press, 1996).

PART IV

The WTO: rules and cases

11. A science-based approach

11.1 INTRODUCTION

The SPS Agreement turns to science as a means of distinguishing between protectionist and legitimate health regulations. Section 11.2 describes the key provisions that require SPS measures to be based on scientific evidence. It also takes note of those GATT provisions that have informed panel and Appellate Body decisions on the more specific provisions of the SPS Agreement. In Section 11.3, it discusses examples of scientific evidence requirements in risk analysis at the domestic level. Section 11.4 then takes stock of the various criticisms that have been levelled at the SPS Agreement's science-based approach. In the final section, a case is made for the validity of the approach.

11.2 THE WTO'S SCIENCE-BASED APPROACH

11.2.1 GATT

In Article XX(b), the GATT provides an exception from its rules for countries to take measures that are necessary to protect human, animal, and plant health, but does not specifically require such measures to be based on science. The exception is subject to compliance with the Article's 'chapeau' which states that such measures must not be 'applied in a manner which would constitute a means of arbitrary or unjustifiable discrimination between countries where the same conditions prevail, or a disguised restriction on international trade'.

11.2.2 The SPS Agreement's Scientific Evidence Requirements

As discussed in Chapter 3, the SPS Agreement contains provisions that relate both to Members' risk assessment activities *and* their risk management activities. Specifically, it *requires* Members to engage in risk assessment, and in doing so, to base their risk management decisions on scientific evidence. Criticisms of the Agreement have focused on these provisions and they will also be the focus of this chapter.

Article 2.1 affirms that Members have the right to enact SPS measures necessary for the protection of human, animal, and plant life or health, provided that such measures are consistent with the provisions of the Agreement. Article 2.2 elaborates on what constitutes consistency with the Agreement by providing that:

> Members shall ensure that any sanitary or phytosanitary measure is applied only *to the extent necessary to protect human, animal or plant life or health*, is based on scientific principles and is not maintained without *sufficient scientific evidence*, except as provided for in paragraph 7 of Article 5. (Emphasis added)

Article 3.1 requires Members to 'base' their SPS measures on international standards, guidelines, and recommendations, where they exist, except as otherwise provided for in the Agreement. Relevant international standards, guidelines, and recommendations are those developed by Codex, OIE, and the IPPC. The Agreement defines 'otherwise' in Article 3.3 which says in part that:

> Members may introduce or maintain sanitary or phytosanitary measures which result in a higher level of sanitary or phytosanitary protection than would be achieved by measures based on the relevant international standards, guidelines or recommendations, if there is a scientific justification, or as a consequence of the level of sanitary or phytosanitary protection a Member determines to be appropriate in accordance with the relevant provisions of paragraphs 1 through 8 of Article 5.

A footnote to Article 3.3 provides that there is a scientific justification if, on the basis of an examination and evaluation of available scientific information in conformity with the relevant provisions of the SPS Agreement, a Member determines that the relevant international standards, guidelines, or recommendations are not sufficient to achieve an appropriate level of sanitary or phytosanitary protection.

Further elaboration is given to the scientific evidence requirements of Articles 2 and 3 in Article 5.1 which requires Members to ensure that their SPS measures are 'based on an assessment, as appropriate to the circumstances, of the risks to human, animal or plant life or health, taking into account risk assessment techniques developed by the relevant international organizations'.

Article 5.2 lists a number of factors that Members *must* take into account in their risk assessments, namely, available scientific evidence; relevant processes and production methods; relevant inspection, sampling, and testing methods; prevalence of specific diseases or pests; existence of pest- or disease-free areas; relevant ecological and environmental conditions; and quarantine or other treatment. It is here that the Agreement gives further meaning to Article 2.2 by actually requiring Members to consider, among other factors, available scientific evidence as part of their risk assessment.

Article 5.7 of the SPS Agreement gives Members the right to take provisional measures in the absence of sufficient scientific evidence. It states that in cases where 'relevant scientific evidence is insufficient, a Member may provisionally adopt sanitary or phytosanitary measures on the basis of available pertinent information . . . in such circumstances, Members shall seek to obtain the additional information necessary for a more objective assessment of risk and review the sanitary or phytosanitary measure accordingly within a reasonable period of time'.

11.3 ASSESSING THE SCIENCE-BASED APPROACH

11.3.1 Criticisms of the Approach

The role of science in the SPS Agreement has been much criticized. Legal scholars have criticized the use of science on a number of grounds, with criticisms sometimes coming from opposite ends of the ideological spectrum. For example, according to Atik, it represents a swing back towards greater national discretion and a move away from the 'monolithic prescriptions' of the world trading system.[1] Walker, on the other hand, argues that the SPS Agreement runs the risk of making the WTO into 'a global meta-regulator'.[2] Somewhere in between these views, Sykes argues that the scientific evidence requirements represent undue hurdles for regulators who sincerely pursue objectives other than protectionism.[3]

Several legal scholars have focused on science's lack of objectivity [4] as a basis for arguments that it offers little hope as a source for neutral principles to resolve disputes between nations.[5] Others argue that the deference paid to science is inappropriate because science is not all-encompassing and other

---

[1] Jeffery Atik, 'Science and International Regulatory Convergence' (1996–7) 17 Nw. J. Int'l L. & Bus. 736 at 740.

[2] Vern Walker, 'Keeping the WTO from Becoming the "World" Trans-science Organization: Scientific Uncertainty, Science Policy, and Fact-finding in the "Growth Hormones Dispute"' (1998) Cornell Int'l L.J. 251 at 255.

[3] Alan O. Sykes, 'Domestic Regulation, Sovereignty, and Scientific Evidence Requirements: A Pessimistic View' (2002) 3:2 Chicago J. Int'l L. 353 at 354.

[4] Robert Hudec, 'Science and "Post-Discriminatory" WTO Law' (2003) 26:2 B.C. Int'l. & Comp. L. Rev. 185 at 189.

[5] Atik, *supra* note 1 at 758. See also Walker who argues that it is a myth that science can be a neutral arbiter. *Supra* note 2 at 228. Winickoff *et al.* argue that the use of a scientific standard falsely presumes a level of consensus on the meaning of sound science. David Winickoff *et al.*, 'Adjudicating the GM Food Wars: Science, Risk, and Democracy in World Trade Law' (2005) 30 Yale J. Int'l L. 81 at 84.

factors should be taken into consideration. These critics find that it represents an intrusion into the regulatory decisions of WTO members. For example, Dayna Nadine Scott argues that WTO rules are based on the misconception that 'the strand of discourse labelled science can be isolated from contextual influences, extracted and imported value-free into WTO disputes'.[6] She argues that by relying on science, the SPS Agreement precludes any consideration of broad social, cultural, and ethical concerns: 'establishing itself as the only accepted discourse through which actors may broker claims about particular technological practice, risk discourse produces a zone of silence within public debate, muting competing critiques of technology framed in social, economic, or cultural terms'.[7] She argues further that the Agreement 'ignores the role of culture in the perception and construction of risks'.[8] With respect to SPS measures to ensure food safety, Echols refers to the cultural, religious, medical, and proscriptive meanings of food which affect people's perceptions about what is safe to eat and argues that the SPS Agreement's science-based framework displaces centuries of food traditions and national attitudes towards food and food safety.[9]

The argument has been made in an NRC report that while the social and life sciences are recognizing that they need to be more holistic and integrative, trade agreements are inappropriately subject to a reductionist perspective by putting complex, multidimensional issues in separate boxes.[10] Scott argues along similar lines. She refers to the *EC – Beef Hormones* case where the Appellate Body found that it was not lawful under the SPS Agreement for a Member to guard against a merely theoretical or hypothetical risk where to do so would affect international trade. She argues that 'until, and unless, the

---

[6] Dayna Nadine Scott, *Nature/Culture Clash: The Transnational Trade Debate over GMOs* (New York: Hauser Global Law School Program, Global Law Working Paper 06/05, 2005) at 42. See also Oren Perez, *Ecological Sensitivity and Global Legal Pluralism* (Oxford: Hart Publishing, 2004) at 118.

[7] Scott, *ibid.* at 43.

[8] *Ibid.*

[9] Marsha A. Echols, *Food Safety and the WTO: The Interplay of Culture, Science and Technology* (New York: Kluwer Law International, 2001) at 3. Echols develops an argument which finds that food is culture and as such, is different from most other subjects of trade. She writes (at 15) that traditions (broadly defined) are more important to most people than the scientific facts about the biological function of food. As such, she finds that the SPS Agreement creates a disjunction between people and governmental actions in the way that it eliminates personal notions of risk from a government's consideration (at 28).

[10] National Research Council, ed., *Incorporating Science, Economics, and Sociology in Developing Sanitary and Phytosanitary Standards in International Trade: Proceedings of a Conference* (Washington DC: National Academy Press, 2000) at 9.

"more than merely theoretical" threshold is reached, only the language of science will resonate in the application of this Agreement'.[11] This is problematic for her, as she goes on to elaborate:

> . . . the world of the SPS Agreement is a world inhabited by experts, in which authority to distinguish right and wrong is the preserve of the 'qualified and respected' scientist. It is a world of technocracy. It is a world in which the contingency of scientific knowledge is denied, and in which the values which enter law through science remain obscured. It is a world in which hypothetical risk must be endured, regardless of the nature of the risk-generating activity and the social worth attaching to it. Context, as well as culture . . . is silenced in this unidimensional world of scientific rationality. This is a world in which law is the servant of science in the name of free trade. A world in which law as an instrument of other values – social order, public confidence, trust, community, rights, democracy or deliberation – has no role. It is a world in which the language of power is science and in which one can only be heard in these terms.[12]

Scott criticizes what she describes as the Appellate Body's unwillingness to countenance the possibility that an SPS measure might be rational below the threshold of 'more than theoretical risk'. The Appellate Body has, she argues, set a threshold which is the same regardless of 'cultural sensibilities or filters which mediate a society's relationship to risk; and regardless of citizen preferences.' She posits that

> in the event that science neither proves the existence of risk, nor proves that there is no risk, there is scope for 'rational' debate as to whether this theoretical risk should be tolerated. But it is a debate which will inevitably transcend scientific rationality, thus shattering the fragile illusion of objectivity and universal commensurability, and reducing the potential for 'scientific universalism . . . [to be] used to overcome the particularism of legal systems'.[13]

---

[11] Joanne Scott, 'On Kith and Kine (and Crustaceans): Trade and Environment in the EU and WTO' in J.H.H. Weiler, ed., *The EU, the WTO and the NAFTA – Towards a Common Law of International Trade?* (New York: Oxford University Press, 2000) at 157.

[12] *Ibid.*

[13] *Ibid.* at 160. See also Skogstad who argues (in the context of food safety regulatory policy) for greater elasticity in the SPS Agreement by means of a broader substantive basis for the legitimacy of national regulatory differences. Such a basis would, she argues, take into account evidence that countries' disparate cultures and experiences enter into their publics' calculations of the hazards posed by various food production and processing methods. Grace Skogstad, 'Internationalization, Democracy, and Food Safety Measures: The (Il)Legitimacy of Consumer Preferences?' (2001) 7 Global Governance 293 at 310.

Peel questions the appropriateness of the scientific evidence criteria, but from a different perspective. She notes a lack of normative standards in the WTO context which could justify resolving the 'balance' between trade and health in any case in favour of a particular standard of health protection.[14] Thus WTO decision-makers reviewing national SPS measures will be operating in a 'normative vacuum' where the only criterion available to guide the 'balance' struck between competing risk regulatory policies of Members is that of science. She argues that the irony of constituting science as a default normative yardstick is that choices about competing risk regulatory policies are thus yielded to a body of knowledge which does not have (or is not purported to have) any normative content. Her contention is that 'science's vision is not one which offers value judgements about whether certain forms of progress are right or wrong'. Her concern is arguably overstated. In the case of human health (and that of wild flora and fauna) the normative void she identifies can be filled by taking note of international human rights and principles of international law, with science essentially occupying a 'gatekeeper' role by requiring some minimum level of rationality as regards finding of a risk.

Most critics of a science-based requirement for justification of trade-restrictive health regulations do not suggest a credible alternative means of sifting out protectionism from genuine health regulations.[15] Perez is one of the exceptions. He argues for replacement of the reliance on science with a strategy of 'active engagement' whereby panels would look to see whether countries have made regulations under conditions of what he calls 'institutional reflexivity' – something he describes as a capacity for self-assessment or self-critique, and an open mind to different types of knowledge claims.[16] He argues that if a risk assessment is not produced under conditions of institutional reflexivity, it cannot warrant an SPS measure. Such an approach is not endorsed here as it is considered that so long as one recognizes the uncertainty and indeterminacy of science, it is quite legitimate to also consider values and other non-scientific factors that go into decision-making in the risk analysis process. However, as discussed below, to abandon science altogether could leave too large a loophole for countries to sneak through protectionist provisions and undermine the balance sought by the WTO.

The lack of alternatives to the science-based approach is likely because there is no obvious or helpful alternative. Howse's view is instructive in this regard. He argues that the SPS Agreement's scientific requirements should be

---

[14] Jacqueline Peel, 'Jean Monnet Working Paper 02/04 – Risk Regulation under the WTO SPS Agreement: Science as an International Normative Yardstick?' (2004) (unpublished, archived at New York University).

[15] Perez, *supra* note 6.

[16] *Ibid.* at 154.

understood not as usurping legitimate democratic choices for stricter regulations, but as enhancing the quality of rational democratic deliberation about risk and its control. As he states:

> There is more to democracy than visceral response to popular prejudice and alarm; democracy's promise is more likely to be fulfilled when citizens, or at least their representatives and agents, have comprehensive and accurate information about risks, and about the costs and benefits associated with alternative strategies for their control. If rational deliberation is an important element in making democratic outcomes legitimate, then providing some role for scientific principles and evidence in the regulatory process may enhance, rather than undermine, democratic control of risk. On the other hand, democracy also requires respect for popular choices, even if different from those that would be made in an ideal democratic environment by scientists and technocrats, if the choices have been made in awareness of the facts, and the manner that they will impact on those legitimately concerned has been explicitly considered.[17]

Howse's argument helps us to answer the question that presents itself in light of the above critiques of the SPS Agreement's science-based requirements. Namely, must we make a choice between an approach that requires scientific evidence to the exclusion of public sentiment, or an approach that opposes the scientific evidence requirement in order to admit public sentiment?[18] It is argued here that a middle ground is both possible and preferable. The scientific evidence requirement does not preclude countries from employing a democratic approach to regulatory decision-making that encourages and pays attention to public input.

It should be noted that there is a distinction between cases where a regulating country's claim of a health risk is supported by public sentiment, and cases where the public supports a trade-restrictive regulation for reasons other than health protection (for example, wanting to restrict cultivation of GM crops because of moral concerns about tampering with nature or concerns about the social effects of agricultural restructuring). Non-health-related regulations

---

[17] Robert Howse, 'Democracy, Science, and Free Trade: Risk Regulation on Trial at the World Trade Organization' (2000) 98 Mich. L. Rev. 2329 at 2330.

[18] Note, however, that Howse's argument does not appear to take account of the cultural cognition theory advanced by Kahan *et al.*, which suggests that an awareness of the facts will do little to change people's risk perceptions. See also Jan Bohanes, 'Risk Regulation in WTO Law: A Procedure-based Approach to the Precautionary Principle' (2002) 40 Columbia Journal of Transnational Law 323 at 362. Bohanes suggests that regulations with a scientific basis are available for public inspection and review and can be critiqued objectively. The public and stakeholders can monitor regulators by conducting experiments of their own or hiring scientists to do the analyses for them, enabling them to rest reasonably assured that the most informed, transparent, and balanced decision is being taken.

may be more properly considered under GATT Article XX(a), which allows measures 'necessary to protect public morals', than under Article XX(b), or the SPS regime, which are both about preserving a country's right to enact measures to *protect health*. This distinction is an important one but also in many cases is likely to be a blurred one. Nevertheless, where the purpose of a measure is ostensibly to protect health, then it must be shown that a risk to health exists.

11.3.2 Mandating the Use of Science Domestically

Given the SPS Agreement's emphasis on science, it is instructive to consider domestic examples of governments mandating the use of science in risk analysis. This is the case in various jurisdictions in the areas of public health and environmental matters. For some time in Europe, the European Court of Justice has accepted the importance of scientific justifications for trade restrictions based on, *inter alia*, health protection under Article 36 of the Treaty of Rome. In the 1987 *Beer Purity* case for example, it recognized a *de facto* scientific evidence requirement, and considered the findings of international research, in particular the Community's scientific committee for food, Codex Alimentarius, and the WHO.[19]

In 1997 the Treaty of Amsterdam introduced provisions which require the European Commission – in proposals that affect the functioning of the common market and that concern health, safety, environmental protection, and consumer protection – to take into account 'any new development based on scientific facts'.[20] While the Commission may adopt measures that apply Community-wide, Article 153 of the Treaty states that Member States are not prohibited from maintaining or introducing more stringent protection measures. Article 95 suggests that such measures must be based on new scientific evidence.[21]

In addition, in making environmental policy (including that which aims to protect human health), the EC Treaty requires that the Community shall

---

[19] Case 178/84, *Commission v. Germany* [1987] ECR 1227. See discussion in Alberto Alemanno, *Science and EU Risk Regulation: The Role of Experts in Decision-Making and Judicial Review* (Milan: Università Commerciale L. Bocconi, 2007) at 3.

[20] Article 95, para 3 EC Treaty.

[21] Article 95, para 5 EC Treaty. A 2005 report recommended that the EU strengthen its policy regarding the use of science in regulatory decision-making, including by having the EU institutions issue a joint Communication affirming that high-quality science will have a principal role in policy-making and decision-making processes. Bruce Ballantine, *Enhancing the Role of Science in the Decision-Making of the European Union* (Brussels: European Policy Centre (EPC), Working Paper No. 17, 2005).

take account of various factors including 'available scientific and technical data'.[22]

In the US, regulators have increasingly embraced the use of science in the policy-making process.[23] In 1993, President Clinton issued Executive Order 12866 which requires all federal regulatory agencies to base their decisions on the 'best reasonably obtainable scientific, technical, economic, and other information concerning the need for, and consequences of, the intended regulation'.[24]

In the US, the use of science by the EPA provides a useful illustrative example of the discipline's limitations as a regulatory decision-making tool. Powell documents a number of these limitations in a 1999 study. He notes widespread perceptions that many EPA decisions do not reflect the best scientific analysis and that science too often is adjusted to fit policy.[25] In this regard, the renowned journal, *Science*, has published a number of editorials alleging regulatory excesses and scientific manipulation at the EPA.[26] For example, it has been argued that decision-makers have sought certainty or political cover from science.[27] Similarly, science can be employed to legitimize or undermine policy choices. Powell cites the example of hearings regarding plans to phase out ozone-depleting CFCs. He reports that while 'equal' time was given to proponents of the phase-out, the presence of the sceptics was disproportionate to the prevailing scientific consensus regarding stratospheric ozone depletion.[28]

Due to lack of scientific data, many regulatory decisions are inevitably based on economic, political, administrative, or technical data.[29] One EPA Science Advisory Board member states that 'the volume of [scientific] involvement doesn't tell you very much about the role of science in the decision'.

---

22 Article 174, para 2.

23 Ragnar Lofstedt and Robyn Fairman, 'Scientific Peer Review to Inform Regulatory Decision Making: A European Perspective' (2006) 26:1 Risk Analysis 25 at 26.

24 Executive Order 12866, 58 Fed. Reg. 51735 (1993).

25 Mark R. Powell, *Science at EPA: Information in the Regulatory Process* (Washington DC: Resources for the Future, 1999) at 1.

26 *Ibid.* at 2.

27 *Ibid.* at 5. Wagner speaks, for example, of a 'science charade' in the regulatory process in the US. She finds that controversial policy decisions may be camouflaged as science, allowing decision-makers to avoid accountability for underlying policy decisions. In doing so, regulators are able to disguise the fact that unpleasant trade-offs are needed. This, she argues, inhibits democratic participation as regulators working in isolation often overlook values that are important to the public. Wendy Wagner, 'The Science Charade in Toxic Risk Regulation' (1995) 95:7 Colum. L. Rev. 1613 at 1617.

28 Powell, *supra* note 25 at 6.

29 *Ibid.* at 3.

Thus, where uncertainties remain, decisions may be based on other non-scientific considerations.[30] Given the inability of science to quantify the health effects of many substances, one commentator has questioned 'whether scientists, in their eagerness for a place at the policy table, inadvertently oversold risk assessment over a number of years as the elixir for ailing environmental policies'. Powell writes that many environmental regulatory choices have been incorrectly framed as issues that can be resolved through science.[31]

Such criticism reflects the realities of regulatory decision-making, namely, that in many instances, regulators will act in their own self-interest and/or are captured by special interests that desire a particular regulatory result. It is crucial that domestic governments recognize the characteristics of science that make it vulnerable to manipulation, including uncertainties and the consequent reliance on value judgments.

The EPA has made use of risk assessment guidelines in recent years that have the objective of ensuring that decision-making is faithful to the principles of 'sound science' in the hope that this will result in a greater level of consistency in the application of science to decision-making. However, Powell suggests that where the science is vague or uncertain, this policy actually increases the burden placed on EPA decision-makers to exercise their science and policy judgment and creates new opportunities for the exercise of administrative discretion.[32]

11.3.3 Validating the Science-Based Approach

This chapter has examined issues relating to the appropriateness of the SPS Agreement's science-based framework and has asked whether such a framework is capable of accommodating consideration of public sentiment in domestic regulatory decision-making. The position taken here is that science *is* a legitimate basis for determining the validity of trade-restrictive health regulations. Despite the documented limitations of science (as exemplified by the experiences of the EPA in the US), it arguably remains the best tool we have to sift out unwanted protectionism from genuine health protection measures. Indeed, at the domestic level in the US, critics have not seriously suggested that the EPA abandon science, but rather, that the organization's practices be strengthened through procedural reforms in order to better harness the benefits of a science-based approach.[33]

---

[30] *Ibid.* at 124.

[31] *Ibid.* at 120.

[32] *Ibid.* at 30.

[33] National Research Council, *Strengthening Science at the US Environmental Protection Agency* (Washington DC: National Academy Press, 2000).

Without science we are left at the mercy of competing unproven claims and perspectives. Even if we accept that public risk perceptions have their own rationality, they cannot be the only guide. As Chapter 9 found, lay people are influenced by many factors in forming their perceptions, some of which may incorporate protectionist agendas. In the words of Cross: 'powerful economic interests cannot change objective truth, but they can change public perception. Money and media are influential.'[34] It may be argued that, from the perspective of trade liberalization, some public risk perceptions are more 'legitimate' than others. Those which are the result of psychological factors or cultural worldviews which are 'internal' to the individual may be seen as more 'legitimate' than those social factors which may encompass protectionist interests. However, even if public sentiment could be determined as arising only from the more 'legitimate' factors, accepting public sentiment as a sole justification for a trade-restrictive regulation would give countries potential to justify almost any chosen regulatory measure. The SPS Agreement's commitment to trade liberalization would be severely constrained if any measure could be justified in the name of a population's risk perceptions, regardless of how arrived at. Therefore it is argued that we need a measure that is capable of being rationalized and assessed, even if that process is not perfect. Science, it is argued, has this capacity.

However, the limitations of science as a decision-making tool must be acknowledged. In particular, they must be acknowledged by panels and the Appellate Body when they are called upon under the SPS Agreement to review a domestic agency's measures to determine whether or not they are based on scientific evidence as required by the SPS Agreement. The next section turns to the question of how panels ought to approach the review of domestic science-based regulations.

11.4 REVIEWING DOMESTIC REGULATORY DECISIONS

The SPS Agreement is silent on the matter of what standard of review panels should adopt when examining a country's SPS measure. As Trebilcock and Soloway note, if too wide a degree of deference is afforded to a Member's regulations, and any remotely plausible explanation can be offered as a rationale for a trade-restrictive health regulation, international trade rules risk being undermined.[35] On the other hand, if the scope for deference is minimal, then

34 Frank B. Cross, 'The Risk of Reliance on Perceived Risk' (1992) 3 Risk 59.

35 Michael Trebilcock and Julie Soloway, 'International Policy and Domestic

a WTO panel would have the power, in effect, to invalidate a Member's regulations that, while trade-restricting, represent a legitimate value choice on the part of domestic consumers,[36] and, in some cases, a valid health protection measure.

In addressing the question of what approach panels should take to reviewing domestic regulatory decisions, it is suggested first, that certain characteristics of science must be noted. In particular, science is vulnerable to manipulation and capture that may defeat the original purpose of the WTO's science-based approach. Manipulation and capture may emanate from various sources, including scientists themselves, as well as from governments, employers, and industry sponsors. As Atik argues, science is vulnerable to becoming nothing more than 'another political ideology explicitly directed by money and power'.[37] As well, lack of scientific data means that many decisions will inevitably be based in part on non-scientific considerations.

Domestically, governments should seek to minimize this problem by ensuring transparency and independence of agencies' risk analysis work. Trebilcock and Soloway have addressed the issues it poses for WTO panels, arguing that the regulating Member ought to bear the burden of proving that its risk regulation reflects at least a credible minority of scientific opinion in terms of the scientific credentials and research methodology of those holding the relevant opinion.[38] This test hints at the need for a threshold of scientific rationality in its mention of research methodology. It is suggested here that this is an appropriate way of proceeding; the discussion below suggests in more detail what role panels should play when reviewing scientific evidence.

Panels should not seek to decide which of two or more differing scientific opinions put forward by the disputing parties is preferable. Rather, their role should be to ensure that the scientific evidence relied upon by an importing party passes a minimum threshold of rationality as determined by the risk assessment procedure followed. As Chapter 10 explained, the discipline of risk assessment is still evolving, with a variety of frameworks used, within and across countries. It would be both unrealistic and unwise to require countries to follow a specific method of evaluation in conducting a risk assessment. This would interfere with national sovereignty, as well as the progress of science itself. As such, WTO panels should not tell regulatory agencies how to

---

Food Safety Regulation: The Case for Substantial Deference by the WTO Dispute Settlement Body under the SPS Agreement' in Daniel Kennedy and James Southwick, eds, *The Political Economy of International Trade* (Cambridge, UK: Cambridge University Press, 2002) at 541.

36 *Ibid.*

37 Atik, *supra* note 1 at 758.

38 Trebilcock and Soloway *supra* note 35 at 551.

perform risk assessments. However, panels should require some level of procedural conformity from countries relying on risk assessments to justify trade-restrictive measures. Panels should ask whether countries have conducted a rigorous risk assessment that is consistent with protocols agreed by the international scientific community. That is, there must be a sense that the process followed is one that is reasonable in the circumstances, given the current state of scientific knowledge. Some common ingredients can be distilled from the myriad of risk assessment frameworks in use. At the highest level of generality these include the need for scientific rigour, consistency, and transparency.

At this point it is helpful to depart from the narrative briefly to explore what is meant by the terms 'minimum threshold of rationality' and 'procedural conformity'. While numerous examples might serve to illustrate the argument, the case of US domestic toxic tort law will be used here. In 1993, in the case of *Daubert v. Merrell Dow Pharmaceuticals, Inc.*, the US Supreme Court decided that district court judges should – in cases involving expert scientific testimony – determine whether that testimony 'represents a sufficiently sound application of scientific principles to a sufficiently robust set of scientific data to justify the expert's scientific conclusions'.[39] This approach departed from the prior test of 'general acceptance in the scientific community' which courts had previously used to admit expert testimony.[40] The *Daubert* test was clarified in *General Electric Co. v. Joiner*[41] where dicta suggested that the trial court's role is to evaluate the 'scientific validity of an expert's conclusions, as opposed to the expert's methodology'.[42]

McGarity writes that *Daubert* and *Joiner* have had a 'profoundly negative' impact on the ability of plaintiffs to hold companies liable for the adverse effects of their products and byproducts on human health and the environment.[43] He finds that courts have largely adopted a 'corpuscular' approach to determining the admissibility of expert scientific evidence in toxic tort cases. This approach involves defendants focusing 'upon flaws in the corpuscles of data underlying the testimony rather than upon the scientific reliability of the expert's overall conclusions'.[44] For example, this approach would allow a defendant to pull apart an epidemiological study by looking at every detail for possible flaws. As McGarity argues, 'given the practical impossibility of

---

39 509 U.S. 579 (1993) at 590 and n. 9, 593–5.

40 *Frye v. United States*, 293 F. 1013 (D.C. Cir. 1923).

41 522 U.S. 136 (1997).

42 See discussion in Thomas O. McGarity, 'On the Prospect of 'Daubertizing' Judicial Review of Risk Assessment' (2003) 66 Law & Contemp. Probs 155.

43 *Ibid.*

44 *Ibid.* at 172.

conducting a perfect epidemiological study, the search is nearly always fruitful'.[45] Further, the determination as to whether or not the study is valid is made by judges who lack scientific expertise. By way of illustration, he refers to the *Joiner* case where the plaintiff had well-documented exposure to the group of chemicals known as polychlorinated biphenyls (PCBs) and relied on expert evidence, which included an animal study. The court held that the animal study relied upon could not validly support the conclusion that PCBs were capable of causing lung cancer because, *inter alia*, the animals were young (whereas the plaintiff was middle-aged), the route of administration was different (direct injection of single doses into the stomach as opposed to continuous dermal and inhalation exposure), the doses the mice received were larger than the plaintiff's exposure, and the mice developed a different form of cancer.[46]

McGarity contrasts the 'corpuscular' approach with the 'weight-of-evidence' approach common in epidemiological science where the focus is on the totality of the scientific information to ask whether a cause–effect conclusion is warranted. Such an approach is appropriate, he argues, given the impossibility of proving with certainty the existence of health risks posed by toxic agents.[47]

There has been debate in the US as to whether or not the *Daubert* approach ought to be extended to the case of courts judicially reviewing agency risk assessments.[48] The approach taken by courts to reviewing decisions of federal agencies is known as the 'hard look' approach. In *Motor Vehicles Manufacturers Association v. State Farm Mutual Automobile Insurance Co.*, the Supreme Court followed the 'hard look' approach. In this case, it was required to consider whether a regulation was arbitrary and capricious. It adopted a two-pronged test, examining first whether the agency had relied on the wrong factors or failed to consider an important aspect of the problem. McGarity refers to this as *prescriptive judicial* substantive review because it oversees the manner in which an agency implements its statutory goals. It includes looking at the analytical methodologies the agency adopts to resolve certain kinds of issues, the criteria it employs, the factors it considers, the range of regulatory options from which it chooses, and the policies it relies upon when existing scientific data do not lead to firm factual conclusions.[49]

---

45 *Ibid.*

46 *Ibid.* at 173.

47 *Ibid.* at 165.

48 For an argument on the merits of such an approach, see Alan Charles Raul and Julie Zampa Dwyer, '"Regulatory *Daubert*": A Proposal to Enhance Judicial Review of Agency Science by Incorporating *Daubert* Principles into Administrative Law' (2003) 66:4 Law & Contemp. Probs 7. For opposition, see generally McGarity, *ibid.* at 156.

49 463 U.S. 29 (1983) at 42.

The second prong involves looking at whether the agency's explanation runs counter to the evidence or is too implausible. McGarity calls this *evaluative* substantive judicial review. This aspect of the test presumes that an agency's concise general statement of basis and purpose contains explanations for how it resolved important scientific, economic, and technical questions. The court's role is to determine whether the agency's technical explanation for how it resolved an important issue, or its response to a critical outside comment, is inconsistent with the evidence in the rulemaking record or is implausible under some appropriate (and judicially ascertainable) measure of plausibility.

McGarity suggests that if the *Daubert* approach was extended to review of agency decisions, these criteria would be intensified as courts would look even more carefully at the agency's decision. McGarity strongly opposes such a move, arguing that judges are not qualified to undertake strict scrutiny of the scientific bases for risk-assessment data and analysis. He also points out that taking a 'Daubert' approach to judicial review would give courts a policy-making role that is inappropriate for a politically unaccountable institution.[50]

The purpose of this discussion of US law is not to engage directly in or comment upon the US debate, but to use it as a useful illustration of what *is* and *is not* meant by the suggestion that WTO panels ought to require a 'minimum threshold of rationality' and 'procedural conformity'. In the *Joiner* case, the court advocated reviewing the scientific validity of an expert's conclusions, as opposed to the expert's methodology. The opposite is being advocated here. WTO panels have neither the expertise, nor the political accountability, to engage in a 'corpuscular' approach to review of domestic risk assessments. As the discussion in Chapter 10 made clear, risk assessments are not an exercise in scientific certainty.[51] It would therefore be inappropriate for WTO panels to adopt a role of intense scrutiny of the minutia of risk assessments. Rather, panels should take an approach more akin to the 'hard look' doctrine, displaying deference to domestic agency decisions, while looking to ensure that agencies have followed a defensible methodology and have not arrived at an arbitrary or manifestly absurd conclusion. This is what is meant here by a procedural focus for review.

How then should panels approach this task? It is suggested here that it would be helpful to use peer review processes as a guide for their role in reviewing countries' compliance with the SPS Agreement's scientific evidence and risk assessment requirements. Peer review is used as a tool by leading scientific journals in ensuring the rationality of the results submitted, and is also used by regulatory agencies who use science as a base for decision-

---

50 McGarity, *ibid* at 156.

51 As McGarity argues, if definitive scientific proof is the goal, risk assessment is the wrong tool. *Ibid*. at 167.

making. Peer review may be described as a critical review of a work product (for example, risk assessment) by experts who are independent of those who developed the product. It is a process designed to ensure that activities are, *inter alia*, technically accurate, competently performed, and properly documented.[52] In the context of the SPS Agreement, panels could draw from the process of peer review in order to ensure that countries' risk assessments have a sound and credible basis. They would ideally do so by employing experts to conduct an in-depth review of the risk assessment, including the assumptions, calculations, extrapolations, alternative interpretations, methodology, acceptance criteria, and conclusions reached.[53] The review would thus be procedurally based, focusing on the underlying scientific merits of the risk assessment, rather than on the actual decision reached.[54] The purpose of the examination would not be to second-guess the scientific judgments made, but to filter out 'junk science' used to justify protectionist interests.[55]

In the language used by Crawford-Brown, Pauwelyn, and Smith, such an approach would enable panels to focus their attention on whether the result of a risk assessment achieves a threshold of rationality.[56] The kinds of matters to be examined by experts will vary according to the specific circumstances (depending on factors such as the amount of information available, and the

---

[52] United States Environmental Protection Agency (EPA), *Science Policy Council Handbook: Peer Review* (Washington DC: EPA, 2000) at para 1.2.3 and Appendix B.3.

[53] *Ibid.* at para 1.2.3.

[54] *Ibid.* at para 3.2.1. Alemanno suggests that panels might also be able to accept a country's evidence that a risk assessment has already been subject to peer review, so long as that prior review was adequate. Alemanno, *supra* note 19 at 25. However, this does not seem wholly desirable as one could foresee a further set of issues that would arise out of the question as to whether or not the country's peer review process was adequate.

[55] See also Trebilcock and Soloway, *supra* note 38. They argue that the purpose of requiring at least a credible minority of scientific opinion is to screen out 'junk science', on the one hand, while avoiding attempts to resolve genuine scientific uncertainty or controversy on the other.

[56] Douglas Crawford-Brown, Joost Pauwelyn and Kelly Smith, 'Environmental Risk, Precaution, and Scientific Rationality in the Context of WTO/NAFTA Trade Rules' (2004) 24:2 Risk Analysis 461 at 467. The authors refer to Bernstein who has argued that the classical view of rationality, based in universal principles agreed upon by the scientific community, should be replaced by *dialogic rationality* that stresses the 'practical, communal, character of this rationality in which there is choice, deliberation, interpretation, judicious weighing and application of universal criteria, and even rational disagreement about which criteria are relevant and most important'. This view places emphasis on dialogue between scientists regarding how to judge data and theories, how to weight lines of evidence, and how to balance these considerations in a judgment of epistemic status.

methodological difficulties encountered) but in each case panels should pose specific questions and concerns to be addressed by the experts as well as inviting general comments concerning the credibility of the risk assessment. These will include the following kinds of issues:[57]

- Whether the results of the risk assessment support the conclusions reached;
- Whether the risk assessment was based on the best available information that is in accord with current scientific thinking;
- The degree of transparency of the process;
- Whether the risk assessment has documented the uncertainties in the data;
- Whether the risk assessment used appropriate models, datasets, and assumptions on which to base a scientifically credible decision;
- Whether the models used, with their associated datasets and assumptions, are able to answer the questions as stated in the risk assessment;
- Whether the modelling approaches used are suitable for assessing the suspected risk to human, animal, or plant health;
- Whether the assessment was based on the most appropriate studies, and whether there are other data or studies that are relevant (that is, useful for the hazard identification or dose response assessment) for the risk assessment; and
- Whether the choice, use, and interpretation of the data employed in the risk assessment is appropriate and scientifically sound.

To conclude, the suggestion here is that WTO dispute panels focus on the procedure followed by countries in their regulatory decision-making processes. Panels should seek to ensure that risk assessments relied upon are conducted in a rigorous manner, and that risk estimates meet a minimum threshold of scientific rationality. Where necessary (which it will be in most cases), they should seek assistance from expert panels familiar with scientific methodology. It is not realistic to suggest that any formula will enable panels to wave a magic wand and answer the question of whether or not a trade-restrictive measure should be justified to protect health. The nature of the world we live in does not allow for such certainties. Rather, by ensuring that a minimum procedural threshold has been reached in terms of risk assessment, panels can allow countries flexibility to make decisions, while seeking to

---

[57] Some of the questions suggested are those asked in example charges for peer reviews as cited in United States Environmental Protection Agency, *supra* note 52 at Appendix C.

prevent speculative or ill-founded regulations that, in essence, are designed to protect local producers.

Where does this conclusion leave public sentiment? As the discussion of risk perception noted, public sentiment may be genuinely concerned with health and should be allowed to play some part in the risk analysis process; indeed, it is arguably important in a democracy that both scientists and regulatory decision-makers take such concerns into account. Such concerns should be accepted as having value (their own 'rationality'), even if they are based on alternative perceptions than those held by experts.[58] In terms of the balancing of health and trade objectives, it is worth noting here Howse's argument that 'if citizens believe they need a certain regulation, however "deluded" such a belief is, their utility will be reduced if they do not get it in the sense that they will believe themselves exposed to a risk they believe to be significant'.[59] In other words, the utility from regulation comes not only from reduced likelihood of an event that one disvalues, but also from the psychological security that results from one's belief about the protection one is receiving.[60]

The SPS Agreement's science-based framework is reconcilable with a role for public sentiment. Such reconciliation rests on the acknowledgement that science incorporates assumptions, judgments, and values, and the argument that the scientific risk assessment process need not (and should not) silence public sentiment as some scholars contend it does. Rather, the scientific risk assessment process is capable of including public debate and alternative views of risk.

In this regard, science's limitations can also be seen as an advantage; its uncertainties lend it a certain malleability as a decision-making tool. This interpretation counters the critique of some authors that science is a technocratic and elitist institution that 'by necessity' excludes the public voice. The public voice need not be excluded; where it is, the fault lies not with science

---

[58] As Cross suggests, 'the people need not be foreclosed from risk determination, but reality (as ascertained through the scientific method) must remain as a check on the powers of government to act on public perceptions. Government systems should be constructed so as not to defer automatically or even presumptively to public perceptions of risk, unchecked by scientific data.' Cross, *supra* note 34.

[59] Howse, *supra* note 17 at 2350. See also European Commission – Health and Consumer Protection Directorate-General, *Final Report on Setting the Scientific Frame for the Inclusion of New Quality of Life Concerns in the Risk Assessment Process* (Brussels, 2003) at 3. This report finds that perception of a risk has a direct impact on well-being by its psychological component as well as having a psychosomatically induced physical health effect.

[60] Howard F. Chang, 'Risk Regulation, Endogenous Public Concerns, and the Hormones Dispute: Nothing to Fear But Fear Itself?' (2004) 77 S. Cal. L. Rev. 743 at 9.

per se (science is, after all, simply a means of obtaining knowledge about the world around us) but with the manner in which its tools are employed. Undoubtedly it can be employed in an elitist and exclusionary manner. However, it is the role of national governments to ensure that it is employed in a democratic manner. After all, we could abolish the use of a scientific benchmark in the SPS Agreement, and governments could continue to ignore public values as they wished.

A regulatory decision-making process that acknowledges and accepts a place for public participation throughout the process will be a more democratic process and is likely to be more widely accepted among critics of the trade system. As Skogstad and Hartley argue, if international science-based regulation is to be widely accepted as legitimate, it will likely have to rest on a more indeterminate understanding of scientific knowledge than has previously been the case. They suggest that such a view of science inevitably gives countries more scope to devise public policies consistent with the preferences of their citizens.[61]

However, incorporation of public sentiment in decision-making should not be allowed to compromise the scientific integrity of the process of risk analysis. Alone, public sentiment is not enough to protect against the spectre of disguised protectionism. As noted above, the social factors that influence public risk perceptions in particular leave open the potential for public opinion to be manipulated by protectionist interests.

It might be questioned whether it is not possible to set out some procedural mechanisms whereby public sentiment could be evaluated and articulated in a systematic way. This question has been considered in the context of calls for safeguards to protect national policies.[62] Pascal Lamy has called for a safeguard clause which would allow nations to enact trade-restrictive measures where necessary to respond to 'collective preferences'.[63] He suggests that in order to rely on such a safeguard clause, governments would have to demonstrate that there was really a coherent underlying social demand and that the measure adopted was consistent with that demand. In order to demonstrate

---

61 Grace Skogstad and Sarah Hartley, 'Science and Policy-Making: The Legitimate Conundrum' (Paper presented to the Public Science in Liberal Democracy: The Challenge to Science and Democracy, The University of Saskatchewan, Saskatoon, 15–16 October 2004).

62 See discussion in Steve Charnovitz, 'An Analysis of Pascal Lamy's Proposal on Collective Preferences' (2005) 8:2 J. Int'l Econ. L. 449 at 453.

63 Lamy defines 'collective preferences' as 'the end result of choices made by human communities that apply to the community as a whole'. Pascal Lamy, *The Emergence of Collective Preferences in International Trade: Implications for Regulating Globalization* (online at http://trade.ec.europa.eu/doclib/docs/2004/september/tradoc_118925.pdf, 2004) at 2.

this, he suggests that countries would have to conduct an internal review of the collective preference in order to find out more about the nature of it and establish whether it was well-founded. As he notes, 'that would entail widespread consultation, study or further scientific research and, in the case of unwarranted collective preferences, educating people with a view to changing their preferences (e.g., where preferences expressed have been radicalized by particular circumstances or where they are a hang-over from a bygone age, with no real *raison d'être*)'.[64] The question in the current context is whether there is any procedural mechanism whereby panels could evaluate public sentiment in a systematic manner. Lamy does not explain exactly how governments could determine whether collective preferences are warranted and/or genuine. Charnovitz suggests as possibilities holding a referendum or poll of public opinion.[65] The discussion in Chapter 9 pointed out that public preferences may be driven by numerous factors, some of which are prone to protectionist influences (social mechanisms), while others (psychological mechanisms and cultural cognition) are less prone to protectionist influence and may therefore be considered more 'genuine'. As noted, the problem therefore arises that it would be extremely difficult and likely impossible to get to the bottom of what drives public opinion in any given case. Indeed, Lamy himself does not provide any suggestions on what to do where there has been untoward government or industry influence in shaping public preferences.[66]

Another problem with finding a systematic means of ascertaining public opinion is that raised by Charnovitz with respect to Lamy's idea of a safeguards clause to protect collective preferences. He points out the problematic nature of the implication that the WTO should show greater respect for broad and/or intense public opinion than for policies chosen through political representation, technocratic expertise, or judicial authority.[67] Noting that the WTO paradigm is that a government speaks for its citizens, he recalls that the WTO does not normally look inside a country to judge the democratic nature of its decisions. In other words, in the WTO, an assertion of a position by a government is conclusive even if the position fails to reflect popular will within that country.[68]

---

[64] Lamy, *ibid.* at 10.

[65] Charnovitz, *supra* note 62 at 457. See for example the case of Switzerland which directly involves the population in its decision-making processes. Franz Xaver Perrez, 'Taking Consumers Seriously: The Swiss Regulatory Approach to Genetically Modified Food' (2000) 8 N.Y.U. Envtl. L.J. 585.

[66] Charnovitz, *ibid.*

[67] *Ibid.* at 455.

[68] *Ibid.* at 456.

In conclusion then, given the difficulties involved in any scheme of systematically ascertaining public preferences, it is argued here that so long as public sentiment – as self-identified by a government – can fit within the bounds of scientific rationality, then it should have a place in domestic risk analysis and panels should recognize the validity of it when they review those processes. That is, so long as scientific rationality is present, the fact that the decision made reflects a view of the evidence that would not have been taken by the complaining party (or indeed, perhaps even by the majority of scientific opinion) should not render the measure illegitimate in WTO law. In some cases, measures may be both rational from a scientific standpoint *and* protectionist. This will happen in the Baptist–Bootlegger cases (that is, mixed motives for health regulations) and the least-trade-restrictive test will need to be applied to resolve the matter.

Further, the SPS Agreement allows Members considerable freedom to set their own acceptable level of protection – and hence, scope for the influence of public sentiment. This freedom is constrained by requirements that measures must be the least-trade-restrictive available and be non-discriminatory. The role that these constraints can play in the balancing equation between the right of nations to regulate to protect health and their obligations under WTO rules not to restrict trade will be examined in Chapter 13.

12. The facts of the health cases

12.1 INTRODUCTION

This chapter provides an overview of cases concerning health regulations decided under the GATT and SPS Agreement and discusses the tensions between health and trade objectives that arise in them. It will be recalled that Chapter 6 sorted cases involving disputes over SPS measures into a typology based on a traffic light analogy, with the colour amber being assigned to cases where health and trade objectives conflict. A number of the health-related disputes that have been adjudicated in the WTO to date can be classified as amber, although some might also be classified as green (serious risk to health) or red (clear case of protectionism). To recap, while cases classified as amber overlap and are not exclusive, at least six possible categories can be identified, namely, cases where: (1) risks are of high probability but not serious; (2) risks are of low probability and not serious; (3) risks are of low probability but are serious; (4) public risk perceptions differ from expert risk perceptions; (5) there is unresolved scientific uncertainty; and (6) there are mixed motives for regulation.

The cases are shown and classified in Table 12.1. The classification is not intended to be definitive and all cases are classified under different headings simultaneously. Cases are classified according to how the facts as set out in the parties' submissions would have looked on their face as this is how panels would have first had to approach their task of adjudication. However, it must be noted that the way in which a complaining Member pleads their case does not necessarily reflect the true difficulty of the issues. For example, in *EC – Asbestos*, Canada claimed that asbestos does not necessarily present a high risk. However, this claim was fairly easily dismissed by the Panel. A discussion of each of the cases follows the table.

12.2 THAILAND – CIGARETTES[1]

This was a GATT case, heard before the implementation of the SPS

[1] *Thailand – Restrictions on Importation of and Internal Taxes on Cigarettes* (1990), WTO Doc. DS10/R-37S/200 (Panel Report) [*Thailand/Cigarettes Panel Report*].

Table 12.1 Cases concerning health regulations

Case	Green: High Probability, Serious Consequences	Amber, e.g., 1: High Probability, Consequences that are not serious	Amber, e.g., 2: Low Probability, Consequences that are not serious	Amber, e.g., 3: Low Probability, Serious Consequences	Amber, e.g., 4: Differing Public Perceptions	Amber, e.g., 5: Scientific Uncertainty	Amber, e.g., 6: Mixed Motives	Red: Protectionism and/or No Health Risk
Thailand – Cigarettes, 1990	√						√	√
EC – Beef Hormones, 1998				√[1]	√	√	√	√
Australia – Salmon, 1998				√[2]			√	√
Japan – Agricultural Products, 1999				√				√[3]
EC – Asbestos, 2001	√			√[4]	√	√		√
Japan – Apples, 2003				√				√[5]
EC – Biotech, 2006		√	√		√	√	√	√
Brazil – Retreaded Tyres, 2007	√							

Notes:

[1] If, indeed, hormones used to promote growth contributed to causing cancer, this would be considered a serious consequence.

[2] According to the argument put forth by Australia.

[3] The Panel Report in this case does not make any mention of alleged protectionism. However, such suggestions were made in the media. For example, the *Seattle Post-Intelligencer* reported statements of various apple producer interests in Washington state that the Japanese requirements were considered a thinly veiled form of protectionism, and that Japan's trade barriers were based on politics and not science. Paul Schukovsky, 'Ruling on Japan Boosts State Apple Growers' *Seattle Post-Intelligencer* (23 February 1999).

[4] Canada raised an argument that when used in a certain manner, asbestos does not present a high risk. However, the green box is checked because the panel found the scientific evidence of a serious health risk to be comprehensive.

[5] See note 1 above.

Agreement in 1994. It proved to be a case where mixed motives were at play. It concerned a serious risk to health in the form of smoking as well as protectionism *vis-à-vis* the local Thai tobacco industry. The case concerned Thailand's *Tobacco Act 1966* which required authorization for the importation of tobacco; such authorization had been granted on only three occasions since 1966, each time to the Thai Tobacco Monopoly. The Act also required payment of an excise tax, a business tax, and a municipal tax on cigarettes. For the excise tax, the ceiling rate for domestic cigarettes was 60 per cent of the retail selling price, while for imported cigarettes, it was 80 per cent of the retail selling price. For the business and municipal tax, section 5 of the *Tobacco Act* provided exemptions for manufacturers and sellers of cigarettes where the cigarettes being sold were made from native tobacco leaves.

The US claimed that the import restrictions and tax provisions were inconsistent with Articles III (national treatment) and XI (quantitative restrictions) of the GATT 1947 and could not be justified under Article XX(b) since, as applied by Thailand, they were not necessary to protect human health. Thailand argued, *inter alia*, that its import restrictions were justified under Article XX(b) due to its public health policy objective of reducing tobacco consumption. It argued that Article XX(b) reflects the recognition that public health protection is a 'basic responsibility of governments'.[2] It contended that chemical and other additives contained in American cigarettes made them more harmful than Thai cigarettes,[3] a claim disputed by the US which argued that American and other foreign cigarettes were in fact less harmful than Thai cigarettes because of their lower tar and nicotine levels.[4] Thailand suggested that the production and consumption of tobacco undermined the objectives set out in the GATT's Preamble, namely, to raise the standard of living, ensure full employment and a large and steadily growing volume of real income and effective demand, develop the full use of the resources of the world and expand the production and exchange of goods. Instead, it argued, 'smoking lowered the standard of living, increased sickness and thereby led to billions of dollars being spent every year on medical costs, which reduced

---

[2] *Ibid.* at para 24.

[3] *Ibid.* at para 28. The Panel argued that US cigarette companies place unknown chemicals in their cigarettes, as well as additives which increase the risks of smoking (for example, cocoa, which according to one study, increased the risk of cancer), and re-add nicotine to the leaves making inhalation easier and absorption of nicotine by the bloodstream and the brain more efficient.

[4] *Ibid.* at para 31. The US also argued that there was no evidence that the additives referred to by Thailand had any adverse effects and that in any event, Thailand's import ban affected all cigarettes and not just those containing additives, many of which such as menthol were also used by the Thai Tobacco Monopoly.

real income and prevented an efficient use being made of resources, human and natural'.[5]

Thailand submitted that it had not banned production of tobacco altogether because this might have led to production and consumption of narcotic drugs even more harmful to human health such as opium, marijuana, and kratom. Historically, the manufacturing of cigarettes in Thailand had been aimed at providing a legal substitute for narcotic products which were themselves outlawed. Cigarette production in Thailand was purportedly a state monopoly because the government believed it was necessary to have total control over a product that could be so harmful to health. A main objective of the *Tobacco Act* was therefore to ensure that cigarettes were produced in a quantity just sufficient to satisfy domestic demand, without increasing demand.[6]

While not disputing the inherent health risks from smoking tobacco, the US argued that Thailand's import ban was not necessary to protect human life or health and further, did not serve the purpose of protecting human health. It based this contention on sales figures for cigarettes in Thailand which showed that (despite the import ban) there was an ongoing substantial increase in the number of cigarettes sold in Thailand, including a well-established demand for foreign cigarettes, met by illegal imports.[7] In addition, it noted that the Thai Tobacco Monopoly produced at least 15 brands of cigarettes, attempted to imitate 'American blend' cigarettes in response to consumer demand, and had an extensive and well-established distribution system.[8]

Advising on technical aspects of the case, the World Health Organization (WHO) submitted that there were substantial differences between Thai and American cigarettes, with US cigarettes being manufactured in a more sophisticated manner that would make the cigarettes easier for people to smoke. It noted this to be of public health concern because groups who did not smoke Thai cigarettes, such as women, might find it easier to smoke American cigarettes, and might fall under the illusion that American brands were safer than local ones. The WHO also suggested that if multinational tobacco companies entered the Thai market, the poorly financed public health programmes would be unable to compete with the marketing budgets of these companies, which would result in higher rates of smoking and consequent illness and death.[9]

---

5 *Ibid.* at para 21.

6 *Ibid.*

7 Thailand disputed this, arguing that smoking levels had actually declined, with increases in cigarette sales being attributable to population increases and a higher standard of living which had encouraged smokers to switch from self-rolled cigarettes and traditional tobacco products to manufactured cigarettes. *Ibid.* at paras 23 and 26.

8 *Ibid.* at para 29.

9 *Ibid.* at para 52.

Having found that Thailand had imposed quantitative restrictions in violation of Article XI:1 of the GATT, the Panel found further that these restrictions were not 'necessary' in terms of Article XX(b) because there were other measures besides an import ban reasonably available to Thailand to control the quality and quantity of cigarettes smoked. These could have included, for example, a regulation on a national treatment basis requiring complete disclosure of ingredients, coupled with a ban on unhealthy substances, as well as legislative bans and other restrictive measures to control the direct and indirect advertising, promotion, and sponsorship of tobacco.[10]

12.3 EC – ASBESTOS[11]

Asbestos[12] is widely known as a toxic material which poses serious health risks. However, it is also an extremely useful material in many industrial and commercial applications, due to its resistance to very high temperatures and to different types of chemical attack. Prior to World War II, asbestos was widely used around the world, in applications as varied as manufacturing brake linings and clutches, insulating cords, and reinforcing cement, plastic, or rubber. Since its serious health impacts have become known, countries have moved to limit as far as possible these negative impacts. However, a lack of available substitute materials means that the use of asbestos has continued in many settings. Asbestos falls into two groups: amphiboles and serpentine. Within the amphibole group, there are five varieties: anthophyllite; amosite (or brown asbestos); crocidolite (or blue asbestos); actinolite; and tremolite. The serpentine group consists of chrysotile (or white asbestos). These varieties have different physical and chemical properties. It is mainly amosite, crocidolite, and chrysotile which are used for industrial and commercial purposes.

In 1996, the French government adopted Decree No. 96-1133 prohibiting the manufacture, processing, sale, importation, exportation, domestic marketing, possession for sale, offer, and transfer of all varieties of asbestos fibre. The purpose of the ban was to protect workers and consumers and was subject to limited exceptions in the case of certain materials containing chrysotile

---

10 *Ibid.* at para 77.

11 *European Communities – Measures Affecting Asbestos and Asbestos-Containing Products* (2000), WTO Doc. WT/DS135/R (Panel Report [*EC/Asbestos Panel Report*]. *European Communities – Measures Affecting Asbestos and Asbestos-Containing Products* (2000), WTO Doc. WT/DS135/AB/R (Appellate Body Report) [*EC/Asbestos AB Report*].

12 See *The Oxford English Dictionary*, 2nd ed., s.v. 'asbestos', describing it as a 'mineral of fibrous texture, capable of being woven into an incombustible fabric'.

fibre when, to perform a similar function, no substitute was available which posed a lesser occupational health risk and which provided the same technical guarantees of safety as asbestos with respect to requirements such as temperature resistance and chemical attack.

Before the ban, France imported 20 000 to 40 000 tonnes of chrysotile fibre from Canada annually. After the French government announced its intention to ban asbestos in July 1996, imports of Canadian chrysotile dropped to under 15 000 tonnes. In 1997, only 17 tonnes were imported. Canada requested a panel in October 1998, challenging the ban of chrysotile fibre and products containing it. It claimed that, unlike amphibole fibre – the asbestos most hazardous to health, which was previously widely used in France – chrysotile fibre can be used without incurring any detectable risk. As such, it considered the Decree to be incompatible with both the TBT Agreement and the GATT 1994.

Canada argued that special attention should be paid to the social and political climate in which the French ban was adopted.[13] Canada claimed that non-scientific and illegitimate factors were at play in France's decision to ban asbestos, including misguided public sentiment as a result of alarmist campaigns by interest groups and in the media. It claimed that 'the ban on asbestos is nothing but a political reaction on the part of the French Government to anti-asbestos propaganda'.[14] Canada also raised the issue of protectionism, suggesting that commercial interests in French industry were moving toward the development of substitute products and fibres, and that the Decree banning asbestos was therefore beneficial to them.[15] It also argued that 'the total ban is both irrational and disproportionate considering the fact that the manufacture and use of modern chrysotile asbestos products do not pose any detectable health risks'.[16]

France rejected Canada's claims, arguing that scientific evidence showed that all types of asbestos were likely to cause asbestosis, lung cancer, and mesothelioma. In response to these public health concerns, many European

---

13 *EC/Asbestos Panel Report*, *supra* note 84 at para 3.26.

14 *Ibid.* at para 3.10.

15 *Ibid.* at para 3.29. It seems, however, that the real protectionist push came from Canada where asbestos is manufactured exclusively in Quebec – a province that for various political reasons has benefited from many industrial assistance and protective measures by the Canadian government. (*EC/Asbestos Panel Report*, *supra* note 84 at para 3.20). See Robert Howse and Elizabeth Tuerk, 'The WTO Impact on Internal Regulation – A Case Study of Canada–EC Asbestos Dispute' in G. de Burca and J. Scott, eds, *The EU and the WTO: Legal and Constitutional Issues* (Oxford: Hart Publishing, 2001) at 291.

16 *EC/Asbestos Panel Report, ibid.* at para 3.11.

ies had introduced legislation restricting and ultimately prohibiting the ting and use of asbestos.[17]

The Panel found that France's measure accorded less favourable treatment to Canadian asbestos and products containing asbestos than to domestic 'like products' and therefore violated GATT Article III.4. However, it found that this violation was justified by Article XX(b) as being necessary to protect human life. The Appellate Body reversed the Panel's finding that the products at issue (asbestos and substitutes for asbestos) were 'like' and therefore reversed the Panel's finding that the measure violated GATT Article III. It also considered Article XX(b) and upheld the Panel's finding that the measure would in any event have been justified under its provisions. Thus, both the panel and Appellate Body considered the threat to health clear and serious enough to justify restricting trade.

12.4 EC – BEEF HORMONES[18]

The *EC/Hormones* dispute arose out of three EC directives enacted in the 1980s that banned from the European market beef produced from animals to which any one or more of six hormones had been administered for the purposes of growth promotion.[19] Three of the hormones were natural: oestradiol, testosterone, and progesterone; while three were synthetic: trenbolone acetate (TBA), zeranol, and melengestrol acetate (MGA).

The hormone ban had its genesis in the 1970s when Italian consumers became concerned about the safety of baby food made from veal. The veal was found to contain residues of a synthetic hormone known as dethylstilboestrol (DES), which was being used illegally in French veal production. Consumers claimed various consequences, including that hormonal irregularities in babies were causing them to grow breasts and menstruate.[20] Significant consumer

---

17 *Ibid.* at para 3.15.

18 *EC – Measures Concerning Meat and Meat Products (Hormones) Complaint by the United States* (1997), WTO Doc. WT/DS26/R/USA (Panel Report) [*EC/Hormones Panel Report (US)*]. *EC – Measures Concerning Meat and Meat Products (Hormones) Complaint by the United States* (1997), WTO Doc. WT/DS48/R/CAN (Panel Report) [*EC/Hormones Panel Report (Canada)*]. *EC – Measures Concerning Meat and Meat Products* (1998), WTO Doc. WT/DS26/AB/R, WT/DS48/AB/R (Appellate Body Report) [*EC/Hormones AB Report*].

19 Council Directive 81/602/EEC of 31 July 1981 ('Directive 81/602'); Council Directive 88/146/EEC of 7 March 1988 ('Directive 88/146'); and Council Directive 88/299/EEC of 17 May 1988 ('Directive 88/299').

20 Anon, 'Brie and Hormones' *The Economist* 310:7584 (1989) 21.

pressure ensued following the DES scandal and the EC ultimately decided that it made sense to ban all hormones, rather than just DES.[21]

In 1981, the European Commission issued Directive 81/602 prohibiting the use of the six above-mentioned hormones for growth promotion purposes. It also prohibited placement on the European market of both domestically produced and imported meat and meat products derived from farm animals to which such substances had been administered. The directive provided exceptions to the prohibition if the hormones were used for therapeutic or zootechnical purposes and administered by a veterinarian or under a veterinarian's responsibility. Directive 81/602 was subsequently extended by further similar directives.

While the US had banned the use of DES in 1979 following a study by the WHO that found it posed health risks, it had continued to permit use of the other hormones for growth promotion purposes. In Canada, only trenbolone was banned. The EC directives were of particular concern to American and Canadian farmers who used the affected hormones to promote faster growth in cattle and thus reduce costs and improve yields.[22] Such regulatory divergence led to the dispute that ultimately became the defining case of the new SPS Agreement, with the US claiming that the EU's ban lacked any scientific basis and was simply a means of protecting the Community's domestic cattle industry.

The *Beef – Hormones* case was complicated by the fact that a number of cultural, social, and political factors had influenced the EC's regulatory decision. Consumer and NGO movements had demanded a ban, taking the approach that 'where there is doubt, there must be a total ban to protect consumers'.[23] Consumer pressure arose in part because of distrust and unwillingness to accept scientifically based assurances that the hormones posed no health risk. This distrust likely had a number of causes, both social and cultural, including the influential role played by NGOs and a recent history of food safety-related scandals (including the DES débâcle) and media coverage

---

21 Anon, 'A Short History of Hormones' *The Economist* 310:7584 (1989) 22. The high level of consumer concern was duly noted by the Disputes Panel. *EC/Hormones Panel Report (US)*, *supra* note 91 at para IV.14. *EC/Hormones Panel Report (Canada)*, *supra* note 18.

22 It was reported in 1989 that use of the hormones – which are typically implanted in cows under the skin behind the ears where small time-release capsules slowly release the hormones over several weeks during key growth phases – can eliminate as many as 21 days of feeding time before the animals reach a target weight of 1000 lbs and save producers approximately $20 per head. Janice Castro, 'Why the Beef over Hormones? (European Ban on Beef From Hormone-Injected US Cattle)' *Time* 133:3 (1989) 44.

23 Castro, *ibid.*

of the same. Political factors were also key, one being the significance of Green parties in the European political context. As Josling, Roberts, and Hassan note, environmental issues in the US are not focused through a single party that can pursue its agenda without compromise; whereas in Europe, Green parties are often coalition partners who gain disproportionate influence by being able to extract single-issue support as a quid-pro-quo for keeping the coalition intact.[24]

Another political factor which impacted the *Beef – Hormones* case is the role of the European Parliament. It has been argued that in the late 1980s and early 1990s, the Parliament saw itself as having a 'social conscience' and keeping a watchdog brief on the Commission, pursuing an agenda which combined idealism and populism. In the case of the Hormones debate, it was accused of pursuing a popular agenda with little respect for the nuances of trade rules. For example, the Parliament established a Committee of Enquiry into the Problem of Quality in the Meat Sector whose report (the 'Pimenta Report') strongly supported the ban on the use of hormones to restore consumer confidence in the meat sector. The Pimenta Report found that successful regulations could not be based solely on scientific information, partly because of scientific uncertainties, but also because the regulatory process had to resolve social and political conflicts that extended beyond scientific considerations.[25]

While public sentiment was considered to be a legitimate influence in Europe, the opposite view was taken in North America. For example, Carter criticizes the EC for citing consumer anxiety over the safety of beef treated with hormones, suggesting that in doing so, it was (wrongly) equating consumer fears over hormones safety with actual public health needs.[26] The US and Canada thus argued that protectionist interests played a role in the hormone ban. Josling, Roberts, and Hassan argue, however, that while the Commission was worried about the state of the beef market in the 1980s, 'blocking a relatively small amount of imports of beef offals for the manufacture of meat pies, along with some high quality restaurant cuts, was hardly a plausible response'. They go on to explain that 'the major problems of the beef market had to do with the surge in cattle from the dairy herd, as quota restrictions limited the number of dairy cows. Hormones added little to that pressure.

---

[24] Tim Josling, Donna Roberts and Ayesha Hassan, *The Beef-Hormones Dispute and its Implications for Trade Policy* (Washington DC: USDA, 1999) at 25.

[25] Report of the Committee of Enquiry into the Problem of Quality in the Meat Sector, 1989.

[26] Michelle D. Carter, 'Selling Science under the SPS Agreement: Accommodating Consumer Preference in the Growth Hormones Controversy' (1997) 6 Minn. J. Global Trade 625 at 627.

Imports from the US were also miniscule relative to the market imbalance. The ban on hormones was undoubtedly instituted as a result of the pressure coming through Parliament to protect the health of consumers . . . The reluctant Commission support for the ban on hormone use is best seen as a capitulation to pressure within the EU not from producers but newly active consumer and environmental groups operating largely through the Parliament.'[27] Noting, however, that regulations often have multiple objectives such as both safety and economic concerns, Chang argues that the EU continued to maintain the hormones ban in part because its beef producers feared beef imports from the US would have a negative impact upon their sales.[28]

Josling, Roberts, and Hassan argue that the hormones ban was a case of the EU regulatory system being both tilted toward producer interests and biased in favour of extremely cautious consumers.[29] Thus the Commission felt obliged by domestic political pressure to set food safety rules for consumption in the EC even though those rules may have incorporated irrational or unscientific motivations.[30]

It has been argued that even if European consumers were wrong about the science, the ban was needed to protect public confidence in beef. EU officials argued that lifting the ban would cause uncertainty among European consumers who would likely lower their beef consumption, which in itself would be a real loss in social welfare.[31] Further, even in the absence of any distortions in behaviour, or where the distortions are too small to justify the regulations, it is arguable that there may have been a benefit to be had from a reduction in fear itself.[32]

In the final result, the Panel ruled against the EC on three grounds. First, it found that the EC's measure was illegal because more permissive international

---

[27] Josling, Roberts and Hassan, *supra* note 24 at 31.

[28] Howard F. Chang, 'Risk Regulation, Endogenous Public Concerns, and the Hormones Dispute: Nothing to Fear But Fear Itself?' (2004) 77 S. Cal. L. Rev. 743 at 769. Chang cites EU officials who expressed concerns that 'lifting the ban would create an over-supply of meat, which could drive rural beef suppliers out of business'. See also Michael J. Smith, 'GATT, Trade, and the Environment' (1993) 23 Envtl L. 533 at 537.

[29] They also argue that the same trend is present in the US. Josling, Roberts and Hassan, *supra* note 24 at 28.

[30] *Ibid*. at 31.

[31] Chang, *supra* note 28 at 748.

[32] See Cass R. Sunstein, 'Probability Neglect: Emotions, Worst Cases, and Law' (2002) 112:61 Yale L.J. 104. Sunstein argues that the reduction of even baseless fear is a social good, given that fear is a real social cost, generating a genuine willingness to pay to eliminate or reduce the risk that is feared.

standards existed for five of the hormones. In making this finding, it interpreted Article 3.1, which says that '[m]embers shall base their sanitary or phytosanitary measures on international standards' as a requirement that SPS measures must 'conform with international standards'.[33] The Appellate Body overruled this finding, finding that the measure only had 'to be based on' the international standard.[34]

Second, the Dispute Panel ruled that the EC measure was not based on a risk assessment as required in Article 5.1. The Appellate Body agreed with this finding. The Panel and Appellate Body found that, for five of the hormones, the EC had obtained assessments of some risks – in particular, a 1982 Report of the EC Veterinary Committee (the 'Lamming Report') and two reports in 1988 and 1989 by the Joint FAO/WHO Expert Committee on Food Additives ('JECFA'). However, the EC measure failed because the EC had not applied risk assessment techniques to the particular risks that the EC claimed were the basis of its SPS measure. The EC did produce risk assessments that found a risk of cancer due to hormone exposure. However, these risk assessments did not examine the risks associated with particular hormones and were not treated as relevant evidence by the Panel, particularly since more specific assessments showed no particular risk.[35] For the sixth hormone, no valid risk assessment existed, and therefore the measure could not have been based on a risk assessment.

Third, the Panel ruled that the EC had violated Article 5.5 of the SPS Agreement by demanding different levels of SPS protection in comparable situations.[36] The EC had allowed carbadox and olaquindox to be used as antimicrobial feed additives that promoted the growth of pigs; yet it had banned the use of hormones as growth promoters in cows although the hormones resulted in similar (or lower) risks to humans. The Appellate Body overturned this aspect of the Dispute Panel's decision, finding that this difference in protection did not amount to treatment that was discriminatory or a disguised restriction on international trade.

---

[33] *EC/Hormones Panel Report (US), supra* note 18 at para 8.72. In 1995, Codex had adopted standards for five of the six hormones in the dispute. These standards did not impose maximum residue levels (MRLs) for the three natural hormones in question because it was considered that naturally produced residues would far exceed the additional residue caused by 'good practice' use of these hormones for promoting growth in cows. For two of the synthetic hormones which mimic the biological activity of natural hormones, the MRLs adopted were far below the residue that would be expected if good veterinary practices were followed. There were no Codex standards for the sixth (synthetic) hormone.

[34] The Appellate Body also affirmed the right of Members to enact SPS measures that are stricter than the relevant international standard.

[35] See *EC/Hormones Panel Report (US), supra* note 18 at para 8.124.

[36] *Ibid.* at para 8.241.

The *EC/Hormones* case has recently been back before a Disputes Panel. In July 1999, the Disputes Settlement Body authorized the US to suspend tariff concessions and other obligations due to the EC's failure to comply with the Appellate Body's ruling. In September 2003, the EC adopted Directive 2003/74/EC which provides that the use for animal growth promotion of one of the six hormones is permanently prohibited while the use of the other five is provisionally forbidden. The EC says that the new Directive is based on a comprehensive risk assessment and is thus fully compliant with the DSB recommendations and rulings. However, the US continues to suspend concessions and related obligations against goods originating in the EC. In November 2004, the EC requested consultations with the US claiming that the US should have removed its retaliatory measures. Canada subsequently joined those consultations. In May 2005, the EC requested the Director General to compose a panel. The case has been heard and the panel report was pending at the time of going to press.

The EC's primary claim is procedural, that the US is in violation of Article 23 of the DSU which requires Members to have recourse to and abide by its rules and procedures when they seek redress of a violation of obligations. Thus the EC argues that the US should have initiated an Article 21.5 claim if it believed that the EC had not complied with the recommendations arising out of the Appellate Body decision.[37] In the alternative, it is arguing that it is in any event in compliance with the recommendations and rulings in that it has now concluded a comprehensive risk assessment which focuses on the potential risks to human health from hormone residues in bovine meat and meat products.

12.5 AUSTRALIA – SALMON[38]

The measure at dispute in *Australia/Salmon* was an import prohibition on fresh, chilled, or frozen (that is, uncooked) salmon from various places, including Canada. The prohibition was found in Quarantine Proclamation 86A (QP86A) which had been in place since February 1975. The basis for the prohibition was that importation of uncooked salmon could result in the introduction of any one

---

[37] Article 21.5 of the DSU provides that where there is disagreement as to the existence or consistency with a covered agreement of measures taken to comply with the recommendations and rulings such dispute shall be decided through recourse to an Article 21 procedure including where possible resort to the original panel.

[38] *Australia – Measures Affecting Importation of Salmon*, WTO Doc. WT/DS18/R (Panel Report) at para 2.24 [*Australia/Salmon Panel Report*]. *Australia – Measures Affecting Importation of Salmon* (1998), WTO Doc. WT/DS18/AB/R [*Australia/Salmon AB Report*].

of 24 different exotic disease agents into Australia, with negative consequences for the health of fish in Australia. None of these disease agents were of concern to human health.

Two of the diseases identified had been listed by OIE in the *International Aquatic Animal Health Code* category of fish diseases that are particularly dangerous threats for spreading and that are considered to be 'of socio-economic and/or public health importance within countries and that are significant in the international trade of aquatic animals and aquatic animal products'.[39] In addition, the OIE also listed four of the diseases in a category of fish diseases that are less well understood but potentially dangerous. The remaining diseases were not listed by OIE.

Following consultations with Canada under the GATT in 1994, Australia conducted a risk analysis regarding the importation of uncooked salmon that had not been heat-treated. While Canada sells five types of uncooked salmon for export, this analysis was limited (by agreement of the parties) to imports of only one type of uncooked salmon, namely, wild, ocean-caught Pacific salmon. The analysis consisted of two Draft Reports, issued in May 1995 and May 1996, and a Final Report issued in December 1996. The first Draft Report found that there had been no evidence of the spread of diseases via fish products for human consumption, despite the 'wide scale movement of salmonid product within and between continents'. It listed a sequence of events of which each event had to occur for the imported salmon to cause an exotic disease to become established in Australia. It noted that although zero risk was not attainable, the risk of disease introduction might be reduced to negligible values if one or more events in the sequence were extremely unlikely to occur or if a number of events in the sequence had a relatively low probability. The Draft Report thus recommended that the importation of wild, ocean-caught Pacific salmon be permitted under certain conditions.[40]

The revised Draft Report issued in May 1996 took into account comments received by the Australian Quarantine and Inspection Service (AQIS) on the May 1995 Draft Report. It made no specific recommendation to permit or not to permit importation but identified a number of risk management options for consideration.

The Final Report was a significant departure from the Draft Reports in that it recommended continuation of the import ban. In doing so it noted that although the probability of establishment of the diseases in question was low, there would be major economic impacts which could seriously threaten the

[39] Office International des Épizooties, *International Aquatic Animal Health Code – 1999*.

[40] *Australia/Salmon Panel Report*, *supra* note 38 at para 2.24.

viability of aquacultural operations and the recreational fishing industries, in addition to adverse environmental impacts. The Final Report took into account comments made by the public as part of a risk communication exercise entered into after publication of the May 1995 Draft Report.

There appear to have been mixed motives present in this case. Canada alleged that Australia's quarantine decisions revealed a protectionist agenda inspired by domestic pressures such as Tasmanian salmon producers who had lobbied against the conclusions in the May 1995 Draft Report.[41] Australia, on the other hand, claimed that it sought only to protect its salmon industry from disease, taking into account the potential environmental impact and the socio-economic impact on recreational fisheries.

Canada requested a panel under the WTO Dispute Settlement Understanding (DSU) in March 1997. It claimed that Australia's measure was an illegal import prohibition under the GATT and that it had not been developed and applied in accordance with the SPS Agreement. It considered the ban arbitrary because Australia did not apply similarly strict quarantine measures against other disease risks, notably, those posed by imports of frozen herring bait fish and live ornamental fish. Canada maintained that these species posed a far greater risk than its salmon. Like the EC in the *EC/Hormones* case, Australia argued that while the risks were low, it could not be certain that the Canadian salmon in question would not spread disease.

The Panel and the Appellate Body ruled against Australia. First, the Appellate Body determined that Australia's ban was not based on an assessment of risks.[42] On this ground, the Appellate Body overruled the Panel's finding that the 1996 Final Report did constitute a risk assessment for the purposes of Article 5.1. Second, the Panel and Appellate Body agreed that the import prohibition was a disguised restriction on trade, in violation of Article 2.3 because Australia did not apply the same high level of protection in other comparable situations (that is, herring bait fish and live ornamental fish).[43]

Third, the Panel decided that the particular SPS measure required by Australia – heat treatment of salmon prior to export to Australia – was more trade-restrictive than necessary and thus violated Article 5.6 of the SPS Agreement because heat treatment would convert fresh or fresh-frozen salmon into less valuable heat-treated fish.[44] The Appellate Body overruled this aspect of the Panel's decision, arguing that the SPS measure at issue was not heat treatment but rather the import prohibition itself.[45] The Appellate Body then

---

41 *Ibid.* at para 8.154.
42 *Australia/Salmon AB Report*, *supra* note 38 at para 135.
43 *Ibid.* at para 177.
44 *Australia/Salmon Panel Report*, *supra* note 38 at para 8.161.
45 *Australia/Salmon AB Report*, *supra* note 38 at para 186.

sought to complete the Panel's analysis in order to determine whether the correct measure at issue (the import prohibition) was 'not more trade restrictive than required' to achieve Australia's appropriate level of protection.[46] However, the Appellate Body was unable to complete the analysis due to a lack of factual information on the record.

12.6 JAPAN – AGRICULTURAL PRODUCTS[47]

This case concerned Japan's long-standing import prohibitions on eight products originating from, among other places, the US. The products at issue were apricots, cherries, plums, pears, quince, peaches (including nectarines), apples, and walnuts. Importation of these products was prohibited on the grounds that they were potential hosts of codling moth, a pest not found in Japan. Codling moth is a pest which invades various fruit crops. Newly hatched larvae of a codling moth may enter into the fruit. In the US, the codling moth is a pest of apples and walnuts; it is also sometimes known to infest nectarines and cherries.

Exporting countries could obtain an exemption from the import prohibition for a particular variety of a product if they were able to propose an alternative measure that would achieve a level of protection equivalent to that achieved by the existing import prohibition. The exporting country had the burden of proving that the proposed alternative would achieve the appropriate level of protection. In reality, the alternative measure had always been disinfestation by using methyl bromide. Guidelines for obtaining an exemption were set out in 1987 by Japan's Ministry of Agriculture, Forestry and Fisheries. They set out procedures to be followed by exporting countries including small-scale dose-mortality tests, followed by large-scale mortality tests and an on-site confirmatory test.[48]

Exemptions were granted for each product on a variety-by-variety basis. Countries had to obtain an exemption separately for each specific variety of the relevant product (for example, a separate permit would have to be obtained

---

[46] *Ibid.* at para 193.

[47] *Japan – Measures Affecting Agricultural Products* (1998), WTO Doc. WT/DS76/R (Panel Report) [*Japan/Agricultural Products Panel Report*]. *Japan – Measures Affecting Agricultural Products* (1999), WTO Doc. WT/DS/76/AB/R (Appellate Body Report) [*Japan/Agricultural Products AB Report*].

[48] The procedures were contained in two sets of guidelines for obtaining an exemption from the ban: (1) the 'Experimental Guideline for Lifting Import Ban – Fumigation' outlined the procedures applicable to the initial lifting of the ban; and (2) the 'Experimental Guide for Cultivar Comparison Test on Insect Mortality – Fumigation' established guidelines for approval of additional varieties.

for Cortland apples and Fuji apples, or for Bing cherries and sour cherries). In other words, a permit for one variety of a product did not extend to other varieties of that product. Japan required that for different varieties of a product, the small-scale dose-mortality test needed to be conducted. If the new variety showed equivalent or superior effectiveness compared to approved varieties, no large-scale mortality test would be necessary. Japan's reasoning for this varietal requirement was that the most effective treatments might vary not only with the characteristics of the fruit/nut but also the season of harvest, which varies between different varieties.

The US contested the validity of Japan's guidelines that required separate testing of the quarantine treatment for different varieties of the same product. It claimed that the guidelines adversely affected exports of US agricultural products, and were inconsistent with Japan's obligations under the SPS Agreement, the GATT 1994, and the Agreement on Agriculture. The key factors at issue were Japan's status as being free of codling moth, and that it is a pest of quarantine significance to Japan.

The Panel found that Japan's testing requirements were inconsistent with the SPS Agreement for three reasons. First, the varietal testing requirement was not based on a risk assessment. The Appellate Body upheld this conclusion, finding that the testing requirement was maintained without sufficient scientific evidence.

Second, the Panel found that the varietal testing requirement was more trade restrictive than necessary and therefore violated Article 5.6 of the SPS Agreement. The Panel suggested another testing method which it considered that Japan could have put in place. However, the Appellate Body overruled this conclusion, finding it to be based on evidence obtained by the Panel itself and not proposed or argued by the complainant. As such, it found that the Panel had over-stepped its authority.

Third, the Panel and Appellate Body both found that Japan had violated the requirement to make its SPS measures transparent, especially the requirement in Article 7 that Members publish their SPS measures.

The Panel and Appellate Body agreed that Japan's violation was not justified by Article 5.7. The Appellate Body confirmed that all of the requirements of Article 5.7 must be fulfilled and that Japan had not sought to obtain the additional information necessary for a more objective risk assessment; rather, the information actually collected by Japan did not 'examine the appropriateness' of the measure and did not address the issue of whether 'varietal characteristics cause a divergency in quarantine efficacy'. Further, the Appellate Body agreed with the Panel that Japan had not reviewed its varietal testing requirement within a reasonable period of time.[49]

---

[49] *Japan/Agricultural Products AB Report*, *supra* note 47 at para 92.

12.7 JAPAN – APPLES[50]

This dispute arose out of Japanese quarantine restrictions that prohibited imports of US apples to protect against the introduction of fire blight (*erwinia amylovora*). Japan's legislation permitted exemptions for some apples on a case-by-case basis if certain criteria were met, including that: importation of apples was prohibited from orchards in which any apple blight was detected; export orchards were inspected three times annually for the presence of fire blight; an orchard would be disqualified from exporting to Japan should fire blight be detected within a 500-metre buffer zone surrounding it; and export apples were treated with chlorine post-harvest.

The US requested a panel in 2002, claiming that Japan's quarantine restrictions were inconsistent with the GATT 1994 and the SPS Agreement. The Panel found that two of the requirements in the measure at issue were maintained without sufficient scientific evidence as required by Article 2.2 of the SPS Agreement, namely: (1) the prohibition on imports where fire blight is detected within a 500 metre buffer zone surrounding an orchard; and (2) the requirement that export orchards be inspected three times a year. The Panel found that these requirements did not bear a rational relationship to the scientific evidence available. On appeal, the Appellate Body upheld these findings.

Japan had claimed that its measures were provisional and therefore justified under Article 5.7. The Panel rejected this claim, finding that the measures, which had been in place for some 48 years, were not provisional, and that there was sufficient scientific evidence available to undertake an objective risk assessment. Again, the Appellate Body upheld these findings on appeal.

Finally, the Panel found that Japan's 1999 Pest Risk Analysis did not meet the requirements of a risk assessment under Article 5.1 of the SPS Agreement. As a result, the measure was not 'based on' a risk assessment as required by Article 5.1. This finding was also upheld by the Appellate Body.

12.8 EC – BIOTECH PRODUCTS[51]

The *EC/Biotech Products*[52] case concerns three complaints by the US,

---

[50] *Japan – Measures Affecting the Importation of Apples* (2003), WTO Doc. WT/DS245/R (Panel Report) [*Japan/Apples Panel Report*]. *Japan – Measures Affecting the Importation of Apples* (2003), WTO Doc. WT/DS245/AB/R (Appellate Body Report) [*Japan/Apples AB Report*].

[51] *European Communities – Measures Affecting the Approval and Marketing of Biotech Products* (2006), WTO Doc. WT/DS291/R, WT/DS292/R, WT/DS293/R (Panel Report) [*EC/Biotech Panel Report*].

[52] The term 'biotech products' is used in the case to refer to 'plant cultivars that

Canada, and Argentina. First, that between October 1998 and August 2003, the EC had imposed a general moratorium on the approval of applications of biotech products, a move which they contended was based on political considerations and not on sound science.[53] Second, that contrary to its WTO obligations, the EU failed to consider for final approval applications concerning certain specified biotech products for which it had commenced approval procedures. The complaints concerned 27 different products. Third, that contrary to WTO obligations, certain EC Member States had adopted and maintained various 'safeguard' measures prohibiting or restricting the marketing of biotech products.

This last issue is of the most relevance to this book as the claims regarding the alleged moratorium did not invoke the SPS Agreement's provisions with respect to finding that a risk to health exists, or the risk management exercise.

The EC regime in question consists of two key legal instruments*: EC Directive 2001/18*[54] which governs 'the deliberate release into the environment of genetically modified organisms'; and *EC Regulation 258/97* which regulates 'novel foods and novel food ingredients'. These legal instruments set out the procedure that must be followed in order to obtain marketing approval for biotech products. Essentially, marketing approval will only be granted upon completion of a case-by-case evaluation of the potential risks to human health and the environment by relevant EC scientific committees.

The safeguard measures at issue were linked to this regime. Directive 2001/18 and Regulation 258/97 allow Member States to provisionally restrict or prohibit the use and/or sale of a product in its territory where it has 'justifiable reasons to consider that a product which has been properly notified and has received written consent . . . constitutes a risk to human health or the environment'.[55] According to Article 23 of Directive 2001/18, justifiable reasons exist where 'as a result of new or additional information made available since the date of the consent and affecting the environmental risk assessment or reassessment of existing information on the basis of new or additional scientific knowledge', a Member State has 'detailed grounds for considering that a GMO . . . constitutes a risk to human health or the environment'. Similar

have been developed through recombinant deoxyribonucleic acid ("recombinant DNA") technology'. *EC/Biotech Panel Report*, *ibid.* at para 2.2.

[53] See for example, Gary G. Yerkey, 'US Looking to Ask EU for Talks in WTO Over Ban on Imports of GMO Food Products', 19 Int'l Trade Rep. (BNA) 1829, 1829 (24 October 2002). Gary G. Yerkey, 'President Bush's High-Profile Criticism of EU over GMOs Seen Exacerbating Trade Dispute', 20 Int'l Trade Rep. (BNA) 916 (29 May 2003).

[54] Its predecessor is also implicated in the dispute, namely, EC Directive 90/220 ('Directive 90/220').

[55] Article 16, Directive 80/220 and Article 23, Directive 2001/18.

provision is made by Article 12 of Regulation 258/97.[56] Safeguard measures enacted pursuant to Directives 90/220 and 2001/18, or Regulation 258/97, can only be applied on a provisional basis. The Member State enacting a safeguard measure must inform the European Commission and other Member States of the measure. Upon notification, the Commission must take a decision as to whether the measure should be allowed, thereby modifying the Community-wide marketing approval, or whether it should be terminated.[57] In making its decision, the Commission is assisted by the Regulatory Committee or the Standing Committee on Foodstuffs.

Each of the products in question had been evaluated as part of the approval process under the deliberate release and novel foods legislation. In each case, the relevant EC scientific committee had found that there was no risk to human health or the environment. However, for the products in question, the individual Member States considered that there was additional information that justified the imposition of the safeguard measure.

The US argued that the EC Member States were not taking a scientific approach and were allowing political factors, including interest group pressure, to influence policy decisions.[58] The EC approval regime and the individual Member State measures were influenced by public opinion which, by the 1990s, had become strongly opposed to biotech products. While scientific evidence suggests that biotech products are safe to humans, the public has perceived enough uncertainty within the scientific community to raise doubt, and in some cases alarm, in their minds.[59] As Perdikis argues, the spectre of scientists arguing about the issues weakens the view that consumers should defer to scientists, and it has indeed been suggested that in Europe consumers are no longer willing to accept scientific evidence used by scientists charged with ensuring human, animal, and plant health as a result of prior food safety scandals where they felt that scientists and government had let them down by denying that there were any risks when it was later revealed that there were, most notably, the mad cow disease crisis.[60]

---

[56] A safeguard measure may be adopted pursuant to Article 12 where 'as a result of new information or a reassessment of existing information', a Member State has 'detailed grounds for considering that the use of a food or a food ingredient complying with this Regulation endangers human health or the environment . . .'.

[57] Article 16(1) of Directive 90/220; Article 23(1) of Directive 2001/18; and Article 12(1) of Regulation 258/97.

[58] N. Perdikis, 'A Conflict of Legitimate Concerns or Pandering to Vested Interests? Conflicting Attitudes Towards the Regulation of Trade in Genetically Modified Goods – The EU and the US' (2000) 1:1 Estey Centre Journal of International Law and Trade Policy 51 at 58.

[59] *Ibid.* at 54.

[60] L.J. Fraver, C. Howard and R. Shepherd, 'Effective Communication about

Concern has revolved around potential risks to both human health and the environment. Europe-wide, in 2002 a survey found that while attitudes varied among nations, majorities in most countries rejected biotech foods which were seen as 'risky' and 'not useful' for society.[61] Ethical considerations have also played a part, with concerns that people should not interfere with nature through biotechnology, as well as concerns over restricting consumers' ability to choose what to eat.[62]

The European legislation is in large part predicated on the precautionary principle which, at the Community level, is explicitly applied in situations where uncertainty as to outcome exists. That is, where risk assessments have been carried out but there are clear limitations to the underlying science, the precautionary principle should apply.[63]

Nine different safeguard measures were at issue in the case:

1. Austria – T25 maize;
2. Austria – Bt-176 maize;
3. Austria – MON810 maize;
4. France – MS1/RF1 oilseed rape (EC-161);
5. France – Topas oilseed rape;
6. Germany – Bt-176 maize;
7. Greece – Topas oilseed rape;
8. Italy – Bt-11 maize (EC-163), MON810 maize; MON809 maize; and T25 maize; and
9. Luxembourg – Bt-176 maize.

Each Member State concerned had notified their intention to enact a safeguard measure to the Commission together with the evidence on which they were relying. For each product, the Commission requested the opinion of the relevant EC scientific committee as to whether the evidence was such as to cause the committee to consider that the product(s) constituted a risk for human health or the environment. Every relevant scientific committee reaffirmed its earlier assessment (or that of another EC scientific committee) that the products

---

Genetic Engineering in Food' (1996) 98(4S) British Food Journal 48–52. Cited in Perdikis, *ibid.* at 59. For a discussion of the mad cow crisis, see Scott C. Ratzan, *The Mad Cow Crisis: Health and the Public Good* (London: UCL Press, 1998).

[61] Pew Initiative on Food and Biotechnology, *US vs. EU: An Examination of the Trade Issues Surrounding Genetically Modified Food* (Richmond, VA: Pew Initiative on Food and Biotechnology, 2005). Citing Eurobarometer 58.0 Europeans and Biotechnology in 2002, http://europa.eu.int/comm/public_opinion/archives/ebs/ebs_177_en.pdf.

[62] Perdikis, *supra* note 58 at 55.

[63] *Ibid.* at 55.

in question did not present any risks to human health or the environment.[64] However, at the date of the panel hearing, no decision had been taken by the Commission as to whether the Community-wide marketing approval for the products should be modified, or whether the safeguard measures should be terminated.[65]

The US, Canada, and Argentina (the 'complaining parties') each challenged a different combination of the above-mentioned safeguard measures.[66] Each of the complaining parties asserted that the EC Member States had failed to base their measures on a risk assessment and on scientific principles pursuant to Articles 5.1 and 2.2 of the SPS Agreement; and that they had applied arbitrary or unjustifiable restrictions on international trade pursuant to Articles 5.5 and 2.3. In addition, Canada and Argentina claimed a violation of Article 5.6 (because the EC's own regulatory regime constitutes another measure, reasonably available, taking into account technical and economic feasibility).

The Panel found violations of Articles 5.1 and 2.2 in respect of each of the measures challenged. Further, in no case did it find that the measure was consistent with Article 5.7. Regarding Canada and Argentina's claims under Articles 5.5, 2.3, and 5.6, it found that there was no need to make a ruling in respect of any of the measures.

12.9 BRAZIL – RETREADED TYRES[67]

In June 2005, the EC complained about Brazil's imposition of measures that adversely affected exports of retreaded tyres from the EC to Brazil. Retreaded tyres are produced by reconditioning used tyres. The measures complained of included a prohibition (pursuant to Portaria 14 of November 2004) on the issuance of import licences for retreaded tyres and a set of measures that banned the importation of used tyres (sometimes used against retreaded

---

64 *EC/Biotech Panel Report*, *supra* note 51 at para 7.2527.

65 *Ibid.*

66 The US challenged all nine safeguard measures. Canada challenged five safeguard measures, namely, Austria – T25 maize; France – MS1/RF1 oilseed rape; France – Topas oilseed rape; Greece – Topas oilseed rape; and Italy – Bt-11 maize (EC-163), MON 809 maize, MON 810 maize, and T25 maize. Argentina challenged six measures, namely, Austria – T25 maize, Austria – Bt-176 maize, Austria – MON810 maize, Germany – Bt-176 maize, Italy – MON810 maize, T25 maize, Bt-11 maize (EC-163); and Luxembourg – Bt-176 maize. *EC/Biotech Panel Report*, *ibid.* at para 2529.

67 *Brazil – Measures Affecting Imports of Retreaded Tyres* (2007), WTO Doc. WT/DS322/R (Panel Report). *Brazil – Measures Affecting Imports of Retreaded Tyres* (2007), WTO Doc. WT/DS322/AB/R (Appellate Body Report).

tyres).[68] The EC also complained about an exemption for imports of retreaded tyres from other MERCOCUR countries.

The EC claimed, *inter alia*, that Brazil's measures violated Article III of the GATT. Brazil responded that its measures were justified by Article XX(b) as being necessary to protect human, animal, and plant health. It said that the measures were necessary to protect human health and life from risks presented by retreaded tyres once they reach the end of their useful life and become waste tyres. It argued that accumulation of waste tyres creates a risk to both humans and animals of mosquito-borne diseases such as dengue and yellow fever because waste tyres create perfect breeding grounds for disease-carrying mosquitoes. It also argued that there is a risk of tyre fires which result in toxic leaching that has substantial adverse effects on human, animal, and plant health.[69] The particular problem with retreaded tyres is that they have a shorter life span than new tyres so by importing retreaded tyres, then there will be a greater accumulation of waste tyres. Therefore, an import ban was considered necessary in order to reduce exposure to the health risks arising from the accumulation of waste tyres. Brazil imposed a ban on import tyres for the same reason; however, the government had faced a number of injunction applications by Brazilian retreaders. While the government had opposed these applications with some success, it had not been able to prevent the continued importation into Brazil of large numbers of used tyres.

The EC alleged that the real reason for Brazil's measures was to protect Brazilian industry. This allegation was rejected by the Panel who found that Brazil's measures were justified under Article XX(b) as necessary to protect human, animal, or plant life and health. This finding was upheld on appeal by the Appellate Body. Turning to the chapeau, the Panel found that the imports of used tyres under court injunctions and the MERCOSUR exemption would only result in the import ban being applied in a manner that constitutes unjustifiable discrimination and a disguised restriction on international trade if such imports had taken place in volumes that significantly undermined the objectives of the import ban. Regarding the court injunctions, there was unjustifiable and arbitrary discrimination, due to the fact that used tyre imports had been taking place under the court injunctions in such amounts that the achievement of Brazil's declared objective was being significantly undermined. Regarding the MERCOSUR exemption, however, there was no unjustifiable or arbitrary discrimination, as volumes of imports from MERCOSUR countries were not significant.

---

68 Other measures complained of were a fine of 400 BRL per unit on the importation, marketing, transportation, storage, or keeping in deposit or warehouses of imported, but not of domestic, retreaded tyres; and various state measures which prohibit the sale of imported retreaded tyres.

69 *Ibid.* at para 7.53.

The Appellate Body rejected the Panel's reliance on quantitative evidence as to whether or not the objectives of the import ban were being undermined. It said that analysis of whether the application of a measure results in arbitrary or unjustifiable discrimination should focus on the cause of the discrimination. It found that both the court injunctions and the MERCOSUR exemption had resulted in the import ban being applied in a manner that constitutes arbitrary or unjustifiable discrimination.

13. Analysis of the health cases

13.1 INTRODUCTION

This section provides an analysis of the foregoing panel and Appellate Body decisions. The examination focuses first on the normative approach taken by panels and the Appellate Body to balancing health and trade objectives when they conflict or are in tension with each other. Second, it analyses the approach taken to balancing these objectives as revealed through interpretation and application of (a) the provisions that require Members to show that there is a risk to health; and (b) the provisions that restrain Members' freedom to make decisions on how to manage risk. Finally, it examines several issues that panels must contend with and that influence the outcome of their analysis in the health cases: determining an appropriate standard of review of countries' regulations; allocation of the burden of proof; and finally, use of expert advice.

13.2 NORMATIVE APPROACH

Chapter 7 suggested that in cases where domestic governments are under an obligation to protect health, panels should strive towards decisions that achieve a socially optimal level of protection. In determining what is 'socially optimal', panels should take an approach that is biased in favour of health objectives. This section looks to see whether such an approach has been followed in the cases to date.

In *Thailand/Cigarettes*, the Panel accepted Thailand's position that smoking was a serious risk to human health and that measures designed to reduce the consumption of cigarettes fell within the scope of Article XX(b). The Panel noted that Article XX(b) clearly allowed countries to give priority to human health over trade liberalization; however, it emphasized that the measure had to be 'necessary'.[1] Given that the existence of a serious risk to health was accepted by the parties, this case offers little in the way of insights into how panels might deal with the balancing of competing interests under the SPS

[1] *Thailand – Restrictions on Importation of and Internal Taxes on Cigarettes* (1990), WTO Doc. DS10/R – 37S/200 (Panel Report), at para 73.

Agreement in amber cases where the risk is less serious or of a lower probability, or where it is only perceived to exist by the public.

More helpful insights can be gleaned from the Appellate Body's decision in *EC/Asbestos*. In that case, the Appellate Body came close to taking judicial notice of human health as a fundamental value.[2] This came about in the context of an examination of the 'likeness' of asbestos and asbestos substitutes (PCG) under the national treatment rule in Article III:4.[3] The Appellate Body relied in part on an open list of four criteria that had been used in previous panel and Appellate Body decisions to determine whether asbestos and PCG were 'like', namely, (i) the properties, nature, and quality of the products; (ii) the end-uses of the products; (iii) whether consumers will view the products as 'like'; and (iv) the tariff classification of the products. While adopting an economic approach to the analysis of 'likeness', the Appellate Body found that broader human interests and values such as health also have a role to play.[4]

One member of the Appellate Body expressed a separate opinion[5] on the issue of the determination of 'likeness', arguing that it is difficult 'to imagine what evidence relating to economic competitive relationships as reflected in end-uses and consumers' tastes and habits could outweigh and set at naught the undisputed deadly nature of chrysotile asbestos fibres, compared with PCG fibres, when inhaled by humans, and thereby compel a characterization of "likeness" of chrysotile asbestos and PCG fibres'.[6] While thus suggesting that in a case as clear cut as that of asbestos, health should trump other considerations in a determination of 'likeness', this member went on to note that under Article III.4, 'not *any* kind or degree of health risk, associated with a particular product, would *a priori* negate a finding of the "likeness" of domestic and imported products. The suggestion is thus a narrow one, limited only to the

---

2 Robert Howse and Elizabeth Tuerk, 'The WTO Impact on Internal Regulation – A Case Study of Canada–EC Asbestos Dispute' in G. de Burca and J. Scott, eds, *The EU and the WTO: Legal and Constitutional Issues* (Oxford: Hart Publishing, 2001) at 301. The authors argue that this could be gleaned from the Appellate Body's finding that the panel had erred in failing to consider evidence of consumer tastes and habits – including their response to the health risks from asbestos – in the context of an examination of 'likeness' under Article III:4.

3 Article III.4 imposes national treatment obligations in the case of 'like' domestic and imported products.

4 Howse and Tuerk, *supra* note 12 at 305.

5 Article 17.11 of the Dispute Settlement Understanding enjoins the Appellate Body to make decisions by consensus but provides that if one member wishes to express a separate opinion it shall be done so anonymously.

6 *European Communities – Measures Affecting Asbestos and Asbestos-Containing Products* (2001), WTO Doc. AB-2000-11, WT/DS135/AB/R (Appellate Body Report), at para 152 [*EC/Asbestos AB Report*].

circumstances of this case, and confined to chrysotile asbestos fibres as compared with PCG fibres.'[7]

The question raised by the Appellate Body's approach is what *kind* or *degree* of health risk is required to negate a finding of 'likeness'. A determination of 'likeness' under the GATT is different from determinations required under the SPS Agreement where the kind or degree of risk is relevant to determining whether an SPS measure is *necessary* to protect health. Nevertheless, it is encouraging that the Appellate Body has recognized the inherent value of human health as there is no reason that this will not be carried forward into decisions under the SPS Agreement.

In the *EC/Hormones* case, the Appellate Body – considering this time the SPS Agreement – said that the interests of promoting international trade and of protecting the life and health of human beings are shared but would sometimes compete with each other.[8] However, despite this recognition and various other references to the objective of balancing the right of nations to protect health and trade liberalization, it did not expressly indicate any principled approach that could be used to establish a balance. While panels and the Appellate Body have recognized the need to balance the competing objectives of health protection and trade liberalization, and have certainly paid attention to the importance of health, there is little sense that they have systematically located the balancing exercise within a broader framework that recognizes the content and values of these dual objectives. For example, they have not explicitly considered the obligation that governments face in the light of health as a fundamental human right or what the implications of this might be.

Perhaps it has been considered that the purpose of health protection and trade liberalization are obvious; however, it is suggested that it would help panels to reach consistently principled decisions if they were to take more explicit note of the policy underpinnings of these dual objectives. Only then can the balancing exercise be addressed in the proper context.

A lack of an overarching principled approach to the balancing exercise required by the SPS Agreement does not mean that panels and the Appellate Body have operated in a normative vacuum. The approach taken to balancing the objectives of health protection and trade liberalization has developed incrementally and can be gleaned from decisions regarding whether or not a risk to health exists and towards the disciplining of risk management decisions. These matters are the focus of the next two sections.

---

7 *Ibid.* at para 153.
8 *Ibid.* at para 177.

13.3 INTERPRETING THE SPS AGREEMENT

13.3.1 Risk Assessment: Existence of a Risk

This section examines interpretation and application of the provisions that deal with the risk assessment phase of risk analysis. That is, the provisions that require Members to show that a risk to health exists. First, it will look at how panels and the Appellate Body have dealt with similar questions under the GATT's Article XX(b) and will then consider the SPS Agreement's scientific evidence requirements. Identification of a risk is critical (both to the GATT and SPS Agreements) because a country can only enact trade-restrictive measures where there is a risk. Under the SPS Agreement, it is only once a country has shown that there is indeed a risk that they are free to nominate their 'appropriate level of protection', thus commencing the risk management process. This is why the question of the legitimacy of incorporating non-scientific factors such as public sentiment into the risk assessment process is critical. It is not disputed that there is a role for the public in deciding on an appropriate level of risk (value judgments being widely recognized to have a role in risk management), but if no risk has been found in the first instance, then there is no basis upon which to move to the risk management phase of determining the appropriate level of protection and the SPS measure that will achieve that protection. However, as discussed in Chapter 9, public sentiment may have a lot to say about the existence or otherwise of a given risk.

13.3.1.1 Lessons from the GATT

Under Article XX(b), a finding of 'necessity' implicitly requires that there be a risk and this raises the question of the relevance of science. In *Thailand/Cigarettes*, the Panel was not required to address directly the question of scientific evidence as the parties to the dispute, and the expert from the WHO assisting the Panel agreed that smoking constituted a serious risk to human health and that measures designed to reduce the consumption of cigarettes fell within the scope of Article XX(b).[9]

In *EC/Asbestos*, Canada claimed that the use of modern chrysotile asbestos products do not pose any detectable health risks and argued that France's import ban on asbestos was both irrational and disproportionate.[10] The Panel noted that under Article XX(b), a measure must (a) fall within the range of policies *designed to protect human life or health*; and (b) be *necessary* to fulfil

---

[9] *Thailand/Cigarettes Panel Report*, *supra* note 11 at para 73.

[10] *European Communities – Measures Affecting Asbestos and Asbestos-Containing Products* (2000), WTO Doc. WT/DS135/R (Panel Report), at para 3.12 [*EC/Asbestos Panel Report*].

the policy objective.[11] Regarding the first requirement, the Panel found that the words 'policies designed to protect' imply the existence of a health risk. As such, it considered its first task to be the determination as to whether chrysotile asbestos posed a risk to human life or health.[12]

The Panel stated that it was required to examine the scientific data in order to determine the existence of a risk to health.[13] While noting the SPS Agreement's detailed provisions on scientific justification, it preferred to confine itself to the provisions of the GATT and to the criteria defined by the practice relating to the application of Article XX.[14] In this context, the Panel stated firmly that it is not its function to settle a scientific debate, 'not being composed of experts in the field of the possible human health risks posed by asbestos'. As such, it refused to set itself up as an arbiter of the opinions expressed by the scientific community.[15] Rather, it saw its role as being to determine whether there is sufficient scientific evidence to conclude that there exists a risk for human life or health and that the measures taken by France were necessary in relation to the objectives pursued. It said that it would base its conclusions with respect to the existence of a public health risk on the scientific evidence put forward by the parties and the comments of the experts consulted within the context of the case.[16]

Taking this approach, the Panel found the evidence tended to show that handling chrysotile-cement products constituted a risk to health. As it stated, 'a decision-maker responsible for taking public health measures might reasonably conclude that the presence of chrysotile-cement products posed a risk because of the risks involved in working with those products'.[17] It is a fine line between arbitrating differing opinions expressed by the scientific community on the one hand, and determining whether there is sufficient scientific evidence on the other. In this case, the Panel's role was simplified by the fact that there was consensus among the scientific experts it consulted. On appeal, the Appellate Body accepted the Panel's finding on the evidence and noted that as there is no requirement under Article XX(b) to actually quantify the risk to human health, a risk may be evaluated in either quantitative or qualitative terms.[18]

---

[11] Citing *United States – Standards for Reformulated and Conventional Gasoline* (1996), WTO Doc. WT/DS2/AB/R (Appellate Body Report [*US/Gasoline AB Report*] at para 6.20.

[12] *EC/Asbestos Panel Report*, *supra* note 110 at para 8.170.

[13] *Ibid.* at para 8.179.

[14] *Ibid.* at para 8.180.

[15] *Ibid.* at para 8.181.

[16] *Ibid.* at para 8.182.

[17] *Ibid.* at para 8.193.

[18] *EC/Asbestos AB Report*, *supra* note 16 at para 167.

In the most recent health-related case in the WTO disputes settlement system, *Brazil/Retreaded Tyres,*[19] the Panel accepted evidence presented by Brazil that the accumulation of waste tyres creates the perfect breeding grounds for disease-carrying mosquitoes and also creates the risk of tyre fires and toxic leaching which has substantial adverse effects on human health and the environment.[20] Evidence regarding tyres as breeding grounds for mosquitoes came from various studies in the US and Australia, WHO fact sheets, as well as the Basel Convention Technical Guidelines on the Identification and Management of Used Tyres which note that 'under certain climatic conditions waste tyre dumps or stockpiles can become the breeding ground for insects, such as mosquitoes, which are capable of transmitting diseases to humans. This is of particular concern in tropical regions.'[21] Regarding the seriousness of the diseases in question (dengue, yellow fever, and malaria), the Panel relied upon the WHO's determination that they are serious diseases of major international public health concern.[22]

As to tyre fires, reliance was again placed on the Basel Convention Technical Guidelines which say that precautions must be taken against the deliberate or accidental igniting of tyre stockpiles.[23] Reports were submitted by Brazil that discussed the potential negative effects of tyre fires – the emission of highly toxic and mutagenic emissions such as a noxious plume with hazardous components including carbon monoxide, dioxins, and furans; pyrolytic oil that contains naphthalene, anthracene, benzene, and various metals; and ash containing heavy metals including lead, arsenic, and zinc.[24] Health problems from emissions include loss of short-term memory, learning disabilities, immune system suppression, cardiovascular problems, while the plume can lead to cancer, premature mortality, reduced lung function, suppression of the immune system, respiratory effects, and heart and chest problems.[25] As common sense would dictate, the Panel considered these various risks serious enough to warrant action by Brazil.

The Panel also accepted that mosquito-borne diseases pose health risks to animals (in relation to dengue fever) while toxic chemicals and heavy metals contained in pyrolytic oil, and hazardous substances contained in toxic plumes

[19] *Brazil – Measures Affecting Imports of Retreaded Tyres* (2007), WTO Doc. WT/DS322/R (Panel Report). *Brazil – Measures Affecting Imports of Retreaded Tyres* (2007), WTO Doc. WT/DS322/R/AB (Appellate Body).

[20] *Brazil – Retreaded Tyres Panel Report, ibid.* at paras 7.71 and 7.77. The Appellate Body accepted these findings of the Panel. *Ibid.*

[21] *Ibid.* at para 7.61.

[22] *Ibid.* at para 7.57.

[23] *Ibid.* at para 7.81.

[24] *Ibid.* at para 7.72.

[25] *Ibid.* at para 7.73.

released from fires, harm animals and plants.[26] In coming to its conclusion, it noted a report submitted by Brazil of a tyre fire in the US state of Ohio that resulted in the death of many fish due to release of pyrolytic oil.[27]

These GATT cases show the willingness of panels to review the scientific evidence presented to them by the parties to determine evidence of risks to health. However, each of the cases to date has been fairly straightforward in this regard, with little need for panels to deal with conflicting evidence. For example, in *Brazil/Retreaded Tyres* the most difficult aspect for the Panel was in determining whether or not tyres actually accumulate in Brazil, and what the chances of fires are likely to be. The actual health risks once the occurrence of these events was determined were clear.

13.3.1.2 The SPS Agreement

Article 2.2 The requirement for measures to be based on scientific evidence
In *EC/Hormones*, the Panel did not examine the claims under Article 2.2, focusing instead on Articles 3 and 5.[28] Despite the Panel's approach, it is helpful to note discussion in the case about the meaning of the term 'scientific principles' as used in Article 2.2. The US submitted that while the SPS Agreement does not define the term, it should at a minimum be interpreted as incorporating 'the scientific method', which represents those principles and practices universally regarded as necessary for scientific investigation, in particular procedures for: (i) the observation of phenomena in nature or under controlled conditions; (ii) the systematic classification of empirical data; (iii) the measurement of empirical quantities and for calculating probability errors and significant deviations; (iv) forming a hypothesis; (v) analysing experimental results using logic and mathematics; and (vi) many other related techniques and processes.[29]

---

26 *Ibid.* at para 7.84.

27 *Ibid.* at para 7.88.

28 *EC – Measures Concerning Meat and Meat Products (Hormones) Complaint by the United States* (1997), WTO Doc. WT/DS26/R/USA (Panel Report), at para 8.271. [*EC/Hormones Panel Report (US)*] *EC – Measures Concerning Meat and Meat Products (Hormones) Complaint by Canada* (1997), WTO Doc. WT/DS48/R/CAN (Panel Report), at para 8.274. [*EC/Hormones Panel Report (Canada)*]. The Appellate Body expressed surprise at this approach, noting that Article 2.2 informs Article 5.1 (and that Article 2.3 informs Article 5.5). However, it suggested that further analysis of the relationship between these articles would have to await another case. *European Communities – Measures Affecting Meat and Meat Products* (1998), WTO Doc. WT/DS26/AB/R, WT/DS48/AB/R (Appellate Body Report), at para 250 [*EC/Hormones AB Report*].

29 *EC/Hormones Panel Report (US), ibid.* at para IV.24.

In response to this submission, the EC argued that the US explanation of the concept of 'scientific principles' was a caricature of the 'scientific method' which could have 'been taken straight from a school textbook circa 1960'. As the Panel described the EC's argument:

> There was absolutely no reason to suppose that the Members of the SPS Agreement had the list presented by the United States in mind when signing it. There were many theories of science and the 'scientific method'; the European Communities relied on biological principles when assessing the risks of using hormones for growth promotion. Measures must be based on scientific principles, as opposed to non-scientific ones such as superstition. If a measure was aimed at reducing or eliminating a risk to health, then it must actually address that risk in a manner which could be scientifically justified. For example, if the measure was aimed at eliminating a pathogenic organism from a food, there were several methods, e.g., heating, salting, pickling, etc which could be scientifically proven to be effective. If, however, a Member required prayers to be said over the food, or a ritual dance to be performed around it, that would not be compatible with the SPS Agreement because such methods could not be scientifically proven to be effective.[30]

The EC further argued that what was important was whether, in the scientific research employed by the EC scientists (or the scientific reports to which they made reference in their reports), the 'minimal attributes of scientific inquiry' were respected.[31] It argued that such an approach is required because what might be an acceptable scientific method for one scientist might not satisfy another, who might be more interested in certain other scientific principles or aspects totally neglected or only partially examined by the first scientist. The EC argued further that it was for this reason that the SPS Agreement only requires 'sufficient' scientific evidence, as opposed to clear or certain scientific evidence.[32] This reasoning led to the EC's position that a Member is entitled to rely on a small or minority part of available scientific evidence.[33]

The EC also argued that by requiring a measure to be based on scientific principles (rather than, for example, requiring that a measure be based on the 'best science' or on the 'weight of evidence') the Agreement recognizes the fact that scientific certainty is rare and many scientific determinations require a judgement among differing scientific views.[34] It is unfortunate that the Panel

---

30 *Ibid.* at para IV.25.

31 *Ibid.* at para IV.26.

32 *Ibid.* at para IV.27.

33 The Appellate Body agreed with this aspect of the EC's argument.

34 See *EC/Hormones Panel Report (US), supra* note 28 at para IV.98. In making this argument, the EC was citing a public document submitted by the United States Trade Representative to Congress. The US objected to reliance on this evidence, see paras IV.100–101.

did not address these submissions as, while directed at Article 2.2, they are broader in their scope and go to the heart of the debate around Article 5.1 and what constitutes a risk assessment. While the EC submission is preferable to the extent that it recognizes there are many scientific methods, it comes close to suggesting that so long as a country says it has used a scientific method then that should be good enough for a panel. This cannot be the case. While the US may have overstated the simplicity of determining whether a scientific method has been followed, as discussed in Chapter 11, there are questions that can be asked to help determine whether a credible method has been followed.

The Appellate Body considered Article 2.2 in *Japan/Agricultural Products* where it agreed with the Panel that the requirement for sufficient scientific evidence in Article 2.2 requires a rational or objective relationship between the SPS measure and the scientific evidence.[35] It said further that 'whether there is a rational relationship between an SPS measure and the scientific evidence is to be determined on a case-by-case basis and will depend upon the particular circumstances of the case, including the characteristics of the measure at issue and the quality and quantity of the scientific evidence'.[36] It found no rational relationship in this case where the US had established that to date not a single instance had occurred in Japan or any other country, where the treatment approved for one variety of a product had to be modified to ensure an effective treatment for another variety of the same product.[37]

The Panel and Appellate Body considered Article 2.2 again in *Japan/Apples*. In this case, the US maintained that there was 'no scientific evidence of any quality' that mature apple fruit as imported from the US had ever transmitted fire blight or was a pathway for the introduction of fire blight.[38] It raised a similar argument as in the *Hormones* case, suggesting that evidence under Article 2.2 must be valid according to the objective principles of the scientific method. Circumstantial evidence should be rejected. Japan responded that indirect evidence should also be taken into account, referring to a variety of published literature which it said established that the bacteria were capable of long-term survival inside, or on the surface, of mature symptomless apple fruit.[39] The Panel recognized a difference between 'direct' and

---

[35] *Japan – Measures Affecting Agricultural Products* (1999), WTO Doc. WT/DS76/AB/R (Appellate Body Report), at para 84 [*Japan/Agricultural Products AB Report*].

[36] *Ibid.* at para 84.

[37] The Panel sought expert advice to assist it in making this determination. *Japan – Measures Affecting Agricultural Products* (1998), WTO Doc. WT/DS76/R (Panel Report), at para 8.42 [*Japan/Agricultural Products Panel Report*].

[38] *Japan – Measures Affecting the Importation of Apples* (2003), WTO Doc. WT/DS245/AB/R (Appellate Body Report), at para 4.49 [*Japan/Apples Panel Report*].

[39] *Ibid.* at paras 4.50 and 8.90.

'indirect' evidence but said that it is to be found in the degree of relationship of the evidence with the facts. It reiterated that the evidence to be considered should be evidence gathered through scientific methods, and that information not acquired through a scientific method should be excluded. Further, it found that in this case, where evidence was available, reliance should have been placed on scientifically produced evidence, rather than on purely circumstantial evidence. At the least, it considered that the circumstantial evidence should have been considered in the light of the body of scientific evidence already available.[40]

In terms of 'sufficiency' of evidence, the Panel referred to the Appellate Body's decision in *Japan/Agricultural Products* and said that the term 'sufficient' implies a 'rational or objective relationship'. Pursuant to this test, it found that Japan's measure was clearly disproportionate to the risk identified on the basis of the scientific evidence available – evidence which it considered to be lacking in both quality and quantity.[41] For the Panel, 'clearly disproportionate' implied that a 'rational or objective relationship' did not exist between the measure and the scientific evidence and therefore the measure was maintained 'without sufficient scientific evidence' within the meaning of Article 2.2. On appeal, the Appellate Body agreed with the Panel's finding. It noted further that the appropriateness of a given approach or methodology in assessing whether a measure is maintained 'without sufficient scientific evidence' depends on the 'particular circumstances of the case' and must be determined on a 'case-by-case' basis. In other words, it suggested that approaches different from that followed by the Panel in this case could also prove appropriate to evaluate whether a measure is maintained without sufficient scientific evidence within the meaning of Article 2.2.

One of Japan's key arguments on appeal was that the Panel had failed to accord a 'certain degree of discretion' to it in the manner in which it chose, weighed, and evaluated scientific evidence.[42] It submitted that the Panel should have made its assessment under Article 2.2 'in the light of Japan's approach to risk and scientific evidence', rather than focusing on the experts' views.[43] The Appellate Body rejected this argument. It found that such an approach would conflict with Article 11 of the DSU which requires panels to make an 'objective assessment of the facts'.[44] Questions remain as to what is

---

40 *Ibid.* at para 8.91.

41 *Ibid.* at paras 8.170 and 8.198.

42 *Japan – Measures Affecting the Importation of Apples* (2003), WTO Doc. WT/DS245/AB/R (Appellate Body Report), at para 166 [*Japan/Apples AB Report*].

43 *Ibid.*

44 *Ibid.* at para 165. The Appellate Body referred to its finding in *EC/Hormones* that Article 11 of the DSU sets out the applicable standard of review that a panel should apply in the assessment of scientific evidence under the SPS Agreement.

'objective'. It should arguably require panels to take an even-handed and unbiased approach, but does not mean that they should substitute their scientific judgment for that of the parties.

As the discussion in Chapter 9 noted, risk assessments are invariably indeterminate. The findings in that chapter support an approach by WTO panels which accepts that within the bounds of procedurally sound scientific methods, different approaches to risk assessment may be taken, and different interpretations of evidence accepted, so long as the result of the assessment fits within the bounds of scientific rationality. Thus, different countries may take different approaches to evidence and risk but so long as a threshold of scientific rationality exists, then it should not be grounds for a panel to set aside a measure because another country's experts would have come to a different conclusion.

Howse argues that the Panel and Appellate Body erred in reading into Article 2.2 a proportionality requirement that is not to be found in the text. He argues that Article 2.2 deals with the sufficiency of *evidence*, not whether a risk is sufficient to regulate. However, the Appellate Body did not expressly address this point; it appeared to read the Panel's language of 'clear disproportion' as merely expressing the conclusion or result of the Panel's application of the 'rational relationship' test, rather than as suggesting a different or additional test for proportionality.[45] Even if the Panel erred, as Howse suggests, it is understandable that it wanted to introduce greater certainty into the interpretation of Article 2.2, given that the Appellate Body's 'rational or objective relationship' test in *Japan/Agricultural Products* provided no concrete guidance as to what constitutes such a relationship. Yet uncertainties remain and amber cases will present particular challenges in this regard. For example, where public sentiment plays a role in a country's regulatory decision, the question arises of 'who's rationality' are we talking about? There may well be a rational relationship in the minds of the public who perceive risks differently from experts. In such a case, whose 'rationality' should a panel consider? Should a panel recognize as legitimate a country's deference to public risk perceptions in deciding whether or not a measure has a rational or objective relationship with the science?

Article 5.1 Risk assessment requirements 'Risk assessment' is defined in paragraph 4 of Annex A to the SPS Agreement (the Annex A(4) definition) as:

> *Risk assessment* – The evaluation of the likelihood of entry, establishment or spread of a pest or disease within the territory of an importing Member according to the

---

[45] Michael J. Trebilcock and Robert Howse, *The Regulation of International Trade*, 3rd ed. (London: Routledge, 2005). *Japan/Apples AB Report*, *ibid.* at para 163.

sanitary or phytosanitary measures which might be applied, and of the associated potential biological and economic consequences; or the evaluation of the potential for adverse effects on human health arising from the presence of additional contaminants, toxins or disease-causing organisms in food, beverages or feedstuffs.

The Annex A(4) definition uses different language depending on whether the risk assessment relates to the entry, establishment, or spread of a pest or disease, *or* to the presence of additional contaminants, toxins, or disease-carrying organisms in food, beverages, or feedstuffs. In the first part, the definition refers to evaluation of 'the likelihood of'. In the second part, the definition refers to 'the potential for adverse effects on . . .'. This distinction seems to place greater importance on risks to human or animal health posed by additives, contaminants, toxins, and disease-causing organisms in food or feedstuffs by basing the evaluation on the *potential* of harm, rather than the harder to prove *likelihood or probability*.[46]

The Panel in *EC/Hormones* interpreted the second part of Annex A(4) to require a two-step process whereby the importing Member should: (i) identify adverse effects on human health (if any); and (ii) if any such adverse effects exist, evaluate the potential or probability of occurrence of such effects.[47] In this interpretation, the Panel spoke of 'probability' as an alternative term for 'potential', thus (in the Appellate Body's view) implying a quantitative dimension to the notion of risk in the case of human health. The Appellate Body considered the ordinary meaning of 'potential' to be similar to 'possibility' and thus different from the ordinary meaning of 'probability'.[48] It found no basis in the Agreement for demonstration of a certain magnitude or threshold level of risk.[49]

In this case, the US argued that the EC had never performed a risk assessment or relied on any risk assessment that could serve as a basis for its ban with respect to the six hormones.[50] It submitted that 'the remarkable charac-

---

[46] Gavin Goh, 'Precaution, Science and Sovereignty: Protecting Life and Health under the WTO Agreements' (2003) 6 Journal of World Intellectual Property 441 at 449.

[47] *EC/Hormones Panel Report (US), supra* note 28 at para 8.98. *EC/Hormones Panel Report (Canada), supra* note 170 at para 8.98.

[48] The Appellate Body stated that: 'The dictionary meaning of "'potential" is 'that which is possible as opposed to actual; a possibility'; L. Brown, ed., *The New Shorter Oxford English Dictionary on Historical Principles*, Vol. 2 (Oxford: Clarendon Press, 1993) at 2310. In contrast, 'probability' refers to 'degrees of likelihood; the appearance of truth, or likelihood of being realized', and 'a thing judged likely to be true, to exist, or to happen'. *EC/Hormones AB Report, supra* note 28 at footnote 164.

[49] *EC/Hormones AB Report, ibid.* at para 186.

[50] *EC/Hormones Panel Report (US), supra* note 28 at para IV.109.

teristic of the public debate in the EC on these hormones was that the "risk" was usually described in terms of consumer anxieties rather than any observable adverse effect on human health'.[51] The EC refuted this argument by stressing that there was nothing in the text of the contested measures, the legislative history, or in any other document to suggest that consumer anxieties was the purpose for which the measures were adopted, although it was likely that consumer concerns had been taken into consideration during the risk management phase, since consumer concerns as to potential risks to human health resulting from the use of hormones were very high at that time.[52]

The EC relied upon a number of reports,[53] arguing that none of them suggested that there is no potential risk from the hormones in question. The Panel confirmed that a risk assessment should be a scientific examination of data and studies, and found that the reports relied upon by the EC appeared to meet these minimum requirements of a risk assessment (in particular the Lamming Report and the 1988 and the JECFA Reports). Further, it noted that the scientists advising the Panel seemed to consider these reports, from a scientific and technical point of view, to be risk assessments. Based on this, the Panel found that the EC had met the burden of demonstrating the existence of a risk assessment carried out in accordance with Article 5.1.[54]

A key point of interpretation in Article 5.1 is that of the phrase 'based on'. In *EC/Hormones*, the Panel ruled that there is a minimum procedural requirement that the respondent actually 'took into account' a risk assessment when it enacted or maintained its measure.[55] In this regard, it found that the EC had not provided any evidence that the studies it referred to had actually been taken into account by the EC when it enacted the directives concerning hormones.[56]

The Panel posited that after this procedural requirement is satisfied, a panel's next task is first to identify the scientific conclusions reached in the risk assessment and the scientific conclusions implicit in the SPS measure; and

---

[51] *Ibid.* at para IV.110.

[52] *Ibid.* at para IV.113.

[53] The reports relied upon included the Lamming Report, Report of the Scientific Group on Anabolic Agents in Animal Production; OIE Scientific Report; JECFA Reports; Pimenta Report – Report of Committee of Enquiry into the Problem of Quality in the Meat Sector; EC Scientific Conference on Growth Promotion in Meat Production (1995); Monographs of the International Agency for Research on Cancer (IARC) on the Evaluation of Carcinogenic Risks to Humans, Supplement 7; and works by individual scientists including Dr Liehr, 'Potential Toxicity of Hormones'.

[54] *Ibid.* at para 8.111.

[55] *Ibid.* at para 8.113. *EC/Hormones Panel Report (Canada), supra* note 28 at para 8.116.

[56] *EC/Hormones Panel Report (US), ibid.*

second, to examine those scientific conclusions to determine whether or not one set of conclusions matches the second set of conclusions.[57] Applying this test, the Panel found that the scientific conclusions implicit in the EC measures (that is, that the use of the hormones in dispute, even in accordance with good practice, posed an identifiable risk to human health) did not conform with any of the scientific conclusions reached in the studies it had submitted as evidence.[58] It found that none of the scientific evidence referred to by the EC which specifically addressed the safety of some or all of the hormones in dispute when used for growth promotion, indicated that an identifiable risk arises for human health from use of those hormones if good practice is followed. All the studies came to the conclusion that the use of the hormones at issue (except for MGA, for which no evidence was submitted) for growth promotion purposes is safe. The Panel noted that the experts it had called confirmed this conclusion.[59]

The Appellate Body rejected the Panel's requirement for the regulating country to have actually taken into account the risk assessment. It found no textual basis for such a requirement and further, said that it inappropriately introduced an element of subjectivity into the question of compliance with Article 5.1. Instead, it ruled that the phrase 'based on' refers to 'a certain *objective relationship* between two elements, that is, an objective situation that persists and is observable between an SPS measure and a risk assessment'.[60] As to the substantive part of the Panel's test, the Appellate Body found that, in principle, the Panel's approach of examining the scientific conclusions implicit in the SPS measure and those yielded by a risk assessment was a useful approach. However, it found that while the relationship between these two sets of conclusions was relevant, it could not be assigned relevance to the exclusion of everything else.[61] It held that Article 5.1 requires the results of the risk assessment to 'sufficiently warrant, that is to say, reasonably support – the SPS measure at stake'. The requirement that an SPS measure be 'based on' a risk assessment is therefore a substantive requirement that there be a rational relationship between the measure and the risk assessment.[62]

The Appellate Body's decision leaves important questions unanswered, including, what constitutes a 'rational relationship', and what is meant by

---

57 *EC/Hormones Panel Report (US), ibid.* at para 8.117. *EC/Hormones Panel Report (Canada), supra* note 28 at para 8.120.

58 *EC/Hormones Panel Report (US), ibid.* at para 8.137. *EC/Hormones Panel Report (Canada), ibid.* at para 8.140.

59 *EC/Hormones Panel Report (US), ibid.* at para 8.124.

60 *EC/Hormones AB Report, supra* note 28 at para 189.

61 *Ibid.* at para 193.

62 *Ibid.*

'reasonably support'? However, the general approach seems correct as it allows Members flexibility which is desirable given the inherent subjectivity of the risk assessment process which would render it inappropriate if panels were to seek to reach their own scientific conclusions. It also discourages panels from seeking to conduct their own risk assessment on the basis of evidence provided by experts or submitted by the parties during the proceedings.[63]

In a critical aspect of its ruling for countries considering the introduction of controversial SPS measures, the Appellate Body stated that Article 5.1 does not require that the risk assessment embody only the view of a majority of the relevant scientific community. Rather, it held that:

> We do not believe that a risk assessment has to come to a monolithic conclusion that coincides with the scientific conclusion or view implicit in the SPS measure. The risk assessment could set out both the prevailing view representing the 'mainstream' of scientific opinion, as well as the opinions of scientists taking a divergent view. Article 5.1 does not require that the risk assessment must necessarily embody only the view of a majority of the relevant scientific community. In some cases, the very existence of divergent views presented by qualified scientists who have investigated the particular issue at hand may indicate a state of scientific uncertainty. Sometimes the divergence may indicate a roughly equal balance of scientific opinion, which may itself be a form of scientific uncertainty. In most cases, responsible and representative governments tend to base their legislative and administrative measures on 'mainstream' scientific opinion. In other cases, equally responsible and representative governments may act in good faith on the basis of what, at a given time, may be a divergent opinion coming from qualified and respected sources. By itself, this does not necessarily signal the absence of a reasonable relationship between the SPS measure and the risk assessment, especially where the risk involved is life-threatening in character and is perceived to constitute a clear and imminent threat to public health and safety.[64]

Despite this concession, the Appellate Body agreed with the Panel that on the facts of this case, the scientific reports cited by the EC did not rationally support its import prohibition.[65] A key factor for the Appellate Body was that much of the scientific evidence related to the carcinogenic potential of entire categories of hormones, or of the hormones at issue in general. They were not

[63] The Panel noted in *EC/Hormones* that it is not for the Panel itself to conduct its own risk assessment on the basis of scientific evidence gathered by the Panel or submitted by the parties during the Panel proceedings. *EC/Hormones Panel Report (US)*, *supra* note 28 at para 8.101.

[64] *EC/Hormones AB Report*, *supra* note 28 at para 194.

[65] The Appellate Body found only one divergent opinion which it did not consider reasonably sufficient to overturn the contrary conclusions reached in other scientific studies referred to by the EC.

specific to carcinogenic effects arising from the presence of the hormones in meat or meat products or residues of the hormones in dispute.[66] It should be noted that in this case, the exercise was relatively straightforward, but it would be more difficult in some cases, for example, where there is minority scientific evidence which the complainant claims is not credible. In such cases, a panel should focus on the procedure followed to arrive at the conclusions relied upon, using expert advice to determine whether accepted scientific protocols were followed and whether there is a threshold of rationality.

Regarding the sixth hormone, MGA, the Appellate Body upheld the Panel's finding that there had been no relevant risk assessment. The EC had referred to assessments that dealt with, *inter alia*, the category of progestins of which the hormone progesterone is a member. The EC argued that these assessments were relevant to MGA because it is an anabolic agent which mimics the action of progesterone. The Panel rejected this argument on the grounds that the assessments did not include any study that demonstrated how closely related MGA is chemically and pharmacologically to other progestins and what effects MGA residues would actually have on humans who ingested them along with meat from cattle to which MGA had been administered for growth promotion purposes.[67] The Appellate Body also upheld the Panel's finding that the EC had not submitted a risk assessment evaluating the level of risk arising from abusive use of hormones and the difficulties of control of the administration of hormones for growth promotion purposes.[68]

Walker criticizes the Appellate Body's approach in *EC/Hormones*, arguing that it did not consider a Member's right to choose its own 'science policy'.[69] He notes the Appellate Body's rejection of a requirement for quantification of a risk to human health, but also notes with some concern its finding that the kind of risk arising from a merely theoretical uncertainty is not enough to justify protective measures. He suggests that the phrase 'probability' (used in relation to risk assessment where risks to animal and plant health from pests or diseases are at issue) can focus one's attention on quantification alone. This, he argues, mischaracterizes the nature of risk which is not only measured by positive knowledge of a quantifiable likelihood, but also by the degree of uncertainty or lack of knowledge about a possible hazard.[70] He finds that on the continuum between merely speculative risk and a conclusively demon-

---

[66] *EC/Hormones AB Report*, *supra* note 28 at para 199.
[67] *EC/Hormones AB Report*, *supra* note 28 at para 201.
[68] *Ibid.* at para 208.
[69] See discussion of science policy in Chapter 10, section 10.4.
[70] Vern Walker, 'Keeping the WTO from Becoming the "World" Trans-science Organization: Scientific Uncertainty, Science Policy, and Fact-finding in the "Growth Hormones Dispute"' (1998) Cornell Int'l L.J. 251 at 305.

strated one is a 'vast stretch of undemonstrated, unquantified, but scientifically plausible risks. Within that zone, the risk of harm is real so long as safety is unproven.'[71] He thus argues that a Member's right under the SPS Agreement to adopt any level of protection is also the right to characterize or assess as a real risk any adverse effect that is 'possible', in the sense of scientifically plausible. Further, he argues that 'a panel should have no discretion to find a Member's measure "not sufficiently warranted" or "not reasonably supported" by the scientific evidence, if the panel finds that there is at least one scientifically plausible set of assumptions under which an adverse effect might occur'.[72] This, he argues, is why the Appellate Body should have recognized the role of science policy, namely, to provide rules that take into account considerations other than science to choose among the plausible alternatives.

Walker's argument is attractive from a standpoint of wanting to provide governments with flexibility to protect the health of their citizens. However, requiring only a 'plausible risk' could open a Pandora's box where countries could justify almost any measure where safety is unproven. His assertion that there is a zone where the risk of harm is real so long as safety is unproven is not particularly helpful given that one of the fundamental characteristics of risk assessment is that it is virtually impossible to prove that something is safe. The Appellate Body's requirement for something more than a theoretical risk is therefore reasonable in light of the competing objectives of health protection and trade liberalization. It is necessary to give meaning to the scientific evidence requirement; otherwise many protectionist measures could be justified by saying that safety is unproven.

Further, the Appellate Body's recognition that quantification of risks is not required arguably provides space within the boundaries of Article 5.1 for Members to implement science policies in the face of uncertainties. Its approach also arguably allows a role for public sentiment in cases where the public view scientific evidence in a light which does not coincide with the way in which some experts see the same evidence, but this has not yet been discussed in a case.

In *Australia/Salmon*, the Appellate Body dealt with the meaning of risk assessment in the first part of Annex A(4). It identified three steps which must be undertaken:[73]

---

[71] *Ibid.*

[72] *Ibid.*

[73] *Australia – Measures Affecting Importation of Salmon* (1998), WTO Doc. AB 1998-5, WT/DS18/AB/R (98-0000) (Appellate Body Report), at para 121 [*Australia/Salmon AB Report*].

1. *Identify* the diseases whose entry, establishment, or spread a Member wants to prevent within its territory, as well as the potential biological and economic consequences associated with the entry, establishment or spread of these diseases;
2. *Evaluate the likelihood* of entry, establishment, or spread of these diseases, as well as the associated potential biological and economic consequences; and
3. *Evaluate the likelihood* of entry, establishment, or spread of these diseases *according to the SPS measures which might be applied.*

The Appellate Body referred to its decision in *EC/Hormones* where it cited the dictionary meaning of 'probability' as 'degrees of likelihood' and 'a thing that is judged likely to be true', for the purpose of distinguishing the terms 'potential' and 'probability'.[74] On this basis, and also relying on the OIE's definition of 'risk' and 'risk assessment', the Appellate Body maintained that for a risk assessment to fall within the meaning of Article 5.1 and the definition in the first part of Annex A(4), it:

> . . . is not sufficient that a risk assessment conclude that there is a possibility of entry, establishment or spread of diseases and associated biological and economic consequences. A proper risk assessment of this type must *evaluate the 'likelihood'*, i.e., the 'probability' of entry, establishment or spread of diseases and associated biological and economic consequences as well as the 'likelihood' i.e., 'probability' of entry, establishment or spread of diseases according to the SPS measures which might be applied.[75] (Emphasis added)

The Appellate Body held that evaluation of the likelihood does not need to be done quantitatively but may be done qualitatively. Further, it held that there is no need for a risk assessment to establish a certain magnitude or threshold level of risk.[76] On the facts of the case, it found that no 'risk assessment' existed as Australia's 1996 Final Report did not meet either the first or second requirement identified.[77] It emphasized that *some* evaluation (as accepted by the Panel) of the likelihood is not enough.

The Appellate Body stressed the importance of distinguishing between the evaluation of risk in a risk assessment and the determination of the appropriate level of protection.[78] The term *'appropriate level of protection'* is defined in paragraph 5 of Annex A as 'the level of protection deemed appropriate by

74 *Ibid.* at para 184.
75 *Ibid.* at para 123.
76 *Ibid.* at para 124. *EC/Hormones AB Report, supra* note 28 at para 186.
77 *Australia/Salmon AB Report, ibid.* at paras 131 and 135.
78 *Ibid.* at para 125.

the Member establishing a sanitary or phytosanitary measure'. The Appellate Body held this to be a *prerogative* of the Member concerned, not of a panel or the Appellate Body.[79] The appropriate level of protection established by a Member and the SPS measure have to be clearly distinguished. They are not one and the same thing. The first is an objective, the second is an instrument chosen to attain or implement that objective.[80] While the Appellate Body emphasized again that the risk evaluated in a risk assessment must be an ascertainable risk, this does not mean that a Member cannot determine its own appropriate level of protection to be 'zero risk'.[81] This point is crucial from a health perspective, as different societies have shown themselves to have very different levels of risk tolerance, even in the light of evidence of a scientifically proven health risk.[82]

In *Japan/Apples*, the Appellate Body upheld the Panel's conclusion that Japan's 1999 Pest Risk Analysis did not satisfy the definition of 'risk assessment' set out in Annex A to the SPS Agreement because it: (1) failed to evaluate the likelihood of entry, establishment, or spread of fire blight through apple fruit;[83] and (2) failed to conduct such an evaluation 'according to the SPS measures which might be applied'.[84] With respect to the latter failure, the Appellate Body noted further that a risk assessment should not be limited to an examination of the measure already in place or favoured by the importing member. That is, it 'should not be distorted by preconceived views on the nature and the content of the measure to be taken; nor should it develop into an exercise tailored to and carried out for the purpose of justifying decisions *ex post facto*'.[85] The Appellate Body was satisfied that Japan had designed and conducted its 1999 Pest Risk Analysis in such a manner that no phytosanitary policy other than the regulatory scheme already in place was considered. As such, it had not properly evaluated the likelihood of entry according to the SPS

[79] *Ibid.* at para 129.

[80] *Ibid.* at para 200.

[81] *Ibid.* at para 125. *EC/Hormones AB Report, supra* note 28 at para 186.

[82] The Appellate Body noted at para 124 that the sixth paragraph of the Preamble and Article 3.3 explicitly recognize the right of Members to establish their own appropriate level of sanitary protection.

[83] The 1999 Pest Risk Analysis made determinations as to the entry, establishment, and spread of fire blight through a collection of various hosts (including apple fruit), but it failed to evaluate the entry, establishment, and spread of fire blight through apple fruit as a 'separate and distinct' vector. *Japan/Apples Panel Report, supra* note 38 at para 8.268. In this regard, the Appellate Body relied on the reasoning of the Appellate Body in *EC/Hormones* where the risk assessment only looked at hormones in general, to conclude that Japan's 1999 Pest Risk Analysis was not sufficiently specific to qualify as a risk assessment under the SPS Agreement.

[84] *Japan/Apples AB Report*, *ibid.* at para 192.

[85] *Ibid.* at para 208.

measures that 'might be applied'.[86] This decision highlights that even if a Member wants to responds to public sentiment, it must ensure that its risk assessment is procedurally sound and considers all possible outcomes.

In the *EC/Biotech* case, the Panel found that it was common ground among the parties to the dispute that the assessments carried out by the lead competent authority of each of the relevant Member States and by the EC scientific committees constituted risk assessments within the meaning of Annex A(4) and Article 5.1 of the SPS Agreement.[87]

The more important inquiry was whether the nine safeguard measures complained of were 'based on' one or more of the risk assessments. With respect to each of the measures, the Panel answered this enquiry in the negative. Noting that it did not matter whether the risk assessment relied upon was performed before or after adoption of the safeguard measure, the Panel stated that it was critical that 'the relevant risk assessment was appropriate to the circumstances existing at the time the Panel was established'.[88] In other words, it found that a change in relevant circumstances could in some cases render a completed risk assessment no longer 'appropriate to the circumstances'.[89] The Panel suggested that the state of scientific knowledge is one example of a circumstance which is relevant to the assessment of risks and which is subject to change over time. That is, evolution of science may result in new and/or better scientific evidence becoming available, and such evidence may have an effect on the continued relevance and validity of the conclusions of an existing risk assessment.[90] The Panel noted that in circumstances where there is little scientific evidence, the phrase 'as appropriate to the circumstances' may provide a measure of flexibility in terms of how the applicable elements of the Annex A(4) definition, including the likelihood evaluation, are satisfied.[91] While the Panel did not discuss the possibility, it is conceivable that in an amber case where public sentiment plays a role, the phrase 'appropriate to the circumstances' allows for the consideration of non-scientific factors that are appropriate to the social, cultural, and political context within which risk must be assessed.

The Panel examined each of the nine safeguard measures complained about by the US, Canada, and/or Argentina. These measures dealt with the two cate-

---

86 *Ibid.* at para 209.

87 *European Communities – Measures Affecting the Approval and Marketing of Biotech Products* (2006), WTO Doc. WT/DS291/R, WT/DS292/R, WT/DS293/R (Panel Report), at para 7.3018 [*EC/Biotech Panel Report*].

88 *Ibid.* at para 7.3025.

89 *Ibid.* at para 7.3023.

90 *Ibid.* at para 7.3022.

91 *Ibid.* at para 7.3044.

gories of risk agents in Annex A(4): first, those for which an evaluation of 'likelihood' must be provided; and second, those for which an evaluation of 'potential' must be provided. The Panel noted the lack of WTO jurisprudence concerning the meaning of the key concepts in the second category. The Panel noted that in *Australia/Salmon* the Appellate Body merely observed that the first and second clauses are substantially different from each other, and that the second clause requires *only* the evaluation of the 'potential' for adverse effects. In this regard, the Panel noted the dictionary definition of 'potential' as 'the possibility of something happening . . . in the future'.[92] In respect of each of the nine safeguard measures at issue, the Panel found that no state had explained, by reference to the original risk assessments, how and why they assessed the risks differently from the community-wide process, and had not provided a revised or supplemental assessment of the risks. Rather, the only risk assessments produced were those carried out at the time when the original Community consent was given, which in each case had found that there was no potential risk to health or the environment from the products in question. Further, the Panel stated that in no case had it been made aware of any divergent views expressed in the risk assessments or any uncertainties or constraints that would justify the safeguards. Finally, the Panel made reference to its experts, none of whom had expressed the view that the potential risks arising from the deliberate release of T25 maize and the other biotech products subject to the dispute could be considered to be risks that are 'life-threatening in character' or that 'constitute a clear and imminent threat to public health and safety'.[93]

Given these findings, the Panel concluded that there was no apparent rational relationship between any of the safeguard measures which, in each case, imposed a complete prohibition, and the risk assessments which found no evidence that the products presented any greater risk to human health or the environment than their conventional counterpart.[94]

---

[92] *Ibid.* at para 7.3039.

[93] *Ibid.* at para 7.3050. The Panel's statement about the experts not finding any risks that were life threatening, serious, or imminent is a reference to a statement by the Appellate Body in *EC/Hormones* where it noted that reliance by a country on a divergent scientific opinion does not necessarily signal the absence of a reasonable relationship between the SPS measure and the risk assessment, especially where the risk involved is life threatening in character and is perceived to constitute a clear and imminent threat to public health and safety. *EC/Hormones AB Report*, *supra* note 28 at para 194. The seriousness or otherwise of the alleged risk to health was not at issue here because there was no scientific evidence, even a divergent opinion, that a risk existed.

[94] *Ibid.*

Article 5.2 Relevant evidence for a risk assessment In *EC/Hormones*, the Panel emphasized that risk assessment is 'a scientific process aimed at establishing the scientific basis for the sanitary measure a Member intends to take'.[95] It held that Article 5 contemplates a distinction between risk assessment (which is not to involve social value judgments), and risk management (which might involve social value judgments and other non-scientific factors). The Appellate Body rejected this distinction, noting that the concept of 'risk management' is not mentioned in any provision of the SPS Agreement and could not be used to sustain a more restrictive interpretation of 'risk assessment' than is justified by the Agreement's actual terms.[96]

The Appellate Body thus took issue with the Panel's purported exclusion from the concept of risk assessment of all matters not susceptible to quantifiable analysis by the empirical or experimental laboratory methods commonly associated with the physical sciences. It did not consider the list of factors in Article 5.2 to be a closed one. In a widely quoted yet rather ambiguous passage, it said that:

> It is essential to bear in mind that the risk that is to be evaluated in a risk assessment under Article 5.1 is not only risk ascertainable in a science laboratory operating under strictly controlled operations, but also risk in human societies as they actually exist, in other words, the actual potential for adverse effects on human health in the real world where people live and work and die.[97]

Given the SPS Agreement's vagueness on this point – Article 5.2 does not explicitly state whether the list it provides is an exclusive one – the Appellate Body's interpretation appears to be a valid one from a treaty interpretation point of view. However, the Appellate Body provides little guidance as to what the Appellate Body envisages as being valid factors, for example, it is not clear whether or not it would include social, cultural, and economic factors.

The Appellate Body's ruling is consistent with current rethinking by scientists and regulatory economists of the soundness of a sharp distinction between risk assessment and risk management. As discussed in Chapter 10, recent literature suggests that it is no longer valid to draw a bright line between risk assessment and risk management.[98] The traditional view suggested that public

---

[95] *EC/Hormones Panel Report (US), supra* note 28 at para 8.107. *EC/Hormones Panel Report (Canada), supra* note 28 at para 8.110.

[96] *EC/Hormones AB Report, supra* note 28 at para 181.

[97] *Ibid.* at para 187.

[98] See for example National Research Council, *Understanding Risk – Informing Decisions in a Democratic Society* (Washington DC: National Academy Press, 1996). For a contrary view, see Jeremy D. Fraiberg and Michael J. Trebilcock, 'Risk Regulation: Technocratic and Democratic Tools for Regulatory Reform' (1998) 43 McGill L.J. 835 at 847.

values and concerns only entered into consideration in the risk management phase which follows from and is procedurally separate from the risk assessment phase. However, current thinking finds that value judgments (including public values and concerns) enter into both risk assessment and risk management.

This book has taken the view that while non-scientific factors should not supplant science, they may well be legitimate considerations in risk assessments where there are uncertainties in the scientific evidence. In such cases, public sentiment may justify a conservative approach in the setting of default assumptions for example. At bottom, given the inevitable uncertainties in risk assessment, it is unrealistic to expect that such factors will have no influence. It thus seems wiser to acknowledge their role in risk assessment, rather than pretending that absolute objectivity is possible. To this end, the Appellate Body's approach in *EC/Hormones* is to be welcomed.

13.3.2 Setting the Appropriate Level of Protection: Risk Management

This section examines interpretation by panels and the Appellate Body of the provisions that restrain Members' freedom to make risk management decisions. In particular, it considers what degree of flexibility they have accorded countries to exercise precaution and to respond to public sentiment in amber cases. An assessment of such flexibility is important in reaching a conclusion regarding the appropriateness of scientific evidence as a benchmark for justifying trade-restrictive SPS measures.

Despite the overlap between risk assessment and risk management, there is also a discernible distinction. It is therefore possible to say something about the approach taken by panels and the Appellate Body towards countries' risk management decisions. Just as risk perceptions differ between countries, so do ideas about the appropriate level of protection. In *EC/Hormones*, the EC argued that differences between governments in their approach to setting a level of protection would arise as a function of their economic priorities and cultural habits. For example, a developed country might find it desirable to set a high level of protection against contamination of foods by waste material from the chemical industry, while a developing country might accord higher priority to encouraging the establishment of a chemical industry rather than worrying about chemical residues in food.[99] Similarly, cultural habits such as eating raw foods might lead a government to set a higher level of protection against some pathogens than would be the case for a government whose population cooks the same food before eating it.[100]

---

99 *EC/Hormones Panel Report (US)*, *supra* note 28 at para IV.57.
100 *Ibid.*

The section will proceed by first examining decisions under the GATT where panels are required to determine whether a health regulation is 'necessary' and to ensure that measures are not 'applied in a manner which would constitute a means of arbitrary or unjustifiable discrimination between countries where the same conditions prevail, or a disguised restriction on international trade'. These GATT decisions may be used to inform decisions under the more specific SPS Agreement where Article 2.1 recognizes the right of Members to take SPS measures that are 'necessary' to protect health. Second, it will examine those provisions of the SPS Agreement which allow panels to review a country's risk management decisions, namely: Articles 2.3 (arbitrary or unjustifiable discrimination), 5.4 (objective of minimizing negative trade effects), 5.5 (arbitrary or unjustifiable distinctions in levels of protection), 5.6 (not more trade-restrictive than necessary), and 5.7 (provisional measures and the precautionary principle).

13.3.2.1 Lessons from the GATT

Is the measure 'necessary'? A GATT panel first interpreted the term 'necessary' as used in Article XX(b) in *Thailand/Cigarettes.* In that case, the Panel suggested that a measure cannot be justified as 'necessary' 'if an alternative measure which it could reasonably be expected to employ and which is not inconsistent with other GATT provisions is available to it'.[101] Here, the Panel found that there were readily available alternative measures consistent with the GATT which would achieve its stated health policy objectives. For example, to ensure the *quality* of cigarettes, Thailand could have implemented a regulation on a national treatment basis requiring complete disclosure of ingredients, coupled with a ban on unhealthy substances.[102]

To address the *quantity* of cigarettes consumed, Thailand could have used tobacco control strategies such as legislative bans and other restrictive measures to control the direct and indirect advertising, promotion, and sponsorship of tobacco.[103] Finally, to restrict the *supply* of cigarettes, Thailand could have maintained a monopoly on the importation and domestic sale of cigarettes in order to regulate the overall supply, prices, and retail availability.[104] However, the Panel did not require any evidence that the alternative measure it suggested would actually work, thus displaying a tendency to fail a

---

[101] *Thailand/Cigarettes Panel Report, supra* note 1 at para 74. The Panel cites the Report of the Panel on *United States – Section 337 of the Tariff Act of 1930* (L/6439, para 5.26, adopted on 7 November 1989).

[102] *Ibid.* at para 77.

[103] *Ibid.* at para 78.

[104] *Ibid.* at para 79.

Member on the necessity test even if there is only a hypothetical less trade-restrictive alternative available, which may or may not be effective in the circumstances.

In the *EC/Asbestos* case, the Panel and Appellate Body considered whether or not France's ban on asbestos was 'necessary' to protect public health within the meaning of Article XX(b).[105] The Panel noted the approach taken in the *Thailand/Cigarettes* case but considered that the assessment of the necessity of a measure could only be focused on the existence of other measures consistent or less inconsistent with the GATT if the parties were agreed on the existence and extent of the health problem in question. Given that the parties did not so agree, it considered that its duty was to determine the existence or otherwise of a health problem.[106] Having found that such a problem existed, in determining the necessity of the measure, the panel took into account the extent of the health problem.[107] This was because if the health problem was not as severe as France alleged, then the import ban might not be necessary. On the facts of the case, the Panel found that the risk due to chrysotile is important to the extent that it can generate lung cancers and mesotheliomas which are very difficult to cure or incurable. As well, the risk existed for a broad sector of the French population.[108]

The Panel noted that in addition to examining the scientific data in order to determine the existence of a risk to health, it was required to do so to determine the existence of other measures consistent or less inconsistent with the GATT 1994 and enabling the same objective of protecting public health to be obtained.[109] In this case, given that France's chosen level of health protection was a 'halt' to the spread of asbestos-related health risks, the Panel found that there was no reasonable alternative to the ban that might be chosen by a decision-maker responsible for developing public health measures.[110] On appeal, Canada argued that in coming to this decision, the Panel did not take into account the risk associated with the use of substitute products without a framework for controlled use. In response to this, the Appellate Body noted its agreement with the Panel that the import ban was 'designed and apt to achieve

---

[105] The Appellate Body refused to interfere with the Panel's factual finding that the evidence before it tended to show that handling chrysotile-cement products constitutes a risk to health rather than the opposite. *EC/Asbestos AB Report*, *supra* note 8 at para 157.

[106] *EC/Asbestos Panel Report*, *supra* note 10 at para 8.174.

[107] *Ibid.* at para 8.176.

[108] *Ibid.* at paras 8.200–201.

[109] *Ibid.* at para 8.179.

[110] *Ibid.* at paras 8.207 and 8.217. The Panel noted that controlled use did not constitute a reasonable alternative, particularly given the difficulties of instituting controls in the building sector.

France's chosen level of health protection'.[111] Further, it said that it was perfectly legitimate for France to seek to halt the spread of a highly risky product while allowing the use of a less risky product in its place.[112]

Canada had argued that 'controlled use' of asbestos was a reasonably available alternative to the Decree banning the product altogether. In considering this argument, the Appellate Body referred to the Panel's decision in *US/Reformulated Gasoline*, in which it held, in essence, that an alternative measure did not cease to be 'reasonably' available simply because the alternative measure involved administrative difficulties for a Member.[113]

The Appellate Body also referred to its earlier decision in *Korea/Beef* in which it addressed the issue of the meaning of the word 'necessary' under Article XX(d) of the GATT 1994.[114] There the Appellate Body said that 'necessary' can have a range of meanings, from 'indispensable' to 'making a contribution to'. In the context of Article XX(d), the Appellate Body found the meaning to be closer to 'indispensable'. The Appellate Body set out a number of factors that influence whether measures are 'necessary' in the context of Article XX(d):

> In sum, determination of whether a measure, which is not 'indispensable', may nevertheless be 'necessary' within the contemplation of Article XX(d), involves in every case a process of weighing and balancing a series of factors which prominently include the contribution made by the [compliance] measure to the enforcement of the law or regulation at issue, the importance of the common interests or values protected by that law or regulation, and the accompanying impact of the law or regulation on imports or exports.

In addition, the Appellate Body observed that 'the more vital or important the common interests or values pursued, the easier it would be to accept as "necessary" measures designed to achieve those ends'.[115] While the Appellate Body's decision must be read in the context of Article XX(d), its statements regarding the meaning of 'necessary' are also of relevance to Article XX(b) and raise the possibility that panels will be more inclined to find that a measure is necessary where vital health interests are at stake.

---

[111] *EC/Asbestos AB Report*, *supra* note 6 at para 168.

[112] *Ibid.*

[113] *US/Gasoline AB Report*, *supra* note 11 at paras 6.26 and 6.28.

[114] Article XX(d) provides an exception to the provisions of the GATT where it is 'necessary to secure compliance with laws or regulations which are not inconsistent with the provisions of this Agreement . . .'.

[115] *Korea – Measures Affecting Imports of Fresh, Chilled and Frozen Beef* (2000), WTO, at para 167.

The Appellate Body stated in *EC/Asbestos* that in determining whether a suggested alternative measure is available, several factors must be taken into account, besides the difficulty of implementation. It referred to two in particular, 'contribution of the measure to the realization of the value pursued' and 'importance of the value pursued'.[116] It concluded with respect to the facts that 'the objective pursued by France's import ban is the preservation of human life and health through the elimination, or reduction, of the well-known, and life-threatening, health risks posed by asbestos fibres. The value pursued is both vital and important in the highest degree.'[117] As to whether or not there was a 'reasonably available' alternative measure that would serve the same end, the Appellate Body found that 'France could not be reasonably expected to employ any alternative measure if that measure would involve a continuation of the very risk that the Decree seeks to "halt". Such an alternative measure would, in effect, prevent France from achieving its chosen level of health protection'.[118] In coming to this conclusion, the Appellate Body appropriately recognized the priority of health protection in this situation. As Howse and Tuerk argue, where imperfect control of a risk through risk management is likely to result in consequences as serious as life-threatening cancer, not to permit an outright ban as the least-trade-restrictive measure would 'impair the very ability of a member to exercise its prerogative (and fulfill its international human rights obligation) to protect the right to life of its citizens'.[119]

In emphasizing the importance of health, the Appellate Body seemed to be giving some priority to health as a value to be promoted by Member governments. It has been questioned whether panels or the Appellate Body have the legitimacy to make such a judgment.[120] Howse and Tuerk suggest in this regard that if the Appellate Body is going to accord some kind of hierarchy to interests where health is at the pinnacle, it must do so following the hierarchies implicit or explicit in international law more generally.[121] Chapter 7 asserted that human health is properly regarded as a fundamental human right and as such, it is argued that the Appellate Body was within the proper bounds of its jurisdiction to make such a judgment. As the Appellate Body stated in *US/Reformulated Gasoline*, WTO Agreements are not to be read 'in clinical isolation' from public international law.[122]

116 *EC/Asbestos AB Report*, *supra* note 6 at para 172.
117 *Ibid.*
118 *Ibid.* at para 174.
119 Howse and Tuerk, *supra* note 2 at 318.
120 Jan Neumann and Elisabeth Tuerk, 'Necessity Revisited: Proportionality in World Trade Organization Law after *Korea-Beef*, *EC-Asbestos* and *EC-Sardines*' (2003) 37:1 J. World Trade 199 at 214.
121 Howse and Tuerk, *supra* note 2 at 326.
122 *US/Gasoline AB Report*, *supra* note 11 at 14.

Neither the Panel nor the Appellate Body said anything in *EC/Asbestos* about how much evidence of a risk to health is needed for a measure to be deemed necessary 'for the protection of health'. Howse and Tuerk argue that there should be a *de minimis* requirement which would set out the 'minimum needed to assert with some plausibility that the measure is directed towards the goal of protecting health'.[123] It is not surprising that the issue did not come up in the *Asbestos/EC* case, given that the evidence regarding the toxicity of asbestos and its negative impact on health was widely accepted.[124] Yet it is likely that under the SPS Agreement, 'amber' cases will emerge in the future which will require panels to grapple more explicitly with the meaning of health. Panels may, for example, have to deal with cases that require them to decide whether 'health' should be given a narrow meaning that focuses on risks to physical health as established by science, or a broader conception that encompasses alternative public perceptions of what constitutes good health including psychological aspects of health.

In *Brazil/Retreaded Tyres*, the Appellate Body stated that a panel must 'consider the relevant factors, particularly the importance of the interests or values at stake, the extent of the contribution to the achievement of the measure's objective, and its trade restrictiveness. If this analysis yields a preliminary conclusion that the measure is necessary, this result must be confirmed by comparing the measure with possible alternatives, which may be less trade restrictive while providing an equivalent contribution to the achievement of the objective. This comparison should be carried out in the light of the importance of the interests or values at stake.'[125]

In this case, the Panel had noted the seriousness accorded to the diseases by the WHO and said that 'protection against such serious diseases is clearly an important objective'.[126] It followed *EC/Asbestos* in its view that 'the objective of protecting human health and life against life-threatening diseases, such as dengue fever and malaria, is both vital and important in the highest degree'. Further, the Panel acknowledged that preservation of animal and plant health constitutes an essential part of protection of the environment and is also an important value.[127] In terms of availability of alternative measures, the Panel noted that a measure would not be reasonably available if it was merely theoretical in nature, where the responding Member was not capable of taking it,

---

123 Howse and Tuerk, *supra* note 12 at 322.

124 For example, the International Labour Organization and the World Health Organization had both determined that governments should take measures to eliminate exposure to asbestos because it represents a grave health risk.

125 *Brazil/Retreaded Tyres Appellate Body Report*, *supra* note 19 at para 178.

126 *Brazil/Retreaded Tyres Panel Report*, *supra* note 19 at para 7.110.

127 *Ibid.* at paras 7.111 and 7.112.

or where it imposes an undue burden on that Member such as prohibitive costs or substantial technical difficulties. As well, the measure must be one that would preserve for the responding Member its right to achieve its desired level of protection with respect to the objective pursued.[128] The Panel held that none of the alternative measures suggested by the EC – measures to reduce the number of waste tyres and to improve the management of waste tyres (such as landfilling, stockpiling, incineration, and material recycling) – would either individually or collectively be such that the risks arising from waste tyres in Brazil would be safely eliminated as is intended by the import ban. Therefore, its import ban could be considered necessary.[129] The Appellate Body upheld the Panel's reasoning in this regard. It noted that the weighing and balancing required of panels is 'a holistic operation that involves putting all the variables of the equation together and evaluating them in relation to each other after having examined them individually, in order to reach a judgment'.[130] The Appellate Body felt that the Panel had been diligent in doing this.

The Appellate Body rejected Brazil's contention that if a Panel found that there were no reasonable alternatives to a measure, then it had to be considered as necessary, regardless of how small its contribution to achieving the objective. Rather, it said that the measure's contribution had to be 'material', not merely marginal or insignificant, especially if the measure is as trade restrictive as an import ban. Importantly, it also recognized that the contributions of some actions may not be immediately observable, but rather that they may only be evaluated with the benefit of time, for example, preventive actions to reduce the incidence of diseases that may manifest themselves only after a certain period of time.[131] This recognition is a welcome one that gives Members the policy space to recognize the importance of health protection and also the reality of illnesses caused by toxins, contaminants, and other agents.

Various uncertainties remain under the GATT and these may also be issues under the SPS Agreement. For example, would a panel hold that a measure was necessary because alternatives were not reasonably available due to factors such as domestic or international political pressures, or public sentiment?[132] This issue may arise in future disputes under the SPS Agreement, given increasing consumer awareness of health and safety issues and consequent demands on

---

[128] *Ibid.* at para 7.158.

[129] *Ibid.* at para 7.214.

[130] *Brazil/Retreaded Tyres Appellate Body Report*, *supra* note 19 at para 182.

[131] *Ibid.* at para 9.

[132] Layla Hughes, 'Limiting the Jurisdiction of Dispute Settlement Panels: The WTO Appellate Body Beef Hormones Decision' (1998) 10 Geo. Int'l Envtl L. Rev. 915 at 935. See also Howse and Tuerk, *supra* note 12 at 328.

politicians to enact regulations. Also, how would necessity be judged in the case of risks that are not as serious as those posed by cigarette smoking, asbestos, mosquito-borne tropical diseases, and burning tyres? As discussed in the following section, the SPS Agreement has a somewhat more relaxed necessity test than that in Article XX, but still requires (in Article 5.6) panels to determine whether another measure is not more trade-restrictive than necessary.

The chapeau Once it is determined that a measure is 'necessary' under GATT Article XX(b), the analysis of a country's risk management decisions turns to the chapeau and whether measures are a means of arbitrary or unjustifiable discrimination between countries where the same conditions prevail, or a disguised restriction on trade. The chapeau is paralleled in part in the Preamble of the SPS Agreement which says that Members are to avoid applying SPS measures in a manner which would 'constitute a means of arbitrary or unjustifiable discrimination', and also in Articles 2.3 and 2.5 (as discussed below).

In *US/Reformulated Gasoline*, the Appellate Body considered that the terms 'arbitrary discrimination', 'unjustifiable discrimination', and 'disguised restriction' may be read side by side; they impart meaning to one another. It considered that 'disguised restriction' may properly be read as 'embracing restrictions amounting to arbitrary or unjustifiable discrimination in international trade taken under the guise of a measure formally within the terms of an exception listed in Article XX'.[133] In other words, 'the kinds of considerations pertinent in deciding whether the application of a particular measure amounts to "arbitrary or unjustifiable discrimination", may also be taken into account in determining the presence of a "disguised restriction" on international trade'.[134] The Panel in *Brazil/Retreaded Tyres* noted that a restriction does not need to be formally 'hidden' or 'dissimulated' in order to constitute a disguised restriction on trade.[135]

In *Brazil/Retreaded Tyres*, the Panel looked to the dictionary meaning of 'arbitrary' which uses, *inter alia*, the phrases dependent on will or pleasure, capricious, unpredictable, inconsistent, and unrestrained in the exercise of will or authority.[136] It did the same for 'unjustifiable', where the dictionary meaning includes not justifiable, indefensible.[137] Previous cases looking at these phrases had found a measure to be arbitrary when it was applied in a rigid and

---

133 *US/Gasoline AB Report, supra* note 11 at 25.
134 *Ibid.*
135 *Brazil/Retreaded Tyres Panel Report,* supra note 19 at para 7.319.
136 *Ibid.* at para 7.257.
137 *Ibid.* at para 7.259.

inflexible manner without inquiring into its appropriateness for the conditions prevailing in the exporting countries.[138] A measure was found to be unjustifiable where it resulted in discrimination that could have been foreseen and that was not 'merely inadvertent or unavoidable'.[139] The Panel took the approach that, in the case of the ban on imported used tyres, while there was no arbitrary discrimination, there was unjustifiable discrimination because, despite the ban, Brazil had nevertheless been taking in imports of used tyres as a result of the court injunctions. This meant that the achievement of Brazil's declared objective of health protection was being significantly undermined and thus the measure was being applied in a manner that constituted a means of unjustifiable discrimination.[140] On appeal, the Appellate Body criticized the Panel's approach. It said that the 'analysis of whether the application of a measure results in arbitrary or unjustifiable discrimination should focus on the cause of the discrimination, or the rationale put forward to explain its existence'.[141] That is, discrimination will be arbitrary or unjustifiable if it is explained by a rationale that bears no relationship to the objective of a measure provisionally justified under one of the paragraphs of Article XX, or goes against that objective.[142] The Appellate Body rejected the Panel's approach of looking at the quantitative impact or effects of the discrimination in order to determine whether or not it amounts to something that can be said to be arbitrary or unjustifiable.[143]

These cases may provide some guidance to panels considering similar concepts under the SPS Agreement. Of particular interest going forward will be the reaction to the differing approaches taken by the Panel and the Appellate Body and whether either approach will be followed in future SPS cases. Both approaches appear to have flaws. The Panel's approach has some merit from a health perspective in that it focuses attention on whether the measure in question is actually achieving the stated objective of health protection. It recognizes the importance of actually achieving the goal of health protection, thus providing a helpful normative framework from which to judge the justifiability or otherwise of discrimination. It may be that discrimination results from an extraneous source (that is, not related to the objective of the measure), but that it does not undermine achievement of the objective. The Appellate Body would find that such a measure violated the chapeau. In such

---

[138] *United States – Import Prohibition of Certain Shrimp and Shrimp Products* (1998), WTO Doc. WT-DS58/AB/R (Appellate Body Report).

[139] *US/Gasoline Appellate Body Report*, *supra* note 11.

[140] *Brazil/Retreaded Tyres Panel Report*, *supra* note 19 at para 7.306.

[141] *Brazil/Retreaded Tyres Appellate Body Report*, *supra* note 19 at para 226.

[142] *Ibid*. at paras 227 and 232.

[143] *Ibid*. at para 229.

a case, however, it seems that a country should be entitled to argue that the discrimination is not unjustified or arbitrary, especially if it is not in a position to apply the measure differently. For example, is Brazil realistically in a position to remove the MERCOSUR exemption? Yet there is some concern from a trade perspective that the Panel's approach might allow a measure that had been applied in a discriminatory or arbitrary manner, so long as it does not undermine the health or environmental objective. For example, a regulation that arbitrarily exempts imports from one country of an otherwise prohibited product may not undermine the objective if the product is expensive and thus unable to penetrate the importing country's market. The Panel's quantitative test would seem to allow such a measure to pass the chapeau test, even although it would surely undermine the chapeau's purpose which seeks to avoid arbitrary or unjustifiable discrimination.

Thus, there are flaws in applying either approach too strictly. Rather, a combination of both approaches would seem to have merit. Indeed, while the Appellate Body rejected the Panel's wholesale reliance on the relationship between the effects of the discrimination and its justifiability, it conceded that, in certain cases, the effects of the discrimination may be a relevant factor for determining whether the discrimination is justifiable.[144] Thus, panels can look to a combination of whether discrimination is related to the objective, and whether the objective has been undermined.

13.3.2.2 The SPS Agreement

Article 2.3 (discrimination/disguised restrictions) and Article 5.5 (distinctions in the level of protection) Article 2.3 requires Members to ensure that their SPS measures do not 'arbitrarily or unjustifiably discriminate between Members where identical or similar conditions prevail' and are not 'applied in a manner which would constitute a disguised restriction on international trade'. This provision is linked to Article 5.5, which requires Members to avoid arbitrary or unjustifiable distinctions in the levels of SPS protection they consider appropriate in different situations, if such distinctions result in discrimination or a disguised restriction on international trade. In *Australia/Salmon*, the Appellate Body held that Australia had acted inconsistently with Article 5.5, and by implication Article 2.3.[145] It also held that a finding of inconsistency with Article 2.3 could be reached independently of a finding under Article 5.5. To date, however, such an independent finding has not been made.

---

144 *Brazil/Retreaded Tyres Appellate Body Report*, *supra* note 19 at para 230.
145 *Australia/Salmon Appellate Body Report*, *supra* note 73 at para 246.

The purpose of Article 5.5 is to detect regulatory inconsistencies national level, since such inconsistencies could indicate unnecessarily tained SPS regulations, which may constitute disguised trade protection.[146]

In *EC/Hormones*, the EC had afforded different levels of protection, depending on the type of hormone (for example, natural or synthetic) and their use (for example, for therapeutic or growth promotion purposes). Also, unlike for beef, the EC imposed no protection for pork imports where the pigs had been administered the anti-microbial agents carbadox and olaquindox.[147] The Appellate Body noted that Article 2.3 provides important context for this provision; however, it declined to consider its decision regarding arbitrary or unjustifiable discrimination under Article XX of the GATT in *US/Reformulated Gasoline*, noting the differing structural context of the GATT Article XX chapeau. It found that the 'degree of difference, or the extent of the discrepancy, in the levels of protection, is only one kind of factor which, along with others, may cumulatively lead to the conclusion that discrimination or a disguised restriction on international trade in fact results from the application of a measure or measures embodying one or more of those different levels of protection'. It went on to state that three elements are required to find a violation of Article 5.5, each of which must be present in order to constitute a violation.[148]

- The Member imposing the measure must have adopted its own appropriate levels of sanitary protection against risks to human life or health in several 'different situations';
- Those levels of protection must exhibit 'arbitrary or unjustifiable' differences in their treatment of the different situations; and
- The arbitrary or unjustifiable differences must result in 'discrimination or a disguised restriction of international trade'.

With respect to the first element, the Panel had said that different situations involving the 'same substance' or the 'same adverse health effect' may be compared to one another.[149] The Appellate Body did not explicitly address this

---

146 Reinhard Quick and Andreas Bluthner, 'Has the Appellate Body Erred? An Appraisal and Criticism of the Ruling in the WTO Hormones Case' (1999) 2 J. Int'l Econ. Law 603 at 620.

147 These agents are used as feed additives that indirectly act as a growth promoter by suppressing the development of bacteria and aiding the intestinal flora of piglets, thereby also exerting preventive therapeutic effects. Carbadox is known to be a genotoxic carcinogen, that is, it induces not merely promotes cancer. *EC/Hormones Appellate Body Report*, *supra* note 28 at para 229.

148 *Ibid*. at paras 214 and 240.

149 *EC/Hormones Panel Report (US)*, *supra* note 28 at para 8.176. *EC/Hormones Panel Report (Canada)*, *supra* note 28 at para 8.179.

statement, noting simply that the 'situations exhibiting differing levels of protection cannot, of course, be compared unless they are comparable, that is, unless they present some common element or elements sufficient to render them comparable'.[150]

With respect to the second element, the Panel had concluded that differences in protection as between naturally occurring and growth promotion hormones were 'arbitrary and unjustifiable'. However, the Appellate Body overruled this finding, explaining that there is a 'fundamental distinction' between 'added hormones' (whether natural or synthetic) and 'naturally-occurring' hormones.[151] The Panel had found it unnecessary to decide whether the difference in the level of protection between administration of hormones for therapeutic or zootechnical purposes and administration for growth promotion purposes constituted a violation of Article 5.5. However, the Appellate Body examined this issue and found that the different levels of protection concerning hormones used for these different purposes were justifiable and therefore not a violation of Article 5.5.[152] The Appellate Body's finding reflects the notion that processes or production methods used to produce a

---

[150] *EC/Hormones AB Report, supra* note 28 at para 217. In this case, the Panel had identified five different situations that could be compared: (i) The level of protection in respect of natural hormones when used for growth promotion; (ii) the level of protection in respect of natural hormones occurring endogenously in meat and other foods; (iii) the level of protection in respect of natural hormones when used for therapeutic or zootechnical purposes; (iv) the level of protection in respect of synthetic hormones (zeranol and trenbolone) when used for growth promotion; and (v) the level of protection in respect of carbadox and olaquindox. *EC/Hormones Panel Report (US), supra* note 28 at para 8.191. *EC/Hormones Panel Report (Canada), supra* note 28 at para 8.194.

[151] *EC/Hormones AB Report, supra* note 28 at para 221. The Appellate Body noted that the EC takes no regulatory action in respect of naturally-occurring hormones in meat and suggested that to require it to prohibit totally the production and consumption of such foods or to limit the residues of naturally occurring hormones in food would entail such a comprehensive and massive governmental intervention in nature and in the ordinary lives of people as to reduce the comparison itself to an absurdity. *EC/Hormones AB Report, supra* note 28 at para 224.

[152] *EC/Hormones AB Report, ibid.* at para 225. In so finding, the Appellate Body accepted the EC's argument that there are two main differences between the administration of hormones for growth promotion purposes and their administration for therapeutic and zootechnical purposes. First, therapeutic use is occasional as opposed to the regular and continuous use that characterizes growth promotion. It is also selective and takes place on a small scale and normally involves cattle intended for breeding and not slaughter; in contrast, the use of hormones for growth promotion purposes occurs on a much larger scale. *EC/Hormones AB Report, ibid.* at para 223. Second, there is much tighter regulation in the EC of the administration of hormones for therapeutic or zootechnical purposes (has to be administered by a veterinarian etc). *EC/Hormones AB Report, ibid.* at para 224.

good are sufficient to distinguish it from another item that is physically identical.[153]

The Panel also found a violation of Article 5.5 in respect of the EC's higher levels of protection concerning hormones for growth promotion as compared to those concerning the anti-microbial agents administered to pigs, finding the difference to be arbitrary and unjustifiable.[154] The Appellate Body agreed with this conclusion.

Moving to the third element, the Panel cited six reasons in support of its conclusion that these differences resulted in 'discrimination or a disguised restriction' on international trade: (1) the 'great difference' in the levels of protection; (2) the absence of a plausible justification for the difference; (3) the nature of the measure as an import ban; (4) the objectives of the measure in reducing barriers to intra-community trade and providing 'favourable treatment' to domestic producers; (5) the fact that before the ban took effect, the percentage of hormone-treated beef was lower in the EC than in Canada and the US; and (6) that the hormones at issue are used in the bovine sector, where the EC wants to limit supplies and is less concerned with international competitiveness, whereas the anti-microbial agents are used in the pork sector, where international competitiveness is a higher priority.[155]

After consideration of these factors, the Appellate Body reversed the Panel's decision, finding that the differences had not resulted in discrimination or a disguised restriction on international trade.[156] It noted that the third element will need to be considered in the light of the circumstances of each case. Looking at the facts at hand, it made, *inter alia*, the following points: (1) the documentation in the Panel record made it clear that there were authentic anxieties concerning the safety of the hormones in question;[157] (2) the necessity for harmonizing measures was part of the effort to establish a common

---

[153] This notion – while historically not accepted by GATT law – is consistent with the approach taken with respect to technical regulations. It remains to be seen how successfully the Appellate Body will be able to apply this notion and whether it will cause undue uncertainty in the law. See generally Ian Sandford, 'Hormonal Imbalance? Balancing Free Trade and SPS Measures after the Decision in Hormones' (1999) 29 V.U.W.L.R. 389.

[154] *EC/Hormones AB Report*, *supra* note 28 at para 235.

[155] *Ibid.* at para 242.

[156] *Ibid.* at para 246.

[157] Such anxieties were caused by results of general scientific studies (showing the carcinogenicity of hormones), the dangers of abuse of hormones and other substances used for growth promotion (highlighted by scandals relating to black-marketing and smuggling of prohibited veterinary drugs in the EC), and the intense concern of consumers in the EC over the quality and drug-free character of the meat available in its internal market. *EC/Hormones AB Report*, *ibid.* at para 245.

internal market for beef; and (3) the Panel's finding was not supported by the 'architecture and structure' of the measures.[158]

An important point here is that the Appellate Body accepted evidence regarding public anxieties and took it into account in finding that the measure did not amount to discrimination or a disguised restriction on trade. It suggested that where a measure is enacted in order to address public anxieties, this may be sufficient to counter a finding of discrimination or disguised protectionism. Kerr and Hobbs suggest that this approach does not take account of the possibility that such public influence may itself constitute a form of protectionism.[159] They suggest that WTO law is largely predicated on the assumption that consumer interests are legitimate, whereas governments are vulnerable to pressure from domestic producers for protection on illegitimate grounds.[160] As such, WTO law only recognizes one source of protectionism, namely, domestic producers. Kerr and Hobbs note that consumers may be manipulated by the media or may be subject to 'a degree of hysteria' regarding food safety issues and conclude that 'pretending that other groups cannot have an interest in protection only reduces the credibility of the WTO and is out of step with the current reality'.[161] As discussed in Chapter 7, the approach suggested in this book is that public sentiment has a legitimate role to play in countries' regulatory decisions in response to risk. The discussion in that chapter noted the potential for public risk perceptions to be manipulated by protectionist interests. This does not, however, justify outright rejection of public perceptions as a means of countering a finding of discrimination – given the cultural and psychological explanations for public risk perception – although it does justify exercise of caution on the part of panels in doing so and supports the SPS Agreement's scientific evidence requirements.

In *Australia/Salmon*, the Appellate Body adopted the three elements necessary for a violation of Article 5.5 that it had set out in *EC/Hormones* and found that they had all been met. With respect to the first element, the existence of different appropriate levels of sanitary protection in several 'different situations', the Panel had found that 'different situations' under the provision exist when the situations have 'common elements' and are therefore 'comparable'. It then set out the four situations it believed were comparable to the import ban on ocean-caught Pacific salmon, namely, the admission of imports of: (1) uncooked herring, certain cod and other products; (2) herring, certain cod,

---

158 *Ibid.*

159 W.A. Kerr and J.E. Hobbs, 'The North American-European Union Dispute over Beef Produced Using Growth Hormones: A Major Test for the New International Trade Regime' (2002) 25:2 The World Economy 283.

160 *Ibid.* at 287.

161 *Ibid.* at 295.

haddock and other products; (3) herring used as bait; and (4) live finfish. The Appellate Body agreed with the Panel that situations could be compared so long as they involved either a risk of entry, establishment or spread of the same or similar disease, *or* a risk of the same or similar 'associated potential biological and economic consequences'.[162] In addition, the Appellate Body observed that it was sufficient for the comparison to have one disease in common, rather than all diseases.[163] Therefore, the Appellate Body upheld the Panel's finding that the import ban on salmon and the four situations involving the import of other fish and fish products were 'different situations' which are comparable under the first element of Article 5.5.[164]

Regarding the second element, the Appellate Body reviewed the Panel's conclusion that the different levels of protection provided to imports of salmon as compared to imports of herring used as bait and finfish were 'arbitrary or unjustifiable'. The Panel had found that these two comparison situations presented at least as high, if not a higher, risk as imports of salmon, yet they were treated much more leniently.[165] On this basis, the Appellate Body upheld the Panel's finding that the differences in levels of protection were arbitrary or unjustifiable.[166]

The Panel had found, with respect to the third element, that the distinctions in levels of protection resulted in a disguised restriction on international trade. In support of this conclusion, the Panel set out three 'warning signals' and three 'additional factors' upon which it had relied. The 'warning signals' were: (1) the arbitrary character of the differences in the levels of protection; (2) the 'rather substantial difference' in the different levels of protection; and (3) the fact that the measure at issue was not 'based on' a risk assessment under Article 5.1.[167]

The 'additional factors' were: (1) the fact that Australia applies two different implementing measures to products which present the same risk suggests that Australia is effectively discriminating between the salmon products at issue (imports of which are prohibited) and herring and finfish (imports of which are allowed); (2) the change in conclusion between the 1995 draft report and the 1996 Final Report 'might well have been inspired by domestic pressures to protect the Australia salmon industry against import competition'; and (3) Australia imposes very strict standards for salmon imports to deal with a

162 *Australia/Salmon AB Report*, *supra* note 73 at para 146.

163 *Ibid.* at para 152.

164 *Ibid.* at para 153.

165 *Australia – Measures Affecting Importation of Salmon* (1998), WTO Doc. WT/DS18/R (Panel Report) at para 8.134. [*Australia/Salmon Panel Report*]

166 *Australia/Salmon AB Report*, *supra* note 73 at para 158.

167 *Australia/Salmon Panel Report, supra* note 165 at para 8.149.

small potential risk, while at the same time it does not appear to apply similarly strict standards to the internal movement of salmon products within Australia.[168] The Appellate Body upheld the Panel's reliance on all three 'warning signals' as well as the second and third 'additional factors' and concluded that the different levels of protection resulted in a disguised restriction on trade. However, it rejected the use of the first 'additional factor', noting that all 'arbitrary or unjustifiable' differences in levels of protection will lead to discrimination between products. This 'additional factor' is therefore no different from the first 'warning signal' and should not be considered a separate factor.[169]

Pauwelyn questions whether the third warning signal (that the measure was not based on a risk assessment) should be read as a departure from the Appellate Body's Article 5.5 findings in the *EC/Hormones* case where, despite a failure by the EC to conduct a risk assessment, the Appellate Body found no discrimination or disguised restriction on trade.[170] It can perhaps be explained by the description of the lack of a risk assessment as a 'warning signal'; that is, it is not a deciding factor. One factor is not enough to make a determination, but in combination with the other factors, may lead to that conclusion. However, it does cause some confusion and provides little guidance as to what weight panels should give to the different factors referred to. Victor notes that in both *EC/Hormones* and *Australia/Salmon* it was relatively easy to identify a comparable situation where SPS risks were substantially different.[171] However, these cases did not shed much light on what is 'comparable', except that in *Australia/Salmon* the Appellate Body rejected a narrow meaning of the phrase when it said that exactly the same sources and types of risks did not need to be at stake.

Article 5.4 Objective of minimizing negative trade effects Article 5.4 of the SPS Agreement provides that Members should, when determining the appropriate level of SPS protection, take into account the objective of minimizing negative trade effects. In *EC/Hormones*, the Panel stated that this provision does not impose an 'obligation' on Members, rather, it is more of a 'best efforts' requirement. It also stated that this objective should be 'taken into

---

[168] *Ibid.* at para 8.153.

[169] *Australia/Salmon AB Report*, *supra* note 73 at para 169.

[170] Joost Pauwelyn, 'The WTO Agreement on Sanitary and Phytosanitary (SPS) Measures as Applied in the First Three SPS Disputes: EC – Hormones, Australia – Salmon, and Japan – Varietals' (1999) 2 J. Int'l Econ. L. 654 at 655.

[171] David G. Victor, 'The Sanitary and Phytosanitary Agreement of the World Trade Organization: An Assessment after Five Years' (2000) 32 N.Y.U.J. Int'l L. & Pol. 865 at 916.

account' in the interpretation of other provisions of the Agreement.[172] To date, no guidance has been provided as to exactly what Article 5.4 means in practice; however, it would seem to bear some relation to the least-trade-restrictive requirement found in Article 5.6.

Article 5.6 Not more trade-restrictive than necessary Article 5.6 stipulates that SPS measures not be 'more trade-restrictive than required to achieve their appropriate level of sanitary or phytosanitary protection, taking into account technical and economic feasibility'. A footnote to this provision says that a measure is more trade-restrictive than required if there is another SPS measure which:

- Is reasonably available taking into account technical and economic feasibility;
- Achieves the Member's appropriate level of protection; and
- Is significantly less restrictive to trade than the SPS measure contested.

Neumann and Tuerk argue that these requirements indicate that the standard of necessity in the SPS Agreement is more relaxed than that under GATT Article XX. First, the reference to technical and economic feasibility is more lenient than the reasonableness test of administrative efforts under Article XX. Second, the fact that an alternative measure would have to be 'significantly' less restrictive gives Members more leeway to choose SPS measures. Third, the reference to the Member's 'appropriate level of protection' ensures that Members retain the right to choose the level of protection they wish to achieve.[173] This reference may be seen as similar to the Appellate Body's consideration in *Korea/Beef* of the importance of the values at stake.[174]

The Appellate Body considered Article 5.6 in *Australia/Salmon* where it confirmed the cumulative nature of the three elements contained in the footnote.[175] The Appellate Body agreed with the Panel that there were alternative measures available to Australia which had been put forward in the 1996 Draft Report.[176] With respect to the first element, the reasonable availability of the

---

172 *EC/Hormones Panel Report (US)*, *supra* note 28 at para 8.166.

173 Neumann and Tuerk, *supra* note 120 at 221.

174 Frank J. Garcia, 'The Salmon Case: Evolution of Balancing Mechanisms for Non-Trade Values in the SPS Agreement' in George A. Bermann and Petros C. Mavroidis, eds, *Trade and Human Health and Safety* (New York: Cambridge University Press, 2006) 133 at 151.

175 *Australia/Salmon AB Report*, *supra* note 73 at 194.

176 *Ibid*. at para 195. Alternative measures might have included evisceration, inspection and grading, or restriction of imports to non-spawning, adult salmoids. See *Australia/Salmon Panel Report*, *supra* note 165 at para 7.146.

alternative measure, the Panel had taken the approach of relying on the Member's own risk assessment. As Garcia notes, such an approach seemed appropriate in this case but would be more difficult in a case where the Member had not considered the alternate measure.[177] He suggests that panels should proceed cautiously in such a case because the risk to the Member's policy autonomy would be substantial if a panel could supplant the Member's regulatory decision on the basis of its evaluation of a hypothetical situation.[178]

Article 5.6 was also considered in *Japan/Agricultural Products* where the Panel rejected a claim by the US that there was an alternative measure within the meaning of Article 5.6, namely, 'testing by product' of the efficacy of the quarantine treatment.[179] The Panel agreed with the US that 'testing by product' would be significantly less restrictive to trade than the varietal testing requirement; however, it concluded that it did not have sufficient evidence to show that it would achieve Japan's appropriate level of protection for any of the products at issue.[180] The US unsuccessfully appealed this finding.[181]

Sykes argues that the least-trade-restrictive test as embodied in Article 5.6 is roughly co-extensive with a crude cost-benefit analysis under conditions of uncertainty.[182] That is, a panel or the Appellate Body does not actually quantify the costs and benefits of alternative regulatory policies in dollars. Rather, decision-makers proceed 'more impressionistically and qualitatively' to assess the effect of alternative policies on trade, administrative difficulties, and resource costs associated with alternative policies, and the regulatory efficacy of those policies. They then weigh these considerations in making a decision.[183] Sykes notes that where the regulatory objective relates to some highly valued interest such as the protection of human life, then the challenged regulation will be upheld if there is any doubt as to the ability of the proposed alternative to achieve the same level of efficacy.[184] The key issue that will arise in 'amber' cases in this regard is how a panel is to assess when a particular health interest is important or vital, such as effectively to raise the threshold for what would be a reasonably available and suitable alternative measure.

Questions of proportionality may be relevant under Article 5.6. Take for example a case where a country prohibits imports of food containing trans-

---

[177] Garcia, *supra* note 174 at 149.
[178] *Ibid.*
[179] *Japan/Agricultural Products Panel Report*, *supra* note 37 at para 8.78.
[180] *Ibid.* at para 8.84.
[181] *Japan/Agricultural Products AB Report, supra* note 35 at para 100.
[182] Alan O. Sykes, 'The Least Restrictive Means' (2003) 70 U. Chicago L. Rev. 403 at 419.
[183] *Ibid.* at 416.
[184] *Ibid.*

fats, which are proven to increase the risk of heart disease, but each food product standing alone will have negligible to minor health consequences. Exporting countries claim that a simple label is a more measured and appropriate response to the nature of the risk. Such a case would fall into the amber classification discussed earlier, being a case where there is a high probability risk but minor consequences. If a proportionality test was introduced, it would not be immediately apparent that there was a rational relationship between the scientific evidence in question, which is not controversial, and the nature of the regulatory response.[185] However, a proportionality test could be used by exporting countries to argue that the measure could not be justified because even if rationally related to the scientific evidence, the measure is out of all proportion to the risk. Such an argument would arguably be problematic because given that the SPS Agreement does not say anything about the level of risk to health required, a panel would have to use its subjective judgment to make a conclusion as to a lack of proportionality. If a complaining party is unable to identify any plausible alternative measure that would protect citizens of the importing state from the health risk in question, then it may be argued — adopting a bias towards health protection – that the measure ought to be allowed, no matter that the risk is not considered very serious by the complaining state. Inevitably though, it would be expected that the less serious the risk, the more likely it will be that suitable alternative less trade-restrictive measures exist to deal with it.

Article 5.7 Provisional measures and the precautionary principle In *EC/Hormones*, the EC argued that its hormone ban was the result of a precautionary approach taken in order to assess the facts. However, maintaining that its import ban was definitive and not provisional, it did not seek to justify its measure under Article 5.7 as that provision is intended to be relied upon only where measures are provisional. In order to make this argument, the EC contended that while Article 5.7 implicitly recognized the precautionary principle, the principle's role ought to be expanded. It argued that the principle has become 'a general customary rule of international law' or at least 'a general principle of law' and that as such it should be used to interpret Articles 5.1 and 5.2 of the SPS Agreement.[186] Thus it suggested that it is not necessary for all scientists around the world to agree on the 'possibility and magnitude' of the

---

[185] Michael Trebilcock and Julie Soloway, 'International Policy and Domestic Food Safety Regulation: The Case for Substantial Deference by the WTO Dispute Settlement Body under the SPS Agreement' in Daniel Kennedy and James Southwick, eds, *The Political Economy of International Trade* (Cambridge, UK: Cambridge University Press, 2002) at 562.

[186] EC appellant's submission, para 91.

risk, nor for all or most of the WTO Members to perceive and evaluate the risk in the same way.[187] It also stressed that Articles 5.1 and 5.2 do not prescribe a particular type of risk assessment and do not prevent Members from being cautious in their risk assessment exercise.[188] This argument is arguably consistent with the literature on risk assessment which suggests that risk assessment and risk management is inherently precautionary.[189] The US and Canada argued that the precautionary principle has not yet been incorporated into customary international law but did not appear to dispute that risk assessments may involve precaution.[190]

The Appellate Body found in *EC/Hormones* that while the precautionary principle is regarded by some as having crystallized into a general principle of customary international environmental law, it is less clear that it has been widely accepted by Members as a principle of general or customary international law.[191] While it did not find it necessary to take a position on the question, the Appellate Body noted some aspects of the relationship of the precautionary principle to the SPS Agreement. First, the principle has not been written into the SPS Agreement as a ground for justifying measures that are otherwise inconsistent with the specific obligations of Members.[192] Second, the precautionary principle is reflected in Article 5.7. The Appellate Body did note, however, that there is no need to assume that Article 5.7 exhausts the relevance of the precautionary principle. Rather, it is also reflected in the sixth paragraph of the Preamble and in Article 3.3 which explicitly recognize the right of Members to establish their own level of sanitary protection. Third, it noted that panels should bear in mind that 'responsible, representative governments commonly act from perspectives of prudence and precaution where risks of irreversible, e.g., life-terminating, damage to human health are concerned'. Finally, the Appellate Body held that the precautionary principle does not, by itself, and without a clear textual directive to that effect, relieve a panel from the duty of applying the normal (that is, customary international law) principles of treaty interpretation in reading the provisions of the SPS

[187] *Ibid.* at para 88.
[188] *Ibid.* at para 94.
[189] See discussion in Chapter 10, section 10.4.2.
[190] US appellee's submission, para 92. Canada's appellee's submission, para 34.
[191] *EC/Hormones AB Report*, *supra* note 28 at para 123.
[192] Prévost writes that while the precautionary principle was becoming increasingly recognized in the environmental context by the end of the Uruguay Round negotiations, it had not been considered in the context of human or animal health. Accordingly, it was not mentioned in the negotiations, nor in the text of the SPS Agreement. Denise Prévost, 'What Role for the Precautionary Principle in WTO Law after *Japan – Apples*?' (2005) 2:4 Journal of Trade and Environment Studies 1 at 2.

Agreement.[193] Overall then, the Appellate Body agreed with the Panel that the precautionary principle does not override the provisions of Article 5.1 and 5.2 of the SPS Agreement which require a risk assessment.

Prévost argues that the Appellate Body based its finding in *EC/Hormones* on a mischaracterization of the nature and role of the precautionary principle. She suggests that the interpretation of the vaguely worded terms such as 'sufficient scientific evidence' in Article 2.2 and 'as appropriate to the circumstances' in Article 5.1, and the application of these requirements to the factual evidence could benefit from guidance provided by the precautionary principle.[194] As noted in Chapter 10, recognition of the precautionary principle would arguably entail giving the benefit of doubt to importing countries in the form of precautionary protection, rather than to the producer where there is any doubt about safety, even if the risks have not been completely assessed.[195]

Yet while it does not say so explicitly, the Appellate Body seems to have left the door open for a country to base a regulation on a risk assessment that reflects a degree of precaution in an 'amber' case such as where the science is uncertain or public perceptions view risk in a different light from experts. It may even be seen as having implied as much in its acknowledgement that a risk assessment does not have to arrive at a 'monolithic' conclusion but may set out both mainstream and divergent scientific views, and further, that Members may base their SPS measures on a divergent opinion from a qualified and respected source.[196] In recognizing the prudence and precaution of governments in cases of risk to health, it seems to be implicitly recognizing the indeterminate nature of science as applied in the discipline of risk assessment. An explicit acknowledgement of this characteristic of risk assessments in future cases would be desirable, helping to focus the determination not on whether countries have been precautionary in their actions, but on *how* precautionary they have been. Precaution is inherent in choosing to accept a minority scientific conclusion that finds existence of a risk, or a scientific conclusion that is uncertain yet still meets a threshold of rationality. Both decisions may be justified under the SPS Agreement. However, where precaution is exercised to protect against a risk which is forecast by a study that does not reach a threshold of scientific rationality as argued for in Chapter 3, then it should not be a justifiable basis for a trade-restrictive SPS measure. Determining when precaution has been taken too far will always be a matter of degree, based on the evidence and the circumstances of the individual case. A recent article

[193] *EC/Hormones AB Report*, *supra* note 28 at para 124.

[194] Prévost, *supra* note 192 at 5.

[195] *EC/Hormones Panel Report (Canada)*, *supra* note 28 at para 4.212. *EC/Hormones Panel Report (US)*, *supra* note 28 at para 4.202.

[196] *EC/Hormones AB Report*, *supra* note 28 at para 193.

argues that a common understanding or rules regarding the issue of precaution would be desirable to allow panels and the Appellate Body to make consistent decisions.[197] The difficulty will be, however, that even if precaution is accepted more formally in WTO law (which is unlikely), countries will continue to have different risk thresholds, leading to the kinds of difficulties discussed here.

In *Japan/Agricultural Products*, the Appellate Body divided Article 5.7 into four requirements which must all be met in order to adopt and maintain a provisional SPS measure. The Member must have:

1. Imposed the measure in respect of a situation where relevant scientific information is insufficient;
2. Adopted the measure on the basis of available pertinent information;
3. Sought to obtain the additional information necessary for a more objective assessment of risk; and
4. Reviewed the . . . measure accordingly within a reasonable period of time.

The Appellate Body stated that, as recognized by Article 2.2, Article 5.7 operates as a qualified exemption from the obligations under Article 2.2 not to maintain SPS measures without sufficient scientific evidence.[198] In this case, Japan had not sought to obtain additional information necessary for a more objective risk assessment nor had it reviewed its varietal testing requirement within a 'reasonable period of time'. The Appellate Body considered that what constitutes a 'reasonable period of time' has to be established on a case-by-case basis and depends on the specific circumstances of each case, including the difficulty of obtaining the additional information necessary for the review and the characteristics of the provisional SPS measure.[199] Here, the Appellate Body considered that it would have been relatively easy for Japan to obtain additional information. This is likely to be a controversial issue in other cases where the science is more complicated and therefore less readily available. In such cases (the case of genetically modified foods is one that comes to mind as a scenario where this might occur), provisional requirements might come to resemble permanent measures if scientific information cannot be obtained. In the medium to long term, this is likely to create tension as to the intended scope of Article 5.7.

Japan/Apples was the first case where a Panel and the Appellate Body addressed the question of what constitutes a situation where relevant scientific

---

[197] United Nations University Institute of Advanced Studies, *Trading Precaution: The Precautionary Principle and the WTO* (Yokohama: UNU IAS, 2005) at 8.

[198] *Japan/Agricultural Products AB Report*, *supra* note 35 at para 80.

[199] *Ibid.* at paras 89 and 93.

evidence is insufficient. After noting that 'the burden is on Japan, as the party invoking Article 5.7 to make a *prima facie* case in support of its position',[200] the Panel found that Article 5.7 was designed to be invoked 'in situations where little, or no, reliable evidence was available on the subject matter at issue'.[201] The Appellate Body upheld the Panel's findings. It concluded that because scientific studies and practical experience on the question of the risk of transmission of fire blight through apple fruit had been accumulating for the past 200 years, Japan's measure could not be justified under Article 5.7 because it was not imposed in respect of a situation where relevant scientific evidence was insufficient.[202] The Appellate Body stated that relevant scientific evidence will be insufficient for the purposes of Article 5.7 if it 'does not allow, in qualitative or quantitative terms, the performance of an adequate assessment of risks as required under Article 5.1'.[203] It also upheld the Panel's decision that Article 5.7 'is triggered *not by the existence of scientific uncertainty*, but rather *by the insufficiency of scientific evidence*' (emphasis added).[204] Notably, the Appellate Body did hold that cases where the available evidence is more than minimal in quantity, but has not led to reliable or conclusive results, fall within the scope of Article 5.7.[205] Thus, according to the Appellate Body's decision in *Japan/Apples*, Article 5.7 may be relied upon either due to the small quantity of evidence on new risks, or due to the fact that accumulated evidence is inconclusive or unreliable. In either case, the insufficiency of the evidence must be such as to render impossible the performance of a risk assessment pursuant to Article 5.1 and must be distinguished from uncertainty.[206]

---

200 *Japan/Apples Panel Report*, *supra* note 38 at para 8.212.

201 *Ibid.* at para 8.215.

202 *Japan/Apples AB Report*, *supra* note 42 at paras 173 and 179. The Panel had observed that in the course of its analysis under Article 2.2, it had come across 'an important amount of relevant evidence' (*ibid.* at para 8.216). The Panel added that a large quantity of high quality scientific evidence on the risk of transmission of fire blight through apple fruit had been produced over the years, and noted that the experts had expressed strong and increasing confidence in the evidence (*ibid.* at para 8.129). Japan challenged the Panel's finding that 'relevant scientific evidence' in Article 5.7 may refer to a particular measure or a particular risk (*ibid.* at para 8.220). However, the Panel held that the term 'insufficient scientific evidence' is meant to refer to evidence *in general* on the phytosanitary question at issue, in this instance the risk of transmission of fire blight through apple fruit (*ibid.* at para 8.218).

203 *Japan/Apples AB Report*, *ibid.* at para 179.

204 In response to Japan's argument that Article 5.7 covers not only situations of 'new uncertainty' (where a new risk is identified) but also 'unresolved uncertainty' (where there is considerable scientific evidence but still uncertainty remains). *Ibid.* at para 184.

205 Noting that the Panel's interpretation did not preclude this. *Ibid.* at para 185.

206 Prévost, *supra* note 192 at 9.

Both *Japan/Apples* and *Japan/Agricultural Products* concerned risk that had potential to cause economic loss from damage to crops. It is arguable that in a case where governments have an obligation to protect health, most notably where human health is at issue, dispute panels ought to interpret Articles 2.2 and 5.7 in a manner that allows countries flexibility to exercise a greater degree of precaution and is therefore more deferential to the substantive regulatory choices of domestic governments.

The most recent decision to consider Article 5.7 is that in *EC – Biotech*. In this case, the Panel departed from the Appellate Body's previous finding in *Japan/Agricultural Products* that Article 5.7 is a qualified exemption from the obligations under Article 2.2. Instead, it held that Article 5.7 should be characterized as an autonomous right.[207] This means that the burden of proof is placed on the complaining party to establish a *prima facie* case that the responding party has failed to comply with the requirements of Article 5.7.

With respect to each of the safeguard measures, the EC argued that the measure met the four requirements of Article 5.7 as set out by the Appellate Body in *Japan/Agricultural Products*. First, it argued that relevant scientific evidence was insufficient. It argued that the concept of 'insufficiency' is relational in that it must refer to the matters of concern to the legislator. That is, 'insufficient' means 'insufficient' for the production of a risk assessment adequate for the purposes of the legislator who must decide whether a measure should be applied, provisionally or otherwise.[208] The EC defined a risk assessment which is adequate as one which has been 'delivered by a reputable source, [which] unequivocally informs the legislator about what the risk is with a sufficient degree of precision, and [which] has withstood the passage of time and is unlikely to be revised'.[209] This argument surely cannot be correct. As discussed in Chapter 10, risk assessments are hardly ever unequivocal and if this was the required standard, the requirements of Article 5.1 would be extremely difficult if not impossible to meet.

The EC also submitted that it is artificial to suppose that there is some kind of magic moment at which the available science becomes sufficient for all purposes. In the EC's view, the actions of a legislator in response to available scientific evidence are a function of what that particular legislator is concerned about. Thus, the higher the level of acceptable risk, the more likely it is that the legislator may

---

[207] *EC/Biotech Panel Report*, *supra* note 87 at para 7.2988.

[208] The Panel noted that the Appellate Body in *Japan/Apples* referred to the insufficiency of available scientific evidence to perform an 'adequate' assessment of risks and suggests that the EC relies on this use of the word 'adequate' in arguing that an 'adequate' assessment of risks is one which is 'adequate' for the purposes of the legislator. *Ibid.* at para 7.3226.

[209] *Ibid.* at para 7.3217.

conclude, within a relatively short period of time, that the scientific evidence is sufficient and that no provisional measure is necessary. The lower the level of acceptable risk, the more likely it may be that the legislator may continue to consider, for a relatively long period of time, that the scientific evidence is insufficient, and that a measure is warranted.[210] The EC is correct that there is no magic moment of saying when evidence is or is not sufficient for the purposes of Article 5.7. As the discussion in Chapter 10 noted, risk assessments may range from being complete, with very little uncertainty, to being riddled with data gaps and uncertainties. Yet even the latter extreme may be procedurally sound and still constitute a valid risk assessment in accordance with accepted scientific protocols.

The EC sought to differentiate the case from *Japan/Apples*, arguing that the technologies in question are still at the 'frontiers of science' and the future consequences (unlike those of fire blight) are highly uncertain and potentially more far-reaching. In the case of fire blight, the matter could be described as having passed into the 'realms of conventional scientific wisdom – an operational hypothesis unlikely to be disturbed other than by some revolutionary and currently totally unforeseeable new scientific discovery'. It argued that the relevant risks are more than the mere theoretical uncertainty that always remains simply because science can never provide absolute certainty that a given substance will never have adverse effects.[211]

The Panel agreed that it must be determined on a case-by-case basis whether the body of available scientific evidence is insufficient to permit the performance of a risk assessment. However, it did not agree that the protection goals of a legislator are relevant to such a determination. In other words, 'there is no apparent link between a legislator's protection goals and the task of assessing the existence and magnitude of potential risks'.[212] It went on to find that 'where a risk assessment has been performed, and that risk assessment meets the standard and definition of Annex A(4), it does not cease to be a risk assessment within the meaning of Annex A(4) merely because a particular Member judges that the risks have not been assessed with a "sufficient" degree of precision, that the assessment has not "withstood" the passage of time, and that it is "likely" that the assessment may need to be revised at some point in the future. If there are factors which affect scientists' level of confidence in a risk assessment they have carried out . . . this may be taken into account by a Member in determining the measure to be applied for achieving its appropriate level of protection from risks'.[213]

---

210 *Ibid.*

211 *Ibid.* at para 7.3221.

212 *Ibid.* at para 7.3229.

213 The Panel noted that a limited body of relevant scientific evidence may be such a factor. *Ibid.* at para 7.3232.

The Panel noted further that the definition of the term 'risk assessment' in Annex A(4) does not indicate that a Member's level of protection is pertinent to an assessment of the existence and magnitude of risks and that scientists do not need to know a Member's 'acceptable level of risk' in order to assess objectively the existence and magnitude of a risk.[214] While scientists do not *need* to know a Member's acceptable level of risk, it seems unrealistic to imagine that a country's protection goals will not play a part in its decision as to whether or not scientific evidence is sufficient to perform a risk assessment for the purposes of Article 5.1. There must always be some factor that triggers a decision that the scientific evidence available is insufficient to perform a risk assessment, and this may be a desire for a high level of protection. The Panel seemed to find a stronger distinction between risk assessment and risk management than actually exists.[215]

The Panel released its interim, confidential, report to the Parties in February 2006. However, the report was leaked to the media which prompted much discussion of the findings among the NGO community. In a letter to the Parties dated 8 May 2006, the Panel sought to clarify certain aspects of its finding which it believed had been misconstrued in these criticisms. One of the points it sought to clarify was that its findings recognize that the notion of 'insufficiency of scientific evidence' as it is used in Article 5.7 of the SPS Agreement includes cases of qualitative insufficiency of scientific evidence. In the current situation, however, the Panel was not persuaded that 'the scientific evidence available at the relevant time was qualitatively (or quantitatively) insufficient' such that a risk assessment as required under Article 5.1 could not be performed.[216]

The Panel found that the sufficiency of scientific evidence should be assessed at the time a provisional measure was adopted.[217] Looking at each of the safeguard measures complained of, it found that the Complaining Parties had established a presumption (which had not in any case been rebutted) that the Member States in question had imposed safeguard measures in respect of a situation where relevant scientific evidence was *not* insufficient. In doing so, it accepted the Complaining Parties' argument that relevant scientific evidence could not have been insufficient to conduct a risk assessment at the time of adoption of the safeguard measures, since a risk assessment had been

---

214 *Ibid.* at para 7.3234.

215 See section 10.4.

216 *European Communities – Measures Affecting the Approval and Marketing of Biotech Products, Report of the Panel, Addendum (Annex 'K')* (2006), WTO Doc. WT/DS291/R/Add.9, WT/DS292/R/Add.9, WT/DS293/R/Add.9 (Panel Report) [*EC/Biotech Annex 'K'*].

217 *EC/Biotech Panel Report, supra* note 87 at para 7.3246.

conducted by the Community on the basis of information provided by the Member States.[218]

On a first reading, the Panel's finding seems to infer that once a country has obtained sufficient scientific evidence to conduct a risk assessment, it is barred from relying upon Article 5.7 in the future. This cannot be correct. If new evidence comes to light that would render the conclusions of the original risk assessment unreliable or inconclusive – but would not necessarily change the result of the risk assessment in a definitive enough manner to allow a completely 'new' measure to be adopted – countries ought to be permitted to enact provisional measures pursuant to Article 5.7 and should not be barred from doing so simply because under earlier circumstances that no longer exist it was able to conduct a risk assessment. The Panel itself appears to have accepted this proposition. In its letter to the Parties dated 8 May 2006, it stated that its findings 'leave room for the possibility that even if at a given point in time relevant scientific evidence is sufficient to perform a risk assessment, a situation might subsequently arise where the relevant scientific evidence could be considered insufficient to perform a risk assessment as required under Article 5.1 and as defined in Annex A(4) of the SPS Agreement. It is conceivable, for instance, that relevant new scientific evidence would negate the validity of the scientific evidence on which an existing risk assessment relied without, however, being sufficient, in quantitative and qualitative terms, to allow the performance of a new risk assessment'.[219]

In making this statement, the Panel has properly recognized that risk assessments do not stand still in time, but are dynamic and continuing. This is a key finding and will be particularly critical in areas of developing scientific knowledge.

13.3.3 Burden of Proof, Standard of Review, and Expert Advice

Three final issues impact how panels review a country's regulatory decision-making under the SPS Agreement: first, how the burden of proof is allocated; second, what standard of review a panel should adopt when reviewing a Member's regulations; and three, what use should be made by panels of expert advice. These three issues are discussed below using both GATT cases where relevant to the SPS Agreement as well as the cases decided under the SPS Agreement.

---

[218] *Ibid.* at para 7.3251.

[219] *EC/Biotech Annex 'K'*, *supra* note 216.

13.3.3.1 Burden of Proof

The Appellate Body considered the *burden of proof* in *EC/Hormones* where it upheld the Panel's finding that the initial burden of proof rests on the complaining party, which must establish a *prima facie* case of inconsistency with a particular provision of the SPS Agreement on the part of the defending party. When that *prima facie* case is made, the burden of proof moves to the defending party, which must in turn counter or refute the claimed inconsistency.[220] However, the Appellate Body reversed the Panel's statement that the SPS Agreement allocates the evidentiary burden to the Member imposing the SPS measure.[221] Instead, it held that the Panel should have begun the analysis of each legal provision by examining whether the US or Canada had presented evidence and legal arguments sufficient to demonstrate that the EC measures were inconsistent with its obligations under the SPS Agreement: '[O]nly after such a prima facie determination has been made by the Panel may the onus be shifted to the EC to bring forward evidence and arguments to disprove the complaining party's claim'.[222] In its consideration of the burden of proof, the Appellate Body emphasized that enacting SPS measures is a right of Members, the exercise of which should not be unjustifiably penalized. This aspect of the Appellate Body's judgment is appropriate in the light of the objectives of the Agreement and the importance of health as proposed by this book as it affirms the right of nations to enact measures to protect health.

In *Japan/Agricultural Products*, the Appellate Body examined the burden of proof in relation to the Panel's finding of a violation of Article 5.6. The Panel had found that Japan had acted inconsistently with Article 5.6 because there was another, less trade-restrictive option open to it. However, the Panel had knowledge of this option, not because it had been argued by the US, but because it had been suggested by the experts advising the Panel.[223] The Appellate Body held that while panels have the right to seek information and consult experts, this information cannot be the basis for a finding of inconsistency with Article 5.6 where the complaining party did not establish a *prima facie* case of inconsistency. In bringing a complaint against Japan, the US had the burden to show a *prima facie* case that there was an alternative measure that meets all three elements under Article 5.6. However, the US had not even argued for the alternative measure so could not have established a *prima facie* case that such measure was an alternative measure within the meaning of

---

[220] *EC/Hormones AB Report*, *supra* note 28 at para 98.
[221] *Ibid.* at para 102.
[222] *Ibid.* at para 109.
[223] *Japan/Agricultural Products Panel Report*, *supra* note 37 at para 8.74. *Japan/Agricultural Products AB Report*, *supra* note 85 at para 125.

Article 5.6.[224] This decision places a significant burden on complaining countries who will not always be in possession of as many facts as the responding Member regarding either the nature of the alleged risk or the social, political, economic, and cultural factors that may have played a role in the risk management decision taken. Nevertheless, it is appropriate in light of the importance of health and consistent with the objectives of the SPS Agreement that the complaining party bear this burden. Article 5.6 requires that the regulating member 'shall ensure' that their measure is not more trade restrictive than necessary but does not require them to lay out all the possible alternative measures that might or might not achieve their desired level of protection. To do so would arguably be an intrusion on their right to enact SPS measures as necessary to protect health.[225]

This type of information imbalance was at issue in *Japan/Apples* where Japan argued that the US, as the exporting country affected by the disease, would 'naturally' have more information on such disease. The Panel rejected this argument, finding no reason why the expertise of the exporting country should justify a different allocation of the burden of proof or the imposition of a heavier burden of proof on that party.[226] It considered that Japan could have sought to perform or commission research on the disease in third countries.[227]

Regarding the burden of proof standard, in *Japan/Apples*, Japan argued that in order for the US to prove its claim under Article 2.2, it 'has to positively prove the insufficiency of scientific evidence'. In response, the US argued that there was 'simply no scientific evidence supporting the measure at issue'. The Panel concluded that in such circumstances, 'we consider that the US should raise a presumption that there are no relevant scientific studies or reports in order to demonstrate that the measure at issue is not supported by sufficient scientific evidence. If Japan submits evidence to rebut that presumption, we would have to weigh the evidence before us.'[228] Similarly, in *Japan/ Agricultural Products*, the Appellate Body required a low degree of

---

[224] *Japan/Agricultural Products AB Report, ibid.* at para 126.

[225] But see Trebilcock and Soloway, *supra* note 185. Trebilcock and Soloway argue that the Appellate Body's decision in *EC/Hormones* was ill-conceived or at least confusing because the importing Member has superior information as to what steps it took to comply with the requirements of the SPS Agreement. However, they do note that complaining Members can invoke Article 5.8 and ask for an explanation of the reasons for the importing country's SPS measures. A failure on the part of the importing Member to supply such particulars should enhance the *prima facie* case of the importing Member. This would effectively shift the burden of proof back to the respondent country.

[226] *Japan/Apples Panel Report, supra* note 38 at para 8.44.

[227] *Ibid.* at para 8.46.

[228] *Ibid.* at para 8.106.

confidence, stating that the complaining Member (the US) was only required to raise a presumption that its products did not pose a risk, while the defending Member must meet a higher burden.[229]

13.3.3.2 Standard of review

As noted, the SPS Agreement is silent on the matter of an appropriate standard of review for panels.[230] It was suggested in Chapter 11 that panels should err on the side of deference to domestic agency decisions, focusing their review on matters of procedural propriety. This section considers the approach that has actually been taken in cases to date.

In *EC/Hormones*, the Appellate Body addressed the question of what was an appropriate standard of review in assessing acts of, and scientific evidence adduced by, the EC.[231] The EC suggested two main alternative approaches panels could take in this regard. The first is '*de novo* review' which would allow a panel complete freedom to come to a different view from the competent authority of the country whose act or determination is being reviewed. A panel would have to 'verify whether the determination by the national authority was correct both factually and procedurally'.[232] The second is described as 'deference'. The EC argued that under this standard, a panel should not seek to redo the investigation conducted by the national authority but instead examine whether the 'procedure' required by the relevant WTO rules had been followed.[233]

The Appellate Body considered that the appropriate standard of review must reflect the balance established in the SPS Agreement between the jurisdictional competences conceded by the Members to the WTO and the jurisdictional competences retained by the Members for themselves.[234] However, the difficulty lies in the fact that the SPS is vague as to where the balance between these competences lies. The Appellate Body placed reliance on Article 11 of the DSU which provides that:

---

229 *Japan/Apples AB Report*, *supra* note 42 at para 137. See also Theofanis Christoforou, 'Settlement of Science-based Trade Disputes in the WTO: A Critical Review of the Developing Case Law in the Face of Scientific Uncertainty' (2000) 8 N.Y.U. Envtl L.J. 622 at 644.

230 *EC/Hormones AB Report*, *supra* note 28 at para 114.

231 On appeal, the EC claimed that the Panel had not accorded sufficient deference to: (i) the EC's decisions in applying a level of sanitary protection higher than that recommended by Codex Alimentarius; (ii) the EC's scientific assessment and management of the risk from the hormones at issue; and (iii) the EC's adherence to the precautionary principle and its aversion to accepting any carcinogenic risk (EC's appellant submissions, para 140 and AB, para 110).

232 EC's appellant submissions, para 122.

233 *Ibid.* at para 123.

234 *EC/Hormones AB Report*, *supra* note 28 at para 115.

> . . . a panel should make an objective assessment of the matter before it, including an objective assessment of the facts of the case and the applicability of and conformity with the relevant covered agreements, and make such other findings as will assist the DSB in making the recommendations or in giving the rulings provided in the covered agreements.

The Appellate Body thus found that the real question was whether the Panel had made an objective assessment of the facts before it. As Trebilcock and Soloway observe, this statement is not particularly helpful.[235] It does not shed any light on exactly what facts are to be objectively assessed. If contested or uncertain scientific facts were 'objectively determined' by panels, this would place panels in the position of a global science court and would be inconsistent with the Appellate Body's own observations on the acceptability of credible minority scientific opinions.[236]

In *Japan/Apples*, the Appellate Body rejected Japan's argument that the Panel should have made its assessment under Article 2.2 in the light of Japan's approach to risk and scientific evidence. It found that Japan's submission that the Panel was obliged to favour Japan's approach to risk and scientific evidence over the views of the experts conflicted with a standard requiring objective assessment of the facts.[237] It also noted that in these cases, the Panel enjoys a margin of discretion in assessing the value of the evidence and weight to be ascribed to it.[238] While this could be interpreted as a rejection of a deferential approach, it can also be seen as specific to the facts of the case, where Japan was in effect asking the Panel to show total deference to and effectively not to review its risk assessment at all.

In summary, the WTO case law to date does little to clarify the issue of what is an appropriate standard of review in SPS cases. Prévost argues that the so-called objective standard allowed the Panel in *EC/Hormones* to essentially substitute its own judgment, however incorrect its appreciation of the scientific evidence before it, for that of the Member government without any real limits.[239] Further, in substituting their own opinion, she argues that panels are not obliged to give as much weight as the regulating Member might desire to

---

235 Trebilcock and Soloway, *supra* note 185 at 564.

236 *Ibid.*

237 *Japan/Apples AB Report*, *supra* note 42 at para 165.

238 *Ibid.* at para 166.

239 Prévost, *supra* note 192 at 11. She notes that in *EC/Hormones*, the Appellate Body agreed with the EC that the Panel had misquoted evidence from an EC expert, Dr Lucien, and that it may have distorted the views of another expert, Dr André. However, the Appellate Body found no reason to overturn the Panel's findings in this regard, provided there was no evidence that the mistake was deliberate'. *EC/Hormones AB Report*, *supra* note 28 at para 138.

statements of caution by scientific experts.[240] On the other hand, too much deference to the scientific conclusions reached in the risk assessment has the potential to turn the scientists who conduct risk assessments into *de facto* fact-finders for the WTO. Trebilcock and Soloway find that Appellate Body decisions show a 'general disposition towards according some significant degree of deference to Members in determining both policy objectives and policy instruments with respect to SPS regulation, [yet] its rulings are not anchored in a coherent conception of an ideal domestic risk regulation process nor in the appropriate scope and limits of supra-national quasi-judicial review of these measures'.[241] Trebilcock and Soloway argue for a relatively deferential form of review of domestic health regulations that focuses on minimum objectively verifiable characteristics of the regulatory process.[242] This book has also argued for a relatively deferential approach, suggesting that panels should focus on procedural propriety when evaluating Members' scientific evidence. To recap, it was suggested in Chapter 11 that Panels should seek to ensure that risk assessments relied upon are conducted in a rigorous manner, and that risk estimates meet a minimum threshold of scientific rationality. In other words, the review should be procedurally based, focusing on the underlying scientific merits of the risk assessment, rather than on the actual decision reached.[243] Panels and the Appellate Body are not qualified to evaluate the merits of scientific evidence or questions of risk tolerance.[244] The purpose of the examination should not be to second-guess the scientific judgments made, but to filter out sham judgments that merely seek to advance protectionist interests.[245]

13.3.3.3 Expert advice

In most amber cases, panels will require assistance from experts familiar with scientific methodology.[246] They may appoint experts pursuant to Article 13.1

---

[240] Prévost, *ibid.* at 12. See also Christoforou, *supra* note 229 at 645.

[241] Trebilcock and Soloway, *supra* note 185 at 573.

[242] *Ibid.* Also Andrew T. Guzman, 'Food Fears: Health and Safety at the WTO' (2004) 45 Va. J. Int'l L. 1.

[243] United States Environmental Protection Agency (EPA), *Science Policy Council Handbook: Peer Review* (Washington DC: EPA, 2000) at para 3.2.1.

[244] Guzman, *supra* note 242 at 1–2.

[245] See also Trebilcock and Soloway, *supra* note 185. They argue that the purpose of requiring at least a credible minority of scientific opinion is to screen out 'junk science', on the one hand, while avoiding attempts to resolve genuine scientific uncertainty or controversy on the other. Guzman argues that panels should leave the evaluation of science, decisions about risk and the relationship between science and SPS measures to domestic governments. Guzman, *supra note 242* at 37.

[246] Christoforou cites a number of scholars who have concluded that non-expert judges cannot verify and judge the scientific bases of competing expert views because

of the Understanding on Rules and Procedures Governing the Settlement of Disputes which allows panels to 'seek information and technical advice from any individual or body which it deems appropriate'. Article 13.2 provides further that panels 'may consult experts to obtain their opinion on certain aspects of the matter. With respect to a factual issue concerning a scientific or other technical matter raised by a party to a dispute, a panel may request an advisory report in writing from an expert review group.' In addition, Article 11.1 of the SPS Agreement provides that 'in a dispute under this Agreement involving scientific or technical issues, a panel should seek advice from experts chosen by the panel in consultation with the parties to the dispute. To this end, a panel may, when it deems it appropriate, establish an advisory technical experts group, or consult the relevant international organizations, at the request of either party to the dispute or on its own initiative'.[247]

In the *EC/Asbestos* case, the Panel saw the role of its appointed experts as being to help it understand and evaluate the evidence submitted and the parties' arguments.[248] In *Japan/Agricultural Products*, the Panel stated that in deciding whether a fact or claim can be accepted, 'we consider that we are called upon to examine and weigh all the evidence validly submitted to us, including the opinions we received from the experts advising the Panel in accordance with Article 13 of the DSU'.[249] This suggests that panels consider themselves capable of verifying the basis of the scientific views and taking a position on the substance of the evidence presented.[250]

In Chapter 7 it was suggested that a key role for experts should be to assist panels to determine whether a country claiming reliance on science can show that they have followed accepted scientific protocols in conducting their risk assessment and whether the result of the risk assessment displays a threshold of scientific rationality.[251] In *EC/Hormones*, the Panel employed experts on an individual basis in order to assist it in determining whether the EC ban was

---

they do not understand the cognitive aims and methods of science and that therefore they must rely on experts to avoid making arbitrary decisions about which of the competing views should be relied upon. Christoforou, *supra* note 229 at 639.

[247] For a discussion of some of the issues surrounding use of experts, including issues concerning selection and appointment, whether individuals or panels should be appointed, and the procedure to be followed during hearings, see Christoforou, *ibid.*

[248] *EC/Asbestos Panel Report*, *supra* note 28 at para 8.182.

[249] *Japan/Agricultural Products Panel Report*, *supra* note 37 at para 7.10.

[250] Christoforou, *supra* note 229 at 644.

[251] Christoforou makes an important point that is relevant in this regard. He notes that the role of experts is to assist tribunals in ferreting out facts, 'not usurping their judicial function or fulfilling the role and duties of the disputing members'. As such, panels must ensure that the appointed experts do not evaluate the risk assessment conducted by the responding party themselves. *Ibid.* at 638.

based on scientific principles and on a risk assessment, and if there was sufficient scientific evidence to support the ban.[252] The Panel asked the experts a total of 35 questions. Some of these appeared to be designed to help it judge the substantive validity or rationality of the scientific conclusions relied upon by the EC. For example, the Panel asked the experts to advise what factors and procedures scientists should consider in establishing an appropriate assessment of the potential adverse effects on human health from the use of the hormones in question.[253] It also asked whether there is any scientific evidence available which demonstrates that a potential for adverse effects on human or animal health arises from the administration of any of the six hormones in dispute, in general, and in particular for animal growth-promotion purposes, if administered in accordance with good animal husbandry practice and/or good veterinary practice.[254] Another question asked whether there is scientific evidence of particular human health effects in countries where meat produced with the use of any of the six hormones in dispute for growth promotion purposes is allowed for consumption as compared to health effects in countries where the use of such hormones is forbidden.[255]

Interestingly, very few questions were directed towards establishing scientific rationality on the basis of the methodology or procedure followed. Further, other questions seemed to indicate that the Panel was to some extent embarking upon its own fact-finding mission. For example, it asked questions concerning the existence of hormone residues in milk products,[256] the difference in health effects of hormones and pesticides,[257] and the effects on growth promotion of the use of female hormones on male animals and vice-versa.[258] It is difficult to see how such questions were necessary to assist the Panel make the required objective inquiry into whether EC regulations were based on scientific evidence.

After reviewing the evidence and the experts' opinions, the Panel found that the scientific conclusions implicit in the EC measures did not conform with any of the scientific conclusions reached in the studies it had submitted as evidence.[259] It found that none of the scientific evidence referred to by the EC which specifically addressed the safety of some or all of the hormones in

---

252 *EC/Hormones Panel Report (US), supra* note 28 at para VI.2.
253 *Ibid.* at section VI.
254 *Ibid.* at question 7.
255 *Ibid.* at question 9.
256 *Ibid.* at question 12.
257 *Ibid.* at question 10.
258 *Ibid.* at question 18.
259 *Ibid.* at para 8.137. *EC/Hormones Panel Report (Canada), supra* note 28 at para 8.140.

dispute when used for growth promotion, indicated that an identifiable risk arose for human health from such use of those hormones if good practice was followed. The Panel noted that the experts it had called confirmed this conclusion.[260]

In *Australia/Salmon*, the Panel moved closer to adopting a more procedural approach. It asked its experts to identify the minimum requirements of a risk assessment generally accepted in the specific area of aquatic animal health and to state whether Australia's 1995 Draft Report and the 1996 Final Report met these requirements.[261] One of the experts, Dr Burmaster, noted that while there are exceptions in some situations (but *not* in the case of this dispute), a risk assessment must use quantitative methods to estimate the probability and the magnitude of desired and adverse consequences. Noting that the requirements might vary depending on the product and/or diseases addressed, Dr Burmaster stated that each risk assessment must, at a minimum, use probabilistic methods to distinguish and quantify the variability and the uncertainty inherent in the problem. In his opinion, because neither the draft risk assessment nor the final risk assessment used quantitative methods, neither met the minimum requirements.[262] Another expert, Dr Rodgers, noted that he was not aware of any prerequisite to use the quantitative method, particularly in view of the lack of data in certain key areas of aquatic animal health.[263] The third expert, Dr Wooldridge, opined that while risk assessments had to evaluate the *probability* of risk, this did not have to be expressed quantitatively, and often could not be.[264]

Thus, while there was some disagreement among the experts as to the need for a quantitative risk assessment, there was agreement on the fundamental issue, namely, that there needs to be an assessment of probability. This, it was agreed, was missing in the present case. Dr Wooldridge identified the following minimum requirements of a risk assessment which she indicated did not vary with the product or disease addressed.[265] (a) It must be transparent, that is, clearly set out, and fully referenced in the report produced. (b) The risk being evaluated must be defined and clearly set out. (c) The hazard/s to be addressed must be defined and clearly set out. If a particular hazard had not been specified in the request for a risk assessment, this would require a hazard identification of appropriate breadth. (d) The potential pathways from the hazards of interest to the outcomes of interest (that is, the sequence of events

260 *EC/Hormones Panel Report (US)*, *ibid.* at para 8.124.
261 *Australia/Salmon Panel Report*, *supra* note 165 at para 6.13.
262 *Ibid.*
263 *Ibid.* at para 6.58.
264 *Ibid.* at para 6.59.
265 *Ibid.* at para 6.23.

necessary) must be elucidated, and clearly set out. Details of any processes incorporated in this pathway (for example, testing for infection) must be fully referenced where appropriate. (e) For each identified step in the pathway, information (data) must be gathered to evaluate the probability of that step occurring. This might vary from hazard to hazard, and therefore must be hazard specific where necessary. This information might be qualitative or quantitative and must be clearly set out, and the source should be fully referenced. (f) The overall probability of the pathway of events from hazard to defined outcome actually occurring should be evaluated either qualitatively or quantitatively for each defined hazard and outcome, using the information obtained. A quantitative assessment should also use appropriate mathematical manipulations. Reasoning, mathematical manipulations (where used) and conclusions must be fully set out. A quantitative assessment must include a time-frame.

Dr Wooldridge found methodological problems with the Final Report. Specifically, the Final Report not only explicitly ruled out the possibility of a quantitative assessment, but the qualitative assessment proceeded 'implicitly on the basis that without quantification, the assessment could only be concerned with the possibility of agent entry, rather than making any attempt to qualitatively ascribe probabilities based on the information available'.[266] In Dr Wooldridge's opinion, looking only at the *possibility* of particular consequences, rather than the *probability*, was neither an appropriate technical method nor an adequate scientific outcome of a risk assessment.[267] In sum, she found that the 1996 Final Report did not meet the minimum methodological requirements of a risk assessment.[268]

Dr Wooldridge identified one of the problems with the 1996 Final Report as being that it lacked clarity and she considered this lack of clarity to be a result of the risk communication process where the results of the risk assessment had been shared with the interested public. She noted that risk communication typically impacts risk assessment by identifying inaccuracies in that process and, further, that it might provide information regarding the acceptable level of risk. In terms of both functions, Dr Wooldridge noted that each point in the documentation of such a consultation process should refer specifically to the place in the initial risk assessment to clearly demonstrate whether, how, and exactly where and why the communication process had modified and altered the initial assessment, conclusions, and resulting policy recommendations.[269] She found that this was not the case with the Final Report which was

---

[266] *Ibid.* at para 6.37.
[267] *Ibid.* at para 6.39.
[268] *Ibid.* at para 6.42.
[269] *Ibid.* at para 6.28.

not as clear or transparent as the Draft Report, and did not clearly distinguish between assessed risk and acceptable risk. Dr Rodgers also addressed the risk communication exercise, noting that, while a valuable exercise, in this case it led to the 1996 Draft Report and the Final Report being less specific as well as more cautious as they gave more weight to the unknown elements of the assessment.[270]

In the *Australia/Salmon* case, the panel appears not to have placed great significance on the opinions of the experts concerning the methodological rigour of Australia's risk assessment. In its decision, it found that the 1996 Final Report did meet the requirements of a risk assessment pursuant to Articles 5.1 and 5.2.[271] This is of concern given the expert testimony that the 1996 Final Report did not meet the minimum methodological requirements of a risk assessment. The Panel appears to have reached its conclusion on the grounds that the Report: (a) identified the diseases whose entry, establishment, and spread Australia wanted to prevent;[272] (b) to some extent evaluated the likelihood of the entry, establishment, or spread of the diseases of concern.[273] However, it found that there was no rational relationship between the measure chosen by Australia (heat treatment) and the risk assessment.[274]

The expert testimony in *Australia/Salmon* highlights the difficulties involved in panels trying to establish with any clarity or certainty a set of requirements which countries must follow pursuant to Article 5.1. The procedures that would constitute a scientifically valid risk assessment will vary depending on the type of risk involved, and even within specific fields, experts will disagree as to the need for quantitative assessment. Clearly there is no formula that can be applied across different disputes. As such, this is where adopting a peer review type procedure would provide a clearer framework for the exercise. It is argued that this is critical, so that panels are consistent in the way in which they use and rely upon expert panels.

A final note. To date, panels have chosen to appoint scientific experts in an individual capacity. Christoforou points out that where a panel appoints five individual experts, it is possible that it may receive five different and possibly conflicting responses to its questions. For example, as noted above, in the *Salmon* case, there was some disagreement among the experts as to the need for a quantitative risk assessment. He suggests that the use of an expert group

---

[270] *Ibid.* at para 6.14.

[271] *Ibid.* at para 8.92.

[272] *Ibid.* at para 8.73.

[273] *Ibid.* at para 8.89.

[274] *Ibid.* at para 8.98. For example, for a number of the diseases in question, the 1996 Final Report mentioned that both freezing and heating may have risk-reduction effects, while one of the diseases was resistant to heat treatment.

would enable the experts to engage in a collective decision-making process which would be more likely to result in consensus.[275] This would mean that panellists would not have to evaluate widely diverging scientific views and Christoforou argues that, as a result, the scientific basis of their decisions may be 'closer to scientific consensus'.[276]

13.4 CONCLUSION

This chapter has sought to shed light on the manner in which WTO panels and the Appellate Body have interpreted GATT Article XX(b) and the relevant provisions of the SPS Agreement in cases where health and trade objectives conflict or are in tension with each other. Here, some brief concluding remarks are provided with respect first, to the normative approach taken by panels and the Appellate Body in light of the discussion in Chapter 7; and second, the analysis of the panel and Appellate Body decisions. Finally, some comments are provided as to what the analysis of the panel and Appellate Body decisions suggests for future 'amber' cases.

Normative approach In terms of the normative approach taken by panels and the Appellate Body, the overriding impression from the decisions is that of a lack of a clearly discernible overriding theoretical approach to the cases. It was argued in Chapter 7 that panels would be assisted in reaching more consistently principled decisions if they took more explicit notice of the substantive content of the values underlying the SPS Agreement's objectives, namely, health protection at a domestic level, and international trade liberalization. This, it was argued, would provide a normative basis for decisions that is lacking where panels appeal solely to the concept of sovereignty in upholding a Member's right to protect health. The Appellate Body perhaps came closest to taking such an approach in *EC/Asbestos* where, in determining the 'necessity' of France's ban on asbestos, it emphasized the importance of the value pursued in health.

Key points from analysis of the cases Some key points can be drawn from the rather lengthy analysis of the case law set out in this chapter. Regarding whether a risk exists, it is argued that panels have not placed enough significance on the procedural aspect of risk assessments (*Australia/Salmon*). They have used expert advice to help them understand and sift through the scientific evidence presented, but have demonstrated a tendency to seek to reach their

---

275 Christoforou, *supra* note 371 at 639.
276 *Ibid.*

own scientific conclusion rather than focusing on whether or not the risk assessment procedure followed was sufficiently rigorous and that the outcome demonstrated a sufficient threshold of scientific rationality. Panels have, however, generally accepted a basic requirement for countries to show that their scientific evidence has been obtained through scientific methods (*EC/Hormones, Japan/Apples,* and *EC/Biotech*), indicating that the suggested approach would not be a significant departure from the approach taken to date.

The Appellate Body has demonstrated its willingness to accept a credible minority scientific opinion (although what constitutes a 'credible' opinion is not clear). In this regard, it is suggested that panels adhere to the procedural approach set out in Chapter 11. Such an approach is desirable, particularly in 'amber' cases where public sentiment may support a minority approach over another and it must be ascertained whether there are scientific grounds to support such an approach.

As to the factors that countries may take into consideration in their risk assessment procedures, the Appellate Body's interpretation of Article 5.2 in *EC/Hormones* seems to give Members significant latitude regarding the factors considered in the risk assessment process, arguably allowing them to include factors other than strict scientific evidence in their risk assessment procedures. Also supportive of a flexible approach to Article 5.2 was the Panel's decision in *EC/Biotech* where it noted that a risk assessment pursuant to Article 5.1 has to be 'appropriate to the circumstances', which may include gaps in the available scientific evidence.

Chapter 11 adopted the position that non-scientific factors including public sentiment have a valid role in the risk assessment process. The Appellate Body's position on this issue is therefore welcomed, although more clarity would be desirable. As noted in Chapter 10, it is not suggested here that public sentiment or other non-scientific factors ought to be the deciding factor in any case. Rather, they should only be admitted as legitimate influencing factors where they fit within the framework of a rigorous scientific methodology, and are compatible with an outcome that meets a minimum threshold of scientific rationality.

Regarding risk management, the approach taken by panels and the Appellate Body to the Article 5.6 least-trade-restrictive test is of particular interest. While this requirement appears to be a fairly relaxed one which the Appellate Body has interpreted so as to pay substantial heed to a Member's 'appropriate level of protection', there remains uncertainty as to how this provision will be applied in some situations. For example, would a panel hold that a measure was necessary because alternatives were not reasonably available due to adverse public opinion?[277] Would the low probability nature of a

[277] Hughes, *supra* note 132 at 935.

risk tend to make panels more amenable to suggestions by the complaining Member that suitable less trade-restrictive alternative measures exist? What would a panel do where the regulating Member had not considered any alternative measure? These kinds of issues are likely to arise more often in the future, given increasing consumer awareness of health issues and demands on politicians to regulate accordingly. Article 5.6 calls for a principled approach which looks at the policy rationale behind health regulations and considers the concept of under- and over-protection in determining what would constitute a reasonably available alternative and in some cases taking a bias towards health protection that allows countries to avoid Type I Error Costs and accept Type II Error Costs.

The precautionary principle remains an area of uncertainty in the area of risk management. It is clear that science will rarely provide absolute certainty. History has shown us that there can be potentially enormous consequences of presuming that a given substance is safe until we have proof that it will cause harm. The precautionary principle is based on this understanding and is equally applicable to human, animal, and plant life and health. Panels and the Appellate Body have not recognized the precautionary principle as such, although the Appellate Body noted in *EC/Hormones* that the principle is reflected in those provisions of the SPS Agreement which recognize the right of Members to establish their own level of SPS protection. It has been noted in this book that it is an accepted fact in public health that precaution is inherent in health regulatory decision-making, due to the indeterminate nature of science as applied in the discipline of risk assessment, as well as the value of human health and the potential irreversibility of adverse consequences. It is also inherent in animal and plant health policy decision-making. An explicit acknowledgement of this characteristic of risk assessments in future cases would be desirable, helping to focus the determination not on whether countries have been precautionary in their actions, but on *how* precautionary they have been. Where precaution is exercised to protect against a risk forecast by a study that does not reach a threshold of scientific rationality, then it should not be a justifiable basis for a trade-restrictive SPS measure. Determining when precaution has been taken too far will always be a matter of degree, based on the evidence and the circumstances of each case.

Overall, panels and the Appellate Body appear to be taking an approach that recognizes the many factors that countries have to take into consideration in enacting health regulations and that allows sufficient flexibility for them to do so without sacrificing democratic participation in regulatory decision-making. The approach arguably allows a role for public sentiment in cases, even where the public view scientific evidence in a light which does not coincide with the way in which experts see the same evidence. However, this remains to be discussed by a panel or the Appellate Body.

Some comments on potential 'amber' cases None of the cases discussed in this chapter have involved conflicts of the kind of conceptual difficulty illustrated in Chapter 6's hypothetical examples. In *EC/Asbestos*, *Thai/Cigarettes*, and *Brazil/Retreaded Tyres*, it was a fairly straightforward matter for the panel to find that a serious risk to health existed. In *EC/Hormones* and *EC/Biotech*, the EC was not able to present any scientific evidence of a risk to health. In *Japan/Apples*, *Japan/Agricultural Products*, and *Australia/Salmon*, the measure was not based on a risk assessment. Chapter 6's hypothetical examples are therefore reproduced here in order to facilitate a brief discussion of the implications of the panel and Appellate Body decisions, as well as the conclusions of previous chapters, for such cases.

Example #1: Risk that is high probability but where the consequences are not serious, such as an import ban on GM nuts because they cause minor allergic reactions in some people In this case, there is conflict or tension between health and trade objectives because the health risks are minor. The allergy in question is not serious or life-threatening. In such a situation, the key questions would likely be whether the import ban was necessary to protect health and whether it was not more trade restrictive than required. Panel and Appellate Body decisions to date suggest that the term 'not more trade restrictive than required' is likely to be interpreted in a fairly flexible manner in future cases, with the country's inferred level of protection being taken as an indication of the importance of the values at stake. However, in a case such as this, it is likely that alternative, less trade-restrictive, measures would have been available and that the regulating country would at some stage have considered them (for example, labelling). It is likely to be a case where the degree of over-protection present is too great to justify adopting a bias towards health that would allow acceptance of Type II Error Costs.

Example #2: Risk that is low probability and where the consequences are not serious, such as an import ban on GM milk where there is a small chance that some people will be intolerant to it due to changes in the enzyme structure In this case, there is conflict or tension between health and trade objectives because the risks to health from the GM milk would be neither very serious nor very likely to eventuate. To date, a panel has not had to consider such a situation. The same issues will arise as in example #1. Alternative measures might also be reasonably available here, such as labelling the milk cartons. The low probability nature of the risk would provide a panel with a strong case to reach a decision that an alternative measure was available, provided that such a measure was available that would achieve the desired health objective. In addition, the low probability nature of the risk might raise a question as to whether there is protectionism. It was suggested in Chapter 7 that in this kind of situation a panel should take into account governmental obligations to protect health and the inherent value of such protection when identifying the socially optimal level of protection. Where over-protection is apparent, this might indicate to a panel that the measure is intended wholly or partly to protect local dairy industries that do not use the GM technique.

Example #3: Risk that is low probability but has serious consequences, such as a ban on imported beef due to an outbreak of 'mad cow disease' in the country of origin In this example, there is conflict or tension between health and trade objectives because the risk, while potentially having a serious negative impact on health, is unlikely to eventuate. The same issues will arise as in the previous example regarding low probability risk. However, because the potential outcome is far more serious from a health perspective – it is a brain wasting disease – the balancing exercise may well come out differently, that is, in favour of maintaining the ban. A panel would have scope to adopt a bias towards health protection and allow for the possibility of Type II Error Costs (assuming of course, that scientific evidence is present with respect to the risk, albeit low probability). The Appellate Body recognized the seriousness of the potential health outcomes of asbestos exposure in *EC/Asbestos* where it found that there was no alternative measure available that would achieve France's desired health protection objective. In this situation, a panel might take the approach of the Appellate Body in *Korea/Beef* that the more vital or important the values pursued, the easier it would be to accept as 'necessary' measures designed to achieve those ends. As Howse and Tuerk argue, restraining a Member from maintaining an import ban where the potential consequences of the risk are as serious as life-threatening cancer would 'impair the very ability of a member to exercise its prerogative (and fulfill its international human rights obligation) to protect the right to life of its citizens'.[278] Finally, a health bias might also be used in identifying the existence of a socially optimal level of protection if the complaining party alleged a disguised restriction on trade.

Example #4: Public risk perceptions that differ from expert risk perceptions of the same risk, such as an import ban on irradiated mangoes in response to public perceptions that a risk to health exists Health and trade objectives are in conflict or tension in this example because it is not clear that there is a literal risk to health that justifies undermining trade liberalization objectives. This example raises a number of questions concerning interpretation of the SPS Agreement, including of Articles 2.2, 5.1, and 5.3. A panel would require that the ban be based on scientific evidence obtained through a risk assessment that

---

[278] Howse and Tuerk, *supra* note 2 at 318.

specifically examined the health effects of irradiated mangoes (or possibly other fruit). In a scenario where there is a consensus of expert opinion that there is no risk from irradiated mangoes, and consequently, the importing country is unable to hold up a risk assessment pointing to a potential risk, then past decisions suggest that a panel would find the ban unjustified under the SPS Agreement. The reasoning of both the panel and the Appellate Body in *EC/Hormones* suggests that this would still be the case even if the public in the importing country had demanded regulation because they feared irradiated mangoes. Such a finding would be in accordance with the conclusions reached by this author as to the appropriate approach in such a situation. While it was concluded that public risk perceptions have value and should be accorded a place in the risk assessment process, if they are completely at odds with the scientific evidence, then they should not justify a trade-restrictive measure on the grounds of health protection (they may induce a country to invoke regulations such as labelling which – if countries considered them to be improperly trade-restrictive – would be considered under the Agreement on Technical Barriers to Trade). Even if consumers' subjective sense of well-being is diminished as a result of the negative weight they attach to the idea of irradiated mangoes due to their particular risk perceptions, it is argued that this cannot outweigh the losses caused as a result of the import ban. If the scenario changed slightly such that there was a lack of consensus as to the existence or otherwise of a risk, or if the science was uncertain, then we would have the situation described in example #5.

Example #5: Scientific uncertainty, such as an import ban on mayonnaise treated using nanotechnology to improve emulsification in response to scientific uncertainty regarding the risk associated with the use of nanotechnology (or a ban on 'energy' drinks with high levels of caffeine, taurine, and glucuronolactone where the effects of these ingredients reacting together in the body is uncertain) Cases involving scientific uncertainty may or may not also involve public sentiment but for the purposes of this example it is assumed that the public fear nanotechnology because they see it as unknown and potentially risky. Such a case sees conflict or tension between health and trade objectives because if the nature and extent of the risk are not known, how can it be determined whether the risk justifies a trade-restrictive measure that negates the potential welfare gains to be had from trade? The panel and Appellate Body decisions to date provide some clues as to approaches that might be taken in a case such as this, but are not definitive. Let us assume that the importing country produces a risk assessment which contains data gaps and uncertainties. Let us also assume that as a result of public input into the risk assessment process, the way in which the risk assessors chose to respond in the face of these uncertainties was conservative. This is a type of situation not yet faced by a panel (in both *EC/Hormones* and *EC/Biotech*, the panel found that there was no risk assessment indicating a risk). In *EC/Hormones*, the Appellate Body gave indications that inclusion of non-scientific factors (here, public sentiment) in a risk assessment would not be inappropriate; however, it is not clear whether it would accept a trade-restrictive measure in the situation described. Taking the approach in this book would have the following result. The public input should be considered valid and valuable in its own right. The panel should investigate the risk assessment process to ensure that it followed acceptable scientific protocol and that the conclusion that there is a risk passes at least a threshold of rationality. If these tests are met, the fact that the conclusion was a result of public sentiment influencing interpretation of scientific data should be accepted by the panel and the measure upheld, subject to there being no reasonably available, less trade-restrictive alternative. Such a result would be consistent with the suggested bias towards health protection and an acceptance of Type II Error Costs if the supposed risk turns out not to exist or is less serious than expected.

Example #6: Some elements of risk to health as described in #1-5, combined with protectionist intent In this scenario, there is conflict or tension because while health is being protected, so is domestic industry to the detriment of trade liberalization. This kind of case requires determination of whether one of the objectives sought ought to override the diminished welfare that will result from the other. This book has argued that where both genuine health protection and protectionism are apparent, the regulation should be allowed so long as it is not more trade-restrictive than necessary and is not inconsistent, arbitrary, or discriminatory in its application. This approach would recognize the value of health as a human right and arguably accords with the SPS Agreement's approach to the question of dealing with conflict or tension between health and trade objectives. It also appears to be in accordance with comments made by panels and the Appellate Body recognizing the value of health. However, recognition by panels and the Appellate Body of health as a human right would be a desirable development to assist in justifying such a conclusion.

PART V

Conclusion

14. Conclusion

At the time of the GATT's implementation in 1944, the main concern of the signatory nations was to liberalize trade in goods by reducing trade barriers imposed at the border, namely, tariffs and quantitative restrictions on imports. The 1994 Uruguay Round Agreements represented the outcome of an intensified effort that began during the Tokyo Round to tackle non-tariff barriers to trade on a multilateral basis. The SPS Agreement is one of a number of WTO Agreements which recognize that domestic regulations may operate as barriers to trade by restricting entry of goods that do not comply with the standards they impose. In doing so, it widens the purview of trade rules and allows exporting countries to 'look behind the border' of importing countries and challenge domestic measures that constitute non-tariff trade barriers.

Extending the reach of international trade rules to domestic regulatory measures has significant implications for countries as they must take these rules into consideration not only when setting tariff rates and quotas, but in any regulatory decision-making exercise that has the potential to impact upon foreign producers. This shift in focus to 'behind the border' measures has been of particular concern to countries in areas where the right to make regulations is regarded as a fundamental exercise of a nation's sovereignty. Nowhere is this concern more pronounced than in the areas impacted by the SPS Agreement, namely regulatory decision-making with respect to protection of human, animal, and plant life and health.

This book has examined the question of whether the SPS Agreement allows an appropriate balance to be struck between conflicting domestic health protection and trade liberalization objectives when they conflict or are in tension with each other. It has reviewed the literature in the area, and sought to move the discussion forward by focusing on difficult situations not yet fully addressed by panels and the Appellate Body.

Part I described the linkage between health and international trade and the extent to which health regulations have conflicted with international trade liberalization – both in the past and currently. It also identified a number of factors that point to likely future conflicts (tariff substitution, food safety issues, public participation in decision-making, regulatory divergence, and North–South conflict). *Part II* then established a normative basis upon which a balance between health and trade objectives can be pursued. *Part III* explored aspects of the global debate around risks. First, it questioned at which level (domestic or international) health regulations are most appropriately made. It then examined how risks are perceived and analysed at the domestic level. Finally, *Part IV* critiqued the rationale for the SPS Agreement's science-based framework, and the panel and Appellate Body's interpretations of its provisions. Here, concluding remarks will address first, on what normative basis ought a balancing of health and trade objectives be approached; second, at what level are regulatory decisions regarding risk most appropriately made; and third, whether the SPS Agreement's science-based framework can be supported based on such an approach.

14.1 A NORMATIVE BASIS UPON WHICH TO BALANCE HEALTH AND TRADE OBJECTIVES

Chapter 6 identified six example categories of 'amber' cases where health and trade objectives may conflict or be in tension with each other under the SPS Agreement. These categories are where:

1. Risks are *high probability but the consequences are not serious*;
2. Risks are *low probability and the consequences are not serious;*
3. Risks are *low probability but have serious consequences;*
4. *Public risk perceptions differ from expert risk perceptions*;
5. There is *scientific uncertainty* concerning the existence of a risk; and/or
6. There are *mixed motives* for a trade-restrictive regulation.

Each of these situations involves conflict or tension between health and trade because it is not clear whether the SPS measure in question is 'necessary' to protect health. As such, panels and the Appellate Body must undertake a balancing exercise to determine whether the alleged domestic welfare gains from the health regulation in question are such that they ought to outweigh the global gains from trade. These situations can be contrasted with those where it is obvious due to the probability and severity of risk that measures are 'necessary' (the 'green' cases) and health is privileged over trade accordingly.

While elements of these 'amber' categories have arisen in WTO cases, it was suggested in Chapter 6 that the more difficult elements have not yet been faced by panels or the Appellate Body. For example, they have not yet had to deal with a case where the disputed SPS measures are based on scientific evidence of a low probability risk with potentially serious consequences *and* where risk-averse public sentiment has influenced interpretation of the scientific evidence.

In considering what would be a desirable approach to balancing health and trade in such a situation, Chapter 7 outlined a perspective that acknowledges the importance of the objectives of both health protection and trade liberalization. It argued that in some amber cases, the importance of health protection may outweigh the case for trade even where it is not apparent that health gains will outweigh the gains from trade or, in other words, where it is not clear that the SPS measure is necessary. In order to support this perspective, it examined the values underlying each objective. Chapter 5 first examined international trade theory and found that while there are numerous objections to the notion of 'free trade', these do not serve to seriously undermine either the validity of trade theory, or the gains to be realized from trade. However, it did find a valid argument that, in some cases, the gains from trade may be exaggerated and that some non-economic values should be given the same, if not more, weight as the wealth maximization goals of trade liberalization.

Second, Chapter 5 provided a normative analysis of domestic regulatory decision-making, examining both economic and non-economic justifications for regulations to protect human, animal, and plant health. Under non-economic justifications, it highlighted factors leading to the conclusion that governments have an obligation under human rights law to protect health, and an obligation under principles of international law to protect wild fauna and flora. It also noted the economic justifications provided by the need to correct market failures. In this regard, it noted that health regulation may be seen as a prerequisite for a functioning market, that is, such regulation is required to ensure acceptability of market mechanisms. Chapter 5 also provided a positive analysis of regulatory decision-making, finding that in many cases, public interest regulation fails to accomplish its goals. Whether this is the result of regulatory capture by interest groups, self-interested actions on the part of decision-makers, perverse administrative processes, or the mistaken (or irrational) judgment of decision-makers and/or consumers, the result may be such that, in some cases, the regulation's claim to be a genuine health protection measure is brought into question.

Having examined trade and regulatory theory, suggestions were made to guide panels in the 'amber' cases. Within the category of 'amber' cases, two situations were identified: (1) where governments have an obligation arising from international law and human rights to protect health, that is, where human health or that of wild flora and fauna is at risk; and (2) the remaining cases where governments have no such obligation. In both situations, panels must ask whether the regulating country has met the SPS Agreement's scientific requirements (*risk assessment*). It must also ask whether, in determining its appropriate level of protection (*risk management*), the country has complied with the provisions requiring Members to take into account the objective of minimizing negative trade effects (Article 5.4), to avoid discrimination or disguised restrictions on trade (Articles 2.3 and 5.5), and to ensure that the measure taken is not more trade restrictive than necessary (Article 5.6). The question of balancing health and trade objectives arises most acutely when assessing a Member's compliance with these constraints on its right to determine its appropriate level of risk protection. It was suggested that for panels to strike an appropriate balance between health and trade in the first type of 'amber' cases, that is, where governments have an obligation to protect health, they ought to adopt a bias towards health protection objectives. This suggestion would have the following two consequences.

First, it would impact decisions concerning whether a country has minimized the negative impacts on trade and whether it has ensured that its measure is not more trade restrictive than necessary. In 'amber' cases where there is no clear indication one way or the other as to the necessity of a measure, it would allow countries to regulate so as to avoid Type I Error Costs (policies made in response to false negatives), but accept Type II Error costs (policies made in response to false positives). Where the use of such a bias might assist panels is, for example, in the case where it is not clear that a measure is necessary due to the risk being of low probability but with potentially serious consequences. In such a case, all other factors being equal, adopting a bias towards health would result in health being privileged over trade.

Second, a bias towards health objectives would impact decisions concerning whether there is discrimination or a disguised trade restriction. It would do so by having panels take into account governmental obligations to protect health and the inherent value of health in identifying the socially optimal level of protection which in turn will aid in determining whether there is a finding of discrimination or a disguised trade restriction. Thus, all other things being equal, a measure to protect against a low probability risk might not be found to constitute a disguised restriction on trade when the value of health protection has been taken into account.

It was argued that in the second situation, that is, where governments are under no obligation to protect health (such as with crop or livestock health), there is no normative argument to be made for the value of health that would justify adopting a bias towards health protection in the manner discussed. Rather, the focus in all enquiries concerning a country's determination of its appropriate level of protection should be on the factors outlined in Article 5.3, namely, the potential damage in terms of loss of production or sales in the event of the entry, establishment, or spread of a pest or disease; the costs of control or eradication in the territory of the importing Member; and the relative cost-effectiveness of alternative approaches to limiting risks.

It must be acknowledged that the notion of a bias towards health objectives immediately presents a concern from a trade liberalization perspective because such an approach could conceivably be abused by a country wishing to protect its producers. The positive analysis of regulation in Chapter 5 demonstrates significant opportunities for regulatory capture by vested interests. Therefore, it has been argued that it is important that panels adopting such a bias have a high level of confidence that the regulating country's professed intent to protect health is genuine. This raises the issue which was explored in Chapter 11, namely, is the SPS Agreement's science-based framework an appropriate means of providing such assurance and thus of justifying trade-restrictive measures?

14.2 SUBSIDIARITY

As noted in Chapter 8, the SPS Agreement reaches into an area that has traditionally been considered core to a country's sovereignty, its regulatory authority over matters of health and safety. In placing controls on the manner in which countries address health and safety issues, the Agreement infringes to a degree on a country's sovereignty. WTO Members have accepted this as part and parcel of what is necessary to ensure that the trade liberalization machine runs smoothly. The balancing exercise we speak of between various objectives is not, after all, something only to be considered by panels and the Appellate Body. Countries negotiated and signed the Uruguay Round Agreements in awareness of what they were giving up in terms of sovereign control over domestic regulation, in return for the benefits of trade liberalization. Yet as so often happens in law and politics, boundaries that may seem clear at the time of negotiation invariably become blurred as time goes on. In the case of the SPS Agreement, one blurred line is that between domestic and international responsibility for standard-setting. The Agreement encourages harmonization. It provides an incentive for countries to base their SPS measures on those set by Codex, OIE, and the IPPC, by providing that if they do, then such standards will be deemed consistent with the Agreement's provisions. However, questions remain as to how much harmonization is desirable or effective. Should WTO Members be aiming to harmonize all of their SPS measures? Surely harmonization would reduce conflict over such measures?

Chapter 8 used the principle of subsidiarity as a lens through which to discuss the desirability and likely effectiveness of greater levels of harmonization. Subsidiarity requires that policy decisions should be made at the lowest level of government capable of effectively addressing the problem at hand. The discussion of regulatory harmonization in Chapter 8 concluded that in the area of health protection, there is only a limited case to be made for international regulatory harmonization and a correspondingly strong case for an emphasis on local decision-making. The benefits of regulatory heterogeneity – recognition of domestic preferences – not only support the 'policed decentralization' approach taken by the SPS Agreement but also militate against further efforts to increase harmonization of standards, and support arguments for significant deference to domestic regulatory decision-making agencies in the exercise of review by WTO panels and the Appellate Body. There is a fine line, however, between an appropriate amount of deference that recognizes a country's domestic preferences, and too much deference that would allow a country to get away with disguised trade restrictions in the name of health. International trade rules must provide some constraint on countries to prevent the latter. This is particularly important in light of findings that even genuine SPS measures often have a significant and negative impact on developing countries. The SPS Agreement's science-based framework is an important tool to help provide such constraint.

14.3 CRITIQUE OF THE SPS AGREEMENT

14.3.1 The Science-based Framework

The SPS Agreement is premised on the notion that requiring scientific justification for standards that deviate from international norms will make it more difficult for countries to shelter domestic industries behind restrictive health regulations than in the absence of such a requirement. This reliance on scientific evidence has drawn much criticism from commentators and scholars. Chapter 1 noted that these critics tend to follow one or other of two strands of thought. The first (minority) strand is concerned with the integrity of the trade liberalization objectives of the SPS Agreement. These critics contend that the use of a scientific benchmark allows Member countries too much discretion in their regulatory decision-making. As such, they allow for the possibility that the scientific evidence requirement fails to achieve the drafters' goal of adequately guarding against protectionism.[1] Reflecting a different view of science, the second strand finds that the scientific evidence requirements impose too much of a straitjacket on governments. For example, Sykes was cited as arguing that the scientific benchmark represents undue hurdles for regulators who sincerely pursue objectives other than protectionism.[2] Other

---

[1] See for example, Robert Hudec, 'Science and "Post-Discriminatory" WTO Law' (2003) 26:2 B.C. Int'l & Comp. L. Rev. 185 at 189. Jeffery Atik, 'Science and International Regulatory Convergence' (1996–7) 17 Nw. J. Int'l L. & Bus. 736 at 758. Vern Walker, 'The Myth of Science as a "Neutral Arbiter" for Triggering Precautions' (2003) 26:2 B.C. Int'l & Comp. L. Rev. 197 at 228.

[2] Alan O. Sykes, 'Domestic Regulation, Sovereignty, and Scientific Evidence Requirements: A Pessimistic View' (2002) 3:2 Chicago J. Int'l L. 353 at 354.

critics are concerned with the apparent exclusion by the SPS Agreement of non-scientific justifications for measures, arguing that reliance on science is misplaced because it precludes any consideration of social, cultural, and ethical concerns and that nations will find their sovereignty diminished if there is no space for consumer anxieties to be respected and domestic politics accommodated.[3]

One of the key questions raised regarding the scientific evidence benchmark is its appropriateness in democracies where public sentiment finds a risk worthy of regulation contrary to the views of experts. Responding to this question required examination of the way in which the public perceives risk. Chapter 9 found that public risk perceptions are influenced by a number of factors: first, *psychological* factors whereby people utilize various heuristics; second, *social factors* whereby people's opinions are influenced by others, including those with protectionist agendas; and third, *cultural cognition* theory which tells us that cultural worldviews may play a role in determining what an individual sees as constituting a risk or not. In relating these conclusions to the SPS Agreement, it may be argued that, from a trade liberalization perspective, some public risk perceptions are more 'legitimate' than others. Those which are the result of internal psychological factors or cultural worldviews may be seen as potentially more 'legitimate' than the social factors which are more vulnerable to being influenced by potentially protectionist interests.

The observation that, in many cases, public risk perceptions do not coincide with 'expert' scientific opinion warranted an examination in Chapter 10 of the relationship between law and science, and science's role in regulatory decision-making. This examination was crucial in enabling further discussion of the appropriateness and usefulness of the scientific evidence requirements. It examined the rationale for science as a benchmark for trade-restrictive regulatory measures, namely, the claim that science is independent and objective. The examination found that the science has serious limitations as a tool for regulatory decision-making and consequently as a means of judging the 'legitimacy' of a trade-restrictive measure. This is because, contrary to the notion of science as a purely technocratic discipline which yields 'correct' and 'objective' answers, any scientific process is in fact highly indeterminate and subjective. Science is vulnerable to manipulation and capture by protectionist interests, and such capture may defeat the original purpose of the SPS Agreement's science-based approach.

---

[3] See for example, Dayna Nadine Scott, *Nature/Culture Clash: The Transnational Trade Debate Over GMOs* (New York: Hauser Global Law School Program, Global Law Working Paper 06/05, 2005) at 42. See also Walker, *supra* note 1 at 255.

Despite its limitations, the position was taken that science *is* a legitimate basis for determining the validity of trade-restrictive measures under the SPS Agreement. It arguably remains the best tool we have to sift out undesirable protectionism from genuine health protection measures because it is capable of being assessed, even if assessment is not perfect. The question then remains of how panels can best conduct an assessment of an SPS measure, given the reality that risk assessments pursuant to Article 5.1 will – in greater or lesser degrees – be dependent on value judgments as to what evidence to believe.

It was suggested that the most desirable and practical way for WTO panels and the Appellate Body to deal with this reality would be to take a procedurally based approach in their review of countries' scientific evidence. It was suggested in Chapter 11 that panels should adopt an approach inspired by peer review when reviewing whether a country has based its SPS measure on scientific evidence. They should ask whether a country has conducted a rigorous risk assessment that is consistent with protocols agreed by the international scientific community. The goal for panels should be to ensure that a country's risk assessment has a sound and credible basis. They would ideally do so by employing experts to conduct an in-depth assessment of the risk assessment, including the assumptions, calculations, extrapolations, alternative interpretations, methodology, acceptance criteria, and conclusions reached.[4] The review could thus be described as being procedurally based, focusing on the underlying scientific merits of the risk assessment, rather than on the actual decision reached.[5] It would pay significant deference to decisions made by domestic agencies. The purpose of the examination would not be to second-guess the scientific judgments made, but to filter out 'junk science' that is used to justify protectionist interests.[6]

Where does this leave the question of public sentiment as a factor in regulatory decision-making? The discussion of risk perception in Chapter 9 argued that public sentiment should be accepted as having value and a legitimate role in domestic regulatory responses to risk. This is important for a number of reasons including that international human rights instruments recognize that

---

4 United States Environmental Protection Agency (EPA), *Science Policy Council Handbook: Peer Review* (Washington DC: EPA, 2000) at para 1.2.3.

5 *Ibid.* at para 3.2.1.

6 See also Michael Trebilcock and Julie Soloway, 'International Policy and Domestic Food Safety Regulation: The Case for Substantial Deference by the WTO Dispute Settlement Body under the SPS Agreement' in Daniel Kennedy and James Southwick, eds, *The Political Economy of International Trade* (Cambridge, UK: Cambridge University Press, 2002). They argue that the purpose of requiring at least a credible minority of scientific opinion is to screen out 'junk science', on the one hand, while avoiding attempts to resolve genuine scientific uncertainty or controversy on the other.

citizens have a right to participate in decisions regarding their health. It is maintained here that the conclusion in favour of the SPS Agreement's science-based framework is reconcilable with a role for public sentiment in both risk assessment and risk management. Such reconciliation rests on the argument that both the risk assessment and risk management processes are capable of including public debate and alternative views of risk. It was suggested that domestic governments would be well advised to consider ways to ensure that the regulatory decision-making process provides room for public participation, including during the risk assessment phase. Currently, evidence suggests that in many cases, across different countries, the public is left with a sense of exclusion from decision-making processes, something which governments should work to avoid if they wish to obtain broader public acceptance of trade liberalization efforts. At the same time, it must be recognized that the extent to which governments wish to include citizens in decision-making processes is in the end a domestic decision for sovereign governments.

In reaching the conclusion that the SPS Agreement's science-based framework is valid, the author accepted Howse's argument that the scientific requirements should be understood not as usurping legitimate democratic choices for stricter regulations, but as enhancing the quality of rational democratic deliberation about risk and its control.[7] This will particularly be the case where public participation is sought within the risk analysis process.

The suggested role for public sentiment was qualified by arguing that it should only be a valid justification for trade-restrictive measures where scientific rationality is also present. In other words, public sentiment should not be allowed to compromise the scientific integrity of the risk analysis process. There is thus a complementary relationship – science should recognize public sentiment yet public sentiment must in some cases cede to scientific rationality.

Finally, it should be noted that where there is no rational scientific evidence to support a measure, countries may still have room to enact regulatory measures that respond to public concerns through means such as labelling rules. Such rules will fall under the jurisdiction of the Technical Barriers to Trade Agreement. As well, where public morals can be invoked, the public morals exception in Article XX(a) of the GATT may assist a country wishing to respond to public sentiment. The extent to which alternatives such as labelling can address public concerns in the 'amber' cases is clearly an important question which warrants further research.

---

[7] Robert Howse, 'Democracy, Science, and Free Trade: Risk Regulation on Trial at the World Trade Organization' (2000) 98 Mich. L. Rev. 2329 at 2330.

2 Interpretation by Panels and the Appellate Body

Chapter 13 examined the manner in which WTO panels and the Appellate Body have interpreted the SPS Agreement to date. The discussion focused on the way in which panels and the Appellate Body have approached: (i) balancing the objectives of health and trade; (ii) interpretation of the scientific evidence requirements; and (iii) disciplining the risk management stage of regulatory decision-making, in particular, the degree of flexibility accorded to countries to exercise precaution and to respond to public sentiment in 'amber' cases.

Regarding the balancing exercise, the review found little evidence that panels or the Appellate Body have located the exercise within a broader framework that recognizes the content and values of the SPS Agreement's sometimes competing objectives. It was suggested that panels would be assisted in reaching consistently principled decisions if they were to take more explicit note of the policy underpinnings of both health protection and trade liberalization to provide context for the balancing exercise. The Appellate Body came closest to taking such an approach in *EC/Asbestos* where, in determining the 'necessity' of France's ban on asbestos, it emphasized the importance of the value pursued in health.

Regarding interpretation of the SPS Agreement's scientific evidence requirements, the Appellate Body has interpreted Article 5.1 as requiring highly specific studies, something which may not always be possible for countries to produce, particularly developing countries. Yet it has also signalled some flexibility in this regard. In *EC/Hormones*, the Appellate Body indicated a willingness to accept a credible minority scientific opinion (without providing any guidance as to what constitutes a 'credible' opinion), rejected the Panel's attempts to introduce a quantitative element into the assessment of potential for adverse effects, and to impose a procedural requirement on Members to actually take into account a risk assessment. These findings seem to give Members a seemingly significant but as yet undefined amount of freedom in enacting SPS measures in the difficult 'amber' situations identified.

Panels have generally accepted the need for responding countries to show that evidence relied upon in their risk assessments has been obtained through scientific methods and have used experts to help ascertain whether in fact this has occurred (for example, *EC/Hormones, Japan/Apples, and EC/Biotech*). In some cases, the questions asked of experts by panels indicate that the panel may have been inappropriately focusing on judging the outcome of the scientific judgments relied upon rather than focusing on procedure. The Appellate Body's interpretation of Article 5.2 in *EC/Hormones* would seem to give Members significant latitude with respect to the risk assessment process, by allowing them to consider factors other than strict scientific evidence in their risk assessment procedures. Its statement in that case that the risk to be evaluated in a risk assessment under Article 5.1 is ' . . . risk in human societies as they actually exist, in other words, the actual potential for adverse effects on human health in the real world where people live and work and die' leaves important unanswered questions concerning the factors that it may accept as justifiable considerations in future disputes, but appears to provide scope for the inclusion of non-scientific factors. This would be consistent with the argument in this book that public sentiment ought to be recognized as a legitimate factor in a country's risk assessment. Yet it would be helpful if it was made clear first, that such factors have a legitimate role in influencing scientific judgment, and second, that nevertheless, they must not be allowed to compromise the integrity of the scientific process. That is, public sentiment must be consistent with an outcome that meets a minimum threshold of scientific rationality and is reached through means of a rigorous scientific methodology.

The third theme of the analysis in Chapter 13 was the approach taken by panels and the Appellate Body to disciplining countries' risk management decisions. Of particular interest in this regard is application of the Article 5.6 least-trade-restrictive means test. While this requirement appears to be a fairly relaxed one which the Appellate Body has interpreted so as to pay substantial heed to a Member's 'appropriate level of protection', there remains uncertainty as to how this provision will be applied in some situations. For example, would a panel hold that a measure was necessary because alternatives were not reasonably available due to adverse public opinion?[8] Would the low probability nature of a risk tend to make panels more amenable to suggestions by the complaining Member that suitable less trade-restrictive alternative measures exist? What would a panel do where the regulating Member had not considered any alternative measure? These kinds of issues are likely to arise more often in the future, given increasing consumer awareness of health issues and demands on governments to regulate accordingly. It was suggested that Article 5.6 calls for a principled approach which looks at the policy rationale behind health regulations and considers the concept of under- and overprotection in determining what would constitute a reasonably available alternative and in some cases taking a bias towards health protection that allows countries to avoid Type I Error Costs and accept Type II Error Costs.

The precautionary principle has been the subject of much discussion in the context of the SPS Agreement. History has shown us that there can be potentially enormous consequences of presuming that a given substance is safe to

---

[8] Layla Hughes, 'Limiting the Jurisdiction of Dispute Settlement Panels: The WTO Appellate Body Beef Hormones Decision' (1998) 10 Geo. Int'l Envtl L. Rev. 915 at 935.

health until we have proof that it will cause harm. The precautionary principle is based on this understanding. It adopts a 'better safe than sorry' approach and as such can be seen as a rejection of the 'wait and see' philosophy that emphasizes scientific certainty as a precondition to adopting protective health measures.[9] It was argued in Chapter 10 that where there is some evidence of a potential risk to human health, panels and the Appellate Body should recognize the concept of precaution, regardless of whether or not it is couched in terms of the 'precautionary principle'. Public health practices typically recognize the importance of precaution and it is arguably inherent in health regulatory decision-making, due to the indeterminate nature of risk assessments, as well as the value of human health and the potential irreversibility of adverse consequences. Similar arguments regarding precaution apply in cases concerning animal and plant health. It was argued that panels and the Appellate Body should focus their enquiries not on whether countries have been precautionary in their actions, but on *how* precautionary they have been. Determining when precaution has been taken too far will always be a matter of degree, based on the evidence and the circumstances of the individual case. The determination should recall the importance and inherent value of health protection. However, if a risk assessment does not reach a threshold of scientific rationality, then it should not be a justifiable basis for a trade-restrictive SPS measure (given that Article 5.7 exists for the situation where there is insufficient scientific evidence to conduct a risk assessment).

Overall, panels and the Appellate Body appear to be taking an approach that recognizes the many factors that countries have to take into consideration in enacting health and safety regulations and that allows sufficient flexibility for them to do so without sacrificing democratic participation in regulatory decision-making. The approach arguably allows a role for public sentiment in cases where the public view scientific evidence in a light which does not coincide with the way in which experts see the same evidence, but this has not yet been explicitly discussed in a case. However, given the overall lack of clarity of drafting in the SPS Agreement, it would be helpful if panels and the Appellate Body sought to provide a coherent overview of its approach and how it fits with the broader context of the conflict between trade and health objectives.

14.4 A FINAL WORD

It has been questioned whether the SPS Agreement is capable in its current

---

9 *Ibid.* at 933.

form of resolving disputes where regulatory differences between nations' health protection regimes arise from differing cultural, political, ethical, and social values.[10] This book has sought to answer this criticism by showing that where health is at issue the SPS Agreement is capable of a flexible interpretation that allows countries significant policy space to regulate in a manner that reflects their unique circumstances. Much lies in the hands of individual countries to manage their regulatory processes in a transparent and inclusive manner and to ensure that they follow accepted scientific protocols. If they do so, indications to date suggest that panels and the Appellate Body are likely to respond with decisions that allow countries flexibility to regulate to protect health. The SPS Agreement arguably gives them the capacity to do so. However, the SPS Agreement is not the appropriate venue where health is not a concern and disputes turn on public morals and concerns about socio-economic factors, such as maintaining a traditional rural sector.

---

10 Perdikis, for example, argues that it cannot. See N. Perdikis, W.A. Kerr and J.E. Hobbs, 'Can the WTO/GATT Agreements on Sanitary and Phyto-Sanitary Measures and Technical Barriers be Renegotiated to Accommodate Agricultural Biotechnology?' in W.H. Lesser, ed., *Transitions in Agbiotech: Economics of Strategy and Policy* (Connecticut: University of Connecticut: Food Marketing Policy Centre, 2000).

Bibliography

INTERNATIONAL AGREEMENTS

Additional Protocol to the American Convention on Human Rights in the Area of Economic, Social and Cultural Rights, Protocol of San Salvador, 17 November 1999 (Ratified 16 November 1999), OAS, Treaty Series, No. 69.

Additional Protocol to the European Social Charter Providing for a System of Collective Complaints, Strasbourg, 9 November 1995, European Treaty Series, No. 158.

African Charter on Human and Peoples Rights, 1981 (Entered into force 1986), OAU Doc. CAB/LEG/67/3 rev.5, 21 I.L.M. 58 (1982), Online at http://www1.umn.edu/humanrts/instree/z1afchar.htm.

Agreement on the Application of Sanitary and Phytosanitary Measures, 15 April 1994, in *The Legal Texts: The Results of the Uruguay Round of Multilateral Trade Negotiations* (Cambridge, UK: Cambridge University Press, 1999).

Agreement on Technical Barriers to Trade, 15 April 1994, in *The Legal Texts: The Results of the Uruguay Round of Multilateral Trade Negotiations* (Cambridge, UK: Cambridge University Press, 1999).

American Declaration on the Rights and Duties of Man, Adopted by the Ninth International Conference of American States, 1948, OAS.

Committee on Economic, Social and Cultural Rights (CESCR) General Comment 14, The right to the highest attainable standard of health, E/C.12/2000/4.

Declaration of Alma Alta, Alma Alta, 1978, available online at the Pan American Health Organization website, http://www.paho.org/English/DD/PIN/alma-ata_declaration.htm.

European Social Charter, Turin, 18 October 1961 (Entered into force 26 February 1965), European Treaty Series, No. 35.

General Agreement on Tariffs and Trade, 30 October 1947, 58 U.N.T.S. 187, Can. T.S. 1947 No. 27 (entered into force 1 January 1948) [GATT].

International Covenant on Civil and Political Rights, UN General Assembly Resolution 217A (III). A/8180 at 71 (1948).

International Covenant on Economic, Social and Cultural Rights, UN General Assembly Resolution 2200A(XXI) A/6316 (1966).

Marrakesh Agreement Establishing the World Trade Organization, 15 April 1994, in *The Legal Texts: The Results of the Uruguay Round of Multilateral Trade Negotiations* (Cambridge, UK: Cambridge University Press, 1999).

North American Free Trade Agreement Between the Government of Canada, the Government of Mexico and the Government of the United States, December 1992, Can. T.S. 1994 No. 2, 32 I.L.M. 289 (entered into force 1 January 1994).

WTO CASES

Australia – Measures Affecting Importation of Salmon (1998), WTO Doc. AB 1998–5, WT/DS18/R (Panel Report).

Australia – Measures Affecting Importation of Salmon (1998), WTO Doc. AB 1998–5, WT/DS18/AB/R (Appellate Body Report).

European Communities – Measures Affecting Asbestos and Asbestos-containing Products (2000), WTO Doc. WT/DS135/R (Panel Report).

European Communities – Measures Affecting Asbestos and Asbestos-containing Products (2001), WTO Doc. AB-2000-11, WT/DS135/AB/R (Appellate Body Report).

European Communities – Measures Concerning Meat and Meat Products (Hormones) Complaint by the United States (1997), WTO Doc. WT/DS26/R/USA (Panel Report).

European Communities – Measures Concerning Meat and Meat Products (Hormones) Complaint by Canada (1997), WTO Doc. WT/DS48/R/CAN (Panel Report).

European Communities – Measures Affecting Meat and Meat Products (1998), WTO Doc. WT/DS26/AB/R, WT/DS48/AB/R (Appellate Body Report).

European Communities – Measures Affecting the Approval and Marketing of Biotech Products (2006), WTO Doc. WT/DS291/R, WT/DS292/R, WT/DS293/R (Panel Report).

European Communities – Measures Affecting the Approval and Marketing of Biotech Products, Reports of the Panel, Addendum (Annex K) (2006), WTO Doc. WT/DS291/R/Add.9, WT/DS292/R/Add.9, WT/DS293/R/Add.9 (Panel Report).

Japan – Measures Affecting Agricultural Products (1998), WTO Doc. WT/DS76/R (Panel Report).

Japan – Measures Affecting Agricultural Products (1999), WTO Doc. WT/DS76/AB/R (Appellate Body Report).

Japan – Measures Affecting the Importation of Apples (2003), WTO Doc. WT/DS245/R (Panel Report).

Japan – Measures Affecting the Importation of Apples (2003), WTO Doc. WT/DS245/AB/R (Appellate Body Report).
Thailand – Restrictions on Importation of and Internal Taxes on Cigarettes (1990), WTO Doc. DS10/R – 37S/200 (Panel Report).
United States – Import Prohibition on Certain Shrimp and Shrimp Products (1998), WTO Doc. WT/DS15/R (Panel Report).
United States – Import Prohibition of Certain Shrimp and Shrimp Products (1998), WTO Doc. WT/DS58/AB/R (Appellate Body Report).
United States – Standards for Reformulated and Conventional Gasoline (1996), WTO Doc. WT/DS2/R (Panel Report).

NON-WTO JURISPRUDENCE

UL Canada Inc. v. Procureur Général du Québec (2003), WTO (J.Q. no. 13505 Report).

SECONDARY MATERIALS: ARTICLES AND REPORTS

Ari Afilalo and Sheila Foster, 'The World Trade Organization's Anti-Discriminatory Jurisprudence: Free Trade, National Sovereignty, and Environmental Health in the Balance' (2003) 15 Geo. Int'l Envtl L.R. 633.
Alberto Alemanno, *Science and EU Risk Regulation: The Role of Experts in Decision-Making and Judicial Review* (Milan: Università Commerciale L. Bocconi, 2007).
Kathleen A. Ambrose, 'Science and the WTO' (2000) 31:3 Law & Pol'y Int'l Bus. 861.
Jason Andrews and Samantha Chaifetz, 'How do International Trade Agreements Influence the Promotion of Public Health? An Introduction to the Issue' (2004) IV:2 Yale Journal of Health Policy, Law, and Ethics 339.
Alessandra Arcuri, 'The Post-Discriminatory Era of the WTO: Toward World-Wide Harmonization of Risk Law?' (2006) [unpublished, archived at Rotterdam].
Judith Asher, *The Right to Health: A Resource Manual for NGOs* (Washington DC: American Association for the Advancement of Science, 2005).
Prema-chandra Athukorala and Sisira Jayasuriya, 'Food Safety Issues, Trade and WTO Rules: A Developing Country Perspective' (2004) 26:9 The World Economy 1395.
Jeffery Atik, 'Science and International Regulatory Convergence' (1996–7) 17 Nw. J. Int'l L. & Bus. 736.

Jeffery Atik, 'Identifying Antidemocratic Outcomes: Authenticity, Self-Sacrifice and International Trade' (1998) 19 U. Pa. J. Int'l Econ. L. 229.
Jeffery Atik, 'The Weakest Link – Demonstrating the Inconsistency of "Appropriate Levels of Protection" in Australia-Salmon' (2004) 24:2 483.
Jeffery Atik and David A. Wirth, 'Science and International Trade – Third Generation Scholarship' (2003) 26:2 B.C. Int'l. & Comp. L. Rev. 171.
Natalie Avery, Martine Drake and Tim Lang, 'Codex Alimentarius: Who is Allowed in? Who is Left out?' (1993) 23:3 The Ecologist 110.
Bruce Ballantine, *Enhancing the Role of Science in the Decision-Making of the European Union* (Brussels: European Policy Centre (EPC), Working Paper No. 17, 2005).
John T. Barcelo, 'Product Standards to Protect the Local Environment – the GATT and the Uruguay Round Sanitary and Phytosanitary Agreement' (1994) 27 Cornell Int'l L.J. 755.
Jan J. Barendregt, Luc Bonneux and Paul J. Van der Maas, 'The Health Care Costs of Smoking' (1997) 337:15 New England Journal of Medicine 1050.
J.H. Barton, 'Biotechnology, the Environment and International Agricultural Trade' (1996) 9 Geo. Int'l Envtl L. Rev. 95.
Philip Bentley, 'A Re-Assessment of Article XX, Paragraphs (b) and (g), of GATT 1994 in the Light of Growing Consumer and Environmental Concern About Biotechnology' (2000) 24 Fordham Int'l L.J. 107.
David Blandford and Linda Fulponi, 'Emerging Public Concerns in Agriculture: Domestic Policies and International Trade Commitments' (1999) 26:3 European Review of Agricultural Economics 409.
M. Gregg Bloche, 'Introduction: Health and the WTO' (2002) 5:4 J. Int'l Econ. L. 821.
M. Gregg Bloche, 'WTO Deference to National Health Policy: Toward an Interpretive Principle' (2002) 5:4 J. Int'l Econ. L. 825.
Jan Bohanes, 'Risk Regulation in WTO Law: A Procedure-based Approach to the Precautionary Principle' (2002) 40 Columbia Journal of Transnational Law 323.
Judith A. Bradbury, 'The Policy Implications of Differing Concepts of Risk' (1989) 14 Science, Technology and Human Values 380.
Richard Braun, 'People's Concerns About Biotechnology: Some Problems and Some Solutions' (2002) 98 Journal of Biotechnology 3.
Geoffrey Brennan and James M. Buchanan, 'Is Public Choice Immoral? The Case for the "Nobel" Lie' (1988) 74 Va. L. Rev. 179.
Gideon Bruckner, *Implementation of OIE Standards in the Framework of the SPS Agreement* (Paris: World Organisation for Animal Health, 73rd General Session, 22–7 May, 2005).
D.E. Buckingham and P.W.B. Phillips, 'Hot Potato, Hot Potato: Regulating Products of Biotechnology by the International Community' (2001) 35:1 J. World Trade 1.

Jean-Cristophe Bureau, Stephan Marette and Alessandra Schiavina, 'Non-tariff Trade Barriers and Consumers' Information: The Case of the EU-US Trade Dispute over Beef' (1998) 25 European Review of Agricultural Economics 437.

Lawrence Busch, 'Virgil, Vigilance, and Voice: Agrifood Ethics in an Age of Globalization' (2003) 16:5 Journal of Agricultural and Environmental Ethics 459.

Lawrence Busch, *et al.*, *Amicus Curiae Brief Submitted to the Dispute Settlement Panel of the World Trade Organization in the Case of EC: Measures Affecting the Approval and Marketing of Biotech Products* (WT/DS291, 292 and 293, 30 April 2004).

Marc L. Busch and Robert Howse, *A (Genetically Modified) Food Fight* (Toronto: C.D. Howe Institute Commentary, 2003).

Jean C. Buzby, *International Trade and Food Safety: Economic Theory and Case Studies* (Washington DC: United States Department of Agriculture, Economic Research Service, 2003).

Paolo G. Carozza, 'Subsidiarity as a Structural Principle of International Human Rights Law' (2003) 97 American Journal of International Law 38.

Michelle D. Carter, 'Selling Science under the SPS Agreement: Accommodating Consumer Preference in the Growth Hormones Controversy' (1997) 6 Minn. J. Global Trade 625.

Damian Chalmers, 'Food for Thought: Reconciling European Risks and Traditional Ways of Life' (2003) 66:4 Modern Law Review 532.

Howard F. Chang, 'Risk Regulation, Endogenous Public Concerns, and the Hormones Dispute: Nothing to Fear But Fear Itself?' (2004) 77 S. Cal. L. Rev. 743.

Steve Charnovitz, 'The Supervision of Health and Biosafety Regulation by World Trade Rules' (2000) 13 Tul. Envtl L.J. 271.

Steve Charnovitz, 'An Analysis of Pascal Lamy's Proposal on Collective Preferences' (2005) 8:2 J. Int'l Econ. L. 449.

Junshi Chen, 'Challenges to Developing Countries After Joining WTO: Risk Assessment of Chemicals in Food' (2004) 198 Toxicology 3.

Sungjoon Cho, 'Linkage of Free Trade and Social Regulation: Moving Beyond the Entropic Dilemma' (2005) 5 Chicago J. Int'l L. 625.

Theofanis Christoforou, 'Settlement of Science-based Trade Disputes in the WTO: A Critical Review of the Developing Case Law in the Face of Scientific Uncertainty' (2000) 8 N.Y.U. Envtl L.J. 622.

R.H. Coase, 'The Problem of Social Cost' (1960) III J.L. & Econ. 1.

Claudio Cocuzzo and Andrea Forabosco, 'Are States Relinquishing their Sovereign Rights? The GATT Dispute Settlement Process in a Globalized Economy' (1996) 4 Tul. J. Int'l & Comp. L. 161.

Codex Alimentarius Commission, *Procedural Manual*, Definitions of Risk Analysis Terms Related to Food Safety, 11.

Shirley A. Coffield, 'The Management and Resolution of Cross Border Disputes as Canada/US Enter the 21st Century: Biotechnology, Food, and Agriculture Disputes or Food Safety and International Trade (Part 1)' (2000) 26 Can.-U.S.L.J. 233.
Commission on Risk Assessment and Risk Management, *Framework for Environmental Risk Management, Presidential/Congressional Commission on Risk Assessment and Risk Management, Final Report, Volume 1* (Washington DC, 1997).
Commission on Risk Assessment and Risk Management, *Risk Assessment and Risk Management in Regulatory Decision-Making, Final Report, Volume 2* (Washington DC, 1997).
Sydney M. Cone, 'The Asbestos Case and Dispute Settlement in the World Trade Organization: The Uneasy Relationship between Panels and the Appellate Body' (2001) 23 Mich. J. Int'l L. 103.
C. Correa, 'Implementing National Public Health Policies in the Framework of the WTO Agreements' (2000) 5 J. World Trade 98.
Nick Covelli and Viktor Hohots, 'The Health Regulation of Biotech Foods Under the WTO Agreement' (2003) 6:4 J. Int'l Econ. L. 773.
Douglas Crawford-Brown, *Scientific Models of Human Health Risk Analysis in Legal and Policy Decisions* (Chapel Hill, NC: Departments of Environmental Sciences and Engineering, University of North Carolina at Chapel Hill, 2001).
Douglas Crawford-Brown, Joost Pauwelyn and Kelly Smith, 'Environmental Risk, Precaution, and Scientific Rationality in the Context of WTO/NAFTA Trade Rules' (2004) 24:2 Risk Analysis 461.
Steven P. Croley, 'Theories of Regulation: Incorporating the Administrative Process' (1998) 98:1 Colum. L. Rev. 1.
Steven P. Croley, 'Public Interested Regulation' (2000) 28:7 Florida State University Law Review 7.
Frank B. Cross, 'The Risk of Reliance on Perceived Risk' (1992) 3 Risk 59.
John Martin Currie, John A. Murphy and Andrew Schmitz, 'The Concept of Economic Surplus and its Use in Economic Analysis' (1971) 81:324 Economic Journal 741.
Allison Marston Danner, 'Enhancing Legitimacy and Accountability of Prosecutional Discretion at the International Criminal Court' (2003) 97 American Journal of International Law 510.
Alan V. Deardoff and Robert M. Stern, 'What You Should Know about Globalization and the World Trade Organization' (2002) 10:3 Review of International Economics 404.
Anne-Célia Disdier, Lionel Fontagné and Mondher Mimouni, *The Impact of Regulations on Agricultural Trade: Evidence From SPS and TBT Agreements* (Paris: Centre d'Etude Prospectives et d'Informations Internationales – Working Paper No. 2007–04, 2007).

Ronald L. Doering, *Margarine Mayhem* (Guelph, Ontario: Food Safety Network, 2004).

Peter Drahos, 'The Regulation of Public Goods' (2004) 7:2 J. Int'l Econ. L. 321.

Marsha A. Echols, 'Food Safety Regulation in the European Union and the United States: Different Cultures, Different Laws' (1998) 4 Colum. J. Eur. L. 525.

Ken Endo, 'The Principle of Subsidiarity: From Johannes Althusius to Jacques Delors' (1994) 44 Hokkaido Law Review 2064.

D.C. Estry and D. Geradin, 'Regulatory Co-opetition' (2000) 3:2 J. Int'l Econ. L. 235.

European Commission, *Communication from the Commission on the Precautionary Principle* (Brussels, 2001, available at http://europa.eu.int/comm/dgs/health_consumer/library/pub/pub07_en.pdf).

European Commission – Health and Consumer Protection Directorate-General, *Final Report on Setting the Scientific Frame for the Inclusion of New Quality of Life Concerns in the Risk Assessment Process* (Brussels, 2003).

Food and Agriculture Organization of the United Nations (FAO) and World Health Organization (WHO), *Understanding the Codex Alimentarius* (Rome: FAO/WHO, 1999).

Food and Agriculture Organization of the United Nations (FAO) and World Health Organization (WHO), *Report of the Evaluation of the Codex Alimentarius and other FAO and WHO Food Standards Work* (Geneva, 2002).

Jeremy D. Fraiberg and Michael J. Trebilcock, 'Risk Regulation: Technocratic and Democratic Tools for Regulatory Reform' (1998) 43 McGill L.J. 835.

S. Frechette, 'Biotechnology, Food, and Agriculture Disputes, or Food Safety and International Trade (Part 2)' (2000) Can.-U.S.L.J. 253.

Heather Berit Freeman, 'Trade Epidemic: The Impact of the Mad Cow Crisis on EU-US Relations' (2002) 25 B.C. Int'l & Comp. L. Rev. 343.

Theresa Garvin, 'Analytical Paradigms: The Epistomological Distances between Scientists, Policy Makers, and the Public' (2001) 21:3 Risk Analysis 443.

General Accounting Office (GAO), *Report to Congressional Requesters: Antibiotic Resistance – Federal Agencies Need to Better Focus Efforts to Address Risk to Humans from Antibiotic Use in Animals* (Washington DC: GAO, 2004).

John L. Gignilliat, 'Pigs, Politics, and Protection: The European Boycott of American Pork, 1879–1891' (1961) 35 Agricultural History 3.

O. Godard, *Integrating Scientific Expertise into Regulatory Decision-Making: Social Decision-Making Under Conditions of Scientific Controversy,*

Expertise and the Precautionary Principle (Florence: European University Institute, EUI Working Paper RSC No. 96/6, 1996).
Gavin Goh, 'Precaution, Science and Sovereignty: Protecting Life and Health under the WTO Agreements' (2003) 6 Journal of World Intellectual Property 441.
Gavin Goh and Andreas R. Ziegler, 'A Real World where People Live and Work and Die: Australian SPS Measures after the WTO Appellate Body's Decision in the Hormones Case' (1998) 32:5 J. World Trade 271.
Bernard Goldstein and Russellyn S. Carruth, 'The Precautionary Principle and/or Risk Assessment in World Trade Organization Decisions: A Possible Role for Risk Perception' (2004) 24:2 Risk Analysis 491.
Lawrence O. Gostin, 'Public Health Law in a New Century: Law as a Tool to Advance the Community's Health' (2000) 283:21 Journal of the American Medical Association 2837.
Lawrence O. Gostin, 'Public Health Law in a New Century — Part III: Public Health Regulation: A Systematic Evaluation' (2000) 283:23 Journal of the American Medical Association 3118.
Willy De Greef, 'Regulatory Conflicts and Trade' (2000) 8 N.Y.U. Envtl L.J. 579.
Andrew J. Green, 'Public Participation and Environmental Policy Outcomes' (1997) 23:4 Can. Pub. Pol'y 435.
U. Grote and S. Kirchhoff, *Environmental and Food Safety Standards in the Context of Trade Liberalization: Issues and Options* (Bonn: Center for Development Research, 2001).
Andrew T. Guzman, 'Food Fears: Health and Safety at the WTO' (2004) 45 Va. J. Int'l L. 1.
Gillian Hadfield, Robert Howse and Michael J. Trebilcock, *Rethinking Consumer Protection Policy* (Prepared for and revised based on the University of Toronto Roundtable on New Approaches to Consumer Law, 1996).
Steve Hathaway, 'Management of Food Safety in International Trade' (1999) 10 Food Control 247.
Steve C. Hathaway and Roger L. Cook, 'A Regulatory Perspective on the Potential Uses of Microbial Risk Assessment in International Trade' (1997) 36 International Journal of Food Microbiology 127.
Dale Hattis and Rob Goble, 'The Red Book, Risk Assessment, and Policy Analysis: The Road Not Taken' (2003) 9 Human and Ecological Risk Assessment 1297.
Caroline Henckels, 'GMOs in the WTO: A Critique of the Panel's Legal Reasoning in EC-Biotech' (2006) Melbourne Journal of International law 7:2.
Spencer Henson and Rupert Loader, 'Impact of Sanitary and Phytosanitary

Standards on Developing Countries and the Role of the SPS Agreement' (1999) 15:3 Agribusiness 355.

Peter Holmes and Alasdair R. Young, 'Trade vs. Public Health Standards: WTO/EU Tensions' (2003) (unpublished, archived at UC Berkeley).

Neal Hooker, 'Food Safety Regulation and Trade in Food Products' (1999) 24:6 Food Policy 653.

Robert Howse, 'Democracy, Science, and Free Trade: Risk Regulation on Trial at the World Trade Organization' (2000) 98 Mich. L. Rev. 2329.

Robert Howse and Petros C. Mavroidis, 'Europe's Evolving Regulatory Strategy for GMOs – The Issue of Consistency with WTO Law: Of Kine and Brine' (2000) 24 Fordham Int'l L.J. 317.

Robert Howse and Donald Regan, 'The Product/Process Distinction – An Illusory Basis for Disciplining "Unilateralism" in Trade Policy' (2000) 11:2 E.J.I.L. 249.

Robert Howse and Ruti G. Teitel, *Beyond the Divide: The Covenant on Economic, Social and Cultural Rights and the World Trade Organization* (Geneva: Friedrich Ebert Stiftung, 2007).

Steve E. Hrudey and William Leiss, 'Risk Management and Precaution: Insights on the Cautious Use of Evidence' (2003) 111:13 Environmental Health Perspectives 1577.

Robert Hudec, 'Science and "Post-Discriminatory" WTO Law' (2003) 26:2 B.C. Int'l & Comp. L. Rev. 185.

Robert E. Hudec, 'GATT/WTO Constraints on National Regulation: Requiem for an "Aim and Effects" Test' (1998) 32 Int'l Law. 619.

W.D. Hueston, 'Science, Politics and Animal Health Policy: Epidemiology in Action' (2003) 60 Preventive Veterinary Medicine 3.

Layla Hughes, 'Limiting the Jurisdiction of Dispute Settlement Panels: The WTO Appellate Body Beef Hormones Decision' (1998) 10 Geo. Int'l Envtl L. Rev. 915.

John P. Huttman, 'British Meat Imports in the Free Trade Era' (1978) 52:2 Agricultural History 247.

Leonardo Iacovone, 'The Analysis and Impact of Sanitary and Phytosanitary Measures' (2005) 22 Integration and Trade 97.

Grant E. Isaac and William A. Kerr, 'Genetically Modified Organisms at the World Trade Organization: A Harvest of Trouble' (2003) 37:6 J. World Trade 1083.

Bernhard Jansen and Maurits Lugard, 'Some Considerations on Trade Barriers Erected for Non-Economic Reasons and WTO Obligations' (1999) J. Int'l Econ. L. 530.

Cindy G. Jardine, *et al.*, 'Risk Management Frameworks for Human Health and Environmental Issues' (2003) Part B:6 Journal of Toxicology and Environmental Health 569.

Christine Jolls, Cass R. Sunstein and Richard Thaler, 'A Behavioural Approach to Law and Economics' (1998) 50 Stan. L. Rev. 1471.
Laura Jones, ed., *Safe Enough? Managing Risk and Regulation* (Vancouver: The Fraser Institute, 2000).
Tim Josling, Donna Roberts and Ayesha Hassan, *The Beef-Hormones Dispute and its Implications for Trade Policy* (Washington DC: USDA, 1999).
F. Kaferstein and M. Abdussalam, 'Food Safety in the 21st Century' (1999) 77:4 Bulletin of the World Health Organization 347.
F.K. Kaferstein, 'Actions to Reverse the Upward Curve of Foodborne Illness' (2003) 14 Food Control 101.
Robert A. Kagan, 'Introduction: Comparing National Styles of Regulation in Japan and the United States' (2000) 22:3&4 L. & Pol'y 225.
Dan M. Kahan and Donald Braman, 'Cultural Cognition and Public Policy' (Yale Law School Public Law Working Paper No. 87, 2005).
Dan M. Kahan and Paul Slovic, 'Cultural Evaluations of Risk: "Values" or "Blunders"?' (2006) 119 Harv. L. R. 166.
Dan M. Kahan, *et al.*, 'Fear and Democracy: A Cultural Evaluation of Sunstein on Risk' (2006) 119 Harv. L.R. 1071.
Justin Kastner and Doug Powell, *Food Safety, International Trade, and History Repeated* (Guelph, Ontario: The Food Safety Network, 2001).
Trish Kelly, 'The WTO, the Environment and Health and Safety Standards' (2003) 26:2 The World Economy 131.
Kevin C. Kennedy, 'Resolving International Sanitary and Phytosanitary Disputes in the WTO: Lessons and Future Directions' (2000) 55 Food & Drug L.J. 81.
William A. Kerr, 'Genetically Modified Organisms, Consumer Scepticism and Trade Law: Implications for the Organization of International Supply Chains' (1999) 4 Supply Chain Management 67.
William A. Kerr, 'International Trade in Transgenic Food Products: A New Focus for Agricultural Trade Disputes' (1999) 22:2 The World Economy 245.
William A. Kerr, 'Science-based Rules of Trade – A Mantra for Some, An Anathema for Others' (2003) 4:2 The Estey Centre Journal of International Law and Trade Policy 86.
W.A. Kerr and J.E. Hobbs, 'The North American-European Union Dispute over Beef Produced Using Growth Hormones: A Major Test for the New International Trade Regime' (2002) 25:2 The World Economy 283.
A.M. Kimbell, K-Y. Wong and K. Taneda, 'An Evidence Base for International Health Regulations: Quantitative Measurement of the Impacts of Epidemic Disease on International Trade' (2005) 24:3 OIE Scientific and Technical Review 825.
Daniel Lee Kleinman and Abby J. Kinchy, 'Boundaries in Science Policy

Making: Bovine Growth Hormone in the European Union' (2003) 44:4 The Sociological Quarterly 577.

Marion Koopmans and Erwin Duizer, 'Foodborne Viruses: An Emerging Problem' (2004) 90 International Journal of Food Microbiology 23.

David Kriebel and Joel Tickner, 'Reenergizing Public Health through Precaution' (2001) 91:9 American Journal of Public Health 1351.

Sheldon Krimsky, 'Risk Assessment and Regulation of Bioengineered Food Products' (2000) 2 International Journal of Biotechnology 231.

Sheldon Krimsky, 'The Weight of Scientific Evidence in Policy and Law' (2005) 95:S1 American Journal of Public Health S129.

Paul R. Krugman, 'Is Free Trade Passé?' (1987) 1:2 The Journal of Economic Perspectives 131.

Pascal Lamy, *The Emergence of Collective Preferences in International Trade: Implications for Regulating Globalization* (online at http://trade.ec.europa.eu/doclib/docs/2004/september/tradoc_118925.pdf, 2004).

Bettina Lange, 'Regulatory Spaces and Interactions: An Introduction' (2003) 12:4 Social and Legal Studies 411.

Lester B. Lave, 'Health and Safety Risk Analyses: Information for Better Decisions' (1987) 236:4799 Science 291.

P. Lee and D. Paxman, 'Reinventing Public Health' (1997) 18:1 Annual Review of Public Health 1.

Peter L. Lindseth, 'Democratic Legitimacy and the Administrative Character of Supernationalism: The Example of the European Union' (1999) 99 Columbia Law Review 628.

Ragnar Lofstedt and Robyn Fairman, 'Scientific Peer Review to Inform Regulatory Decision Making: A European Perspective' (2006) 26:1 Risk Analysis 25.

Frank Loy, 'Statement on Biotechnology: A Discussion of Four Important Issues in the Biotechnology Debate' (2000) 8 N.Y.U. Envtl L.J. 607.

F. MacMillan and M. Blakeney, 'Regulating GMOs: Is the WTO Agreement on Sanitary and Phytosanitary Measures Hormonally Challenged? Part I' (2000) 6:4 Int'l Trade L. Reg. 131.

Fiona MacMillan and Michael Blakeney, 'Labelling of GMOs under the Rules of the World Trade Organization' (2000/1) 4:1 Bio-Science Law Review 3.

Gabrielle Marceau and Joel P. Trachtman, 'The Technical Barriers to Trade Agreement, the Sanitary and Phytosanitary Measures Agreement, and the General Agreement on Tariffs and Trade: A Map of the World Trade Organization Law of Domestic Regulation of Goods' (2002) 36:5 J. World Trade 811.

David L. Markell, 'Slack in the Administrative State and its Implications for Governance: The Issue of Accountability' (2005) 84 Oregon Law Review 1.

David L. Markell and Tom Tyler, 'Using Empirical Research to Design Government Citizen Participation Processes: A Case Study of Citizens' Roles in Environmental Compliance and Enforcement' (FSU College of Law, Public Law Research Paper No. 270, 2nd Annual Conference of Empirical Legal Studies Paper, FSU College of Law, Law and Economics Paper No. 07-014).

Jerry L. Mashaw, 'Law and Engineering: In Search of the Law-Science Problem' (2003) 66:4 Law & Contemp. Probs 135.

Larry A. Di Matteo, *et al.*, 'The Doha Declaration and Beyond: Giving a Voice to Non-trade Concerns within the WTO Trade Regime' (2003) 36:1 Vand. J. Transnat'l L. 95.

Walter Mattli, 'The Politics and Economics of International Institutional Standards Setting: An Introduction' (2001) 8:3 Journal of European Public Policy 328.

Walter Mattli and Tim Büthe, 'Setting International Standards: Technological Rationality or Primacy of Power?' (2003) 56 World Politics 1.

Graham Mayeda, 'Developing Disharmony? The SPS and TBT Agreements and the Impact of Harmonisation on Developing Countries' (2003) 7:4 J. Int'l Econ. L. 737.

Patrick McAuslan, 'Public Law and Public Choice' (1988) 51:6 Mod. L. Rev. 681.

Mathew McCubbins, Roger Noll and Barry Weingast, 'Administrative Procedures as Instruments of Political Control' (1989) 3:2 J. Law Econ. Org. 243.

Thomas O. McGarity, 'Substantive and Procedural Discretion in Administrative Resolution of Science Policy' (1979) 67 Geo. L.J. 729.

Thomas O. McGarity, 'On the Prospect of "Daubertizing" Judicial Review of Risk Assessment' (2003) 66 Law and Contemp. Probs 155.

A.J. McMichael and R. Beaglehole, 'The Changing Global Context of Public Health' (2000) 356:9228 Lancet 495.

Paul S. Mead, *et al.*, 'Food-Related Illness and Death in the United States' (1999) 5:5 Emerging Infectious Diseases 607.

Richard A. Merrill, 'Foreword: Symposium on Science in the Regulatory Process' (2003) 66:4 Law & Contemp. Probs 1.

David Michaels and Celeste Monforton, 'Manufacturing Uncertainty: Contested Science and the Protection of the Public's Health and Environment' (2005) 95:S1 American Journal of Public Health S39.

Lydia Miljan, 'Unknown Causes, Unknown Risks' in Laura Jones, ed., *Safe Enough? Managing Risk and Regulation* (Vancouver: The Fraser Institute, 2000) 31.

Robert M. Millimet, 'The Impact of the Uruguay Round and the New Agreement on Sanitary and Phytosanitary Measures' (1995) 5 Transnational Law and Contemporary Problems 443.

Chantal Millon-Delsol, 'L'Etat Subsidaire: Ingérence et non-ingérence de l'etat: Le principle de subsidarité aux fondements de l'histoire européenne' (1992).

Erik Millstone and Patrick van Zwanenberg, 'The Evolution of Food Safety Policy-making Institutions in the UK, EU and Codex Alimentarius' (2002) 36:6 Social Policy and Administration 593.

Yasmine Motarjemi and Frtiz Kaferstein, 'Food Safety, Hazard Analysis and Critical Control Point and the Increase in Foodborne Diseases: A Paradox?' (1999) 10 Food Control 325.

Sean D. Murphy, 'Biotechnology and International Law' (2001) 42 Harv. Int'l L.J. 47.

National Foreign Trade Council Inc., *Looking Behind the Curtain: The Growth of Trade Barriers that Ignore Sound Science* (Washington DC: National Foreign Trade Council, Inc, 2003).

Dorothy Nelkin, Philippe Sands and Richard B. Stewart, 'Foreword: The International Challenge of Genetically Modified Organism Regulation' (2000) 8 N.Y.U. Envtl L.J. 523.

Jan Neumann and Elisabeth Tuerk, 'Necessity Revisited: Proportionality in World Trade Organization Law after *Korea-Beef*, *EC-Asbestos* and *EC-Sardines*' (2003) 37:1 J. World Trade 199.

Roger G. Noll and James E. Krier, 'Some Implications of Cognitive Psychology for Risk Regulation' in Cass R. Sunstein, ed., *Behavioural Law and Economics* (Cambridge, UK: Cambridge University Press, 2000).

Julie A. Nordlee, *et al.*, 'Identification of a Brazil-Nut Allergen in Transgenic Solutions' (1996) 334:11 New England Journal of Medicine 688.

Alessandra Nucara, 'Precautionary Principle and GMOs: Protection or Protectionism?' (2003) 9:2 Int'l Trade L. Reg. 47.

OECD, *Measuring the Trade Effects of the SPS Agreement*, COM/AGR/TD/WP(2002)71/FINAL (Paris: OECD, 2003).

Matthias Oesch, 'Standards of Review in WTO Dispute Resolution' (2003) 6:3 J. Int'l Econ. L. 635.

Tsunehiro Otsuki, John S. Wilson and Mirvat Sewadeh, 'Saving Two in a Billion: Quantifying the Trade Effect of European Food Safety Standards on African Exports' (2001) 26 Food Policy 495.

T. Ademola Oyejide, E. Olawale Ogunkola and S. Abiodun Bankole, *Quantifying the Trade Impact of Sanitary and Phystosanitary Standards: What is Known and Issues of Importance for Sub-Saharan Africa* (Washington DC, 2000).

John Paterson, 'Trans-Science, Trans-Law and Proceduralization' (2003) 12:4 Social and Legal Studies 525.

Joost Pauwelyn, 'The WTO Agreement on Sanitary and Phytosanitary (SPS) Measures as Applied in the First Three SPS Disputes: EC – Hormones, Australia – Salmon, and Japan – Varietals' (1999) 2 J. Int'l Econ. L. 654.

Jacqueline Peel, 'Jean Monnet Working Paper 02/04 – Risk Regulation under the WTO SPS Agreement: Science as an International Normative Yardstick?' (2004) (unpublished, archived at New York University).

N. Perdikis, 'A Conflict of Legitimate Concerns or Pandering to Vested Interests? Conflicting Attitudes towards the Regulation of Trade in Genetically Modified Goods – The EU and the US' (2000) 1:1 Estey Centre Journal of International Law and Trade Policy 51.

N. Perdikis, W.A. Kerr and J.E. Hobbs, 'Reforming the WTO to Defuse Potential Trade Conflicts in Genetically Modified Organisms' (2001) 24:3 The World Economy 379.

Franz Xaver Perrez, 'Taking Consumers Seriously: The Swiss Regulatory Approach to Genetically Modified Food' (2000) 8 N.Y.U. Envtl. L.J. 585.

Pest Management Risk Agency, *Technical Paper: A Decision Framework for Risk Assessment and Risk Management in the Pest Management Risk Agency* (Ottawa: Pest Management Risk Agency, Health Canada, 2000).

Pew Initiative on Food and Biotechnology, *US vs. EU: An Examination of the Trade Issues Surrounding Genetically Modified Food* (Richmond, VA: Pew Initiative on Food and Biotechnology, 2005).

Richard A. Posner, 'Theories of Economic Regulation' (1974) 5:2 The Bell Journal of Economics and Management Science 335.

Diahanna L. Post, 'The Precautionary Principle and Risk Assessment in International Food Safety: How the World Trade Organization Influences Standards' (2006) 26:5 Risk Analysis 1259.

Denise Prévost, 'What Role for the Precautionary Principle in WTO Law after *Japan–Apples*?' (2005) 2:4 Journal of Trade and Environment Studies 1.

Reinhard Quick and Andreas Bluthner, 'Has the Appellate Body Erred? An Appraisal and Criticism of the Ruling in the WTO Hormones Case' (1999) 2 J. Int'l Econ. Law 603.

John Quinn and Philip Slayton, eds, *Non-Tariff Barriers after the Tokyo Round* (Montreal: The Institute for Research on Public Policy, 1980).

Sara Pardo Quintillan, 'Free Trade, Public Health Protection and Consumer Information in the European and WTO Context' (1999) 33:6 J. World Trade 147.

Jeffrey J. Rachlinski and Cynthia R. Farina, 'Getting Beyond Cynicism: New Theories of the Regulatory State Cognitive Psychology and Optimal Government Design' (2002) 87 Cornell L. Rev. 549.

Paul Raeburn, 'Clamor over Genetically Modified Foods Comes to the United States' (2000) 8 N.Y.U. Envtl L.J. 610.

Alan Charles Raul and Julie Zampa Dwyer, '"Regulatory *Daubert*": A Proposal to Enhance Judicial Review of Agency Science by Incorporating *Daubert* Principles into Administrative Law' (2003) 66:4 Law and Contemp. Probs.

Donald H. Regan, 'Further Thoughts on the Role of Regulatory Purpose under Article III of the General Agreement on Tariffs and Trade' (2003) 37:4 J. World Trade 737.

D. Roberts, 'Preliminary Assessment of the Effects of the WTO Agreement on Sanitary and Phytosanitary Trade Regulation' (1998) J. Int'l Econ. L. 377.

Donna Roberts, 'Analyzing Technical Trade Barriers in Agricultural Markets: Challenges and Priorities' (1999) 15:3 Agribusiness 335.

Donna Roberts, Timothy E. Josling and David Orden, *A Framework for Analyzing TBTs in Agricultural Markets, USDA Technical Bulletin No. 1876* (Washington DC: United States Department of Agriculture, Economic Research Service, 1999).

Donna Roberts and Barry Krissoff, 'Regulatory Barriers in International Horticultural Markets' (Washington DC: United States Department of Agriculture, 2004).

David Robertson, 'Civil Society and the WTO' (2000) 23:9 The World Economy

B.J. Rothstein, 'Bringing Science to Law' (2005) 95:S1 American Journal of Public Health S4.

Gene Rowe and Lynn J. Frewer, 'A Typology of Public Engagement Mechanisms' (2005) 30:2 Science, Technology and Human Values 251.

Gene Rowe and Lynn J. Frewer, 'Public Participation Methods: A Framework for Evaluation' (2000) 25:1 Science, Technology and Human Values 3.

Gene Rowe and George Wright, 'Differences in Expert and Lay Judgments of Risk: Myth or Reality?' (2001) 21:2 Risk Analysis 341.

William Ruckelshaus, 'Risk in a Free Society' (1984) 4 Risk Analysis 157.

William D. Ruckelshaus, 'Science, Risk, and Public Policy' (1983) 221:4 Science 1026.

Holly Saigo, 'Agricultural Biotechnology and the Negotiation of the Biosafety Protocol' (2000) 12 Geo. Int'l Envtl. L.R. 229.

Ian Sandford, 'Hormonal Imbalance? Balancing Free Trade and SPS Measures after the Decision in Hormones' (1999) 29 V.U.W.L.R. 389.

Pierre Sauve and Americo Beviglia Zampetti, 'Subsidiarity Perspectives on the New Trade Agenda' (2000) 3:1 Journal of International Economic Law 83.

M. van Schothorst, 'Microbiological Risk Assessment of Foods in International Trade' (2002) 40 Safety Science 359.

Dayna Nadine Scott, *Nature/Culture Clash: The Transnational Trade Debate over GMOs* (New York: Hauser Global Law School Program, Global Law Working Paper 06/05, 2005).

Hansjorg Seiler, 'Harmonised Risk Based Regulation – A Legal Viewpoint' (2002) 40 Safety Science 31.

Ellen R. Shaffer, 'Global Trade and Public Health' (2005) 95:1 American Journal of Public Health 23.

Ira S. Shapiro, 'Treating Cigarettes as an Exception to the Trade Rules' (2002) XXII:1 SAIS Review 87.

Rajeev Sharma, 'WTO Beware: The Newest Global Non-Tariff Barriers' (2002) 1 Int. T.L.R. 1.

Bruce A. Silverglade, 'The WTO Agreement on Sanitary and Phytosanitary Measures: Weakening Food Safety Regulations to Facilitate Trade?' (2000) 55 Food and Drug L.J. 517.

Lennart Sjoberg, 'Principles of Risk Perception Applied to Gene Technology' (2004) 5: Special Issue EMBO Reports 547.

Grace Skogstad, 'Internationalization, Democracy, and Food Safety Measures: The (Il)Legitimacy of Consumer Preferences?' (2001) 7 Global Governance 293.

Grace Skogstad, 'The WTO and Food Safety Regulatory Policy Innovation in the European Union' (2001) 39:3 Journal of Common Market Studies 485.

Grace Skogstad, 'Contested Political Authority, Risk Society, and the Transatlantic Divide in Genetic Engineering Regulation' (2005) (unpublished, archived at University of Toronto).

Marco M. Slotbloom, 'Do Public Health Measures Receive Similar Treatment in European Community and World Trade Organization Law?' (2003) 37:3 J. World Trade 553.

P. Slovic, 'Perception of Risk' (1987) 236:4799 Science 280.

Paul Slovic, 'Going Beyond the Red Book: The Sociopolitics of Risk' (2003) 9 Human and Ecological Risk Assessment 1.

Michael J. Smith, 'GATT, Trade, and the Environment' (1993) 23 Envtl L. 533.

Alan Sokal, 'A Plea for Reason, Evidence and Logic' (1997) 6:2 New Politics 126.

Alan D. Sokal, 'Transgressing the Boundaries: An Afterword' (1996) 43:4 Dissent 93.

C. Starr, 'Risk Management, Assessment, and Acceptability' (1985) 5 Risk Analysis 97.

Terence P. Stewart and David S. Johanson, 'The SPS Agreement of the World Trade Organization and International Organizations: The Roles of the Codex Alimentarius Commission, the International Plant Protection Convention, and the International Office of Epizootics' (1998) 26 Syracuse J. Int'l L. & Com. 27.

Terence P. Stewart and David S. Johanson, 'Policy in Flux: The European Union's Laws on Agricultural Biotechnology and their Effects on International Trade' (1999) 4 Drake J. Agric. L. 243.

Terence P. Stewart and David S. Johanson, 'The SPS Agreement of the World Trade Organization and the International Trade of Dairy Products' (1999) 54 Food & Drug L.J. 55.

George J. Stigler, 'The Theory of Economic Regulation' (1971) 1 Bell Journal of Economics 3.

Cass R. Sunstein, 'Book Review – The Laws of Fear' (2001) 115 Harv. L. Rev. 1119.

Cass R. Sunstein, 'Probability Neglect: Emotions, Worst Cases, and Law' (2002) 112: 61 Yale L.J.

Cass R. Sunstein, 'Misfearing: A Reply' (2006) 119 Harv. L. Rev. 1110.

Steve Suppan, *Consumers International's Decision-Making in the Global Market: Codex Briefing Paper* (Minneapolis: Institute for Agriculture and Trade Policy, 2004).

Alan Swinbank, 'The Role of the WTO and the International Agencies in SPS Standard Setting' (1999) 15:3 Agribusiness 323.

Alan O. Sykes, 'Comparative Advantage and the Normative Economics of International Trade Policy' (1998) 1 J. Int'l Econ. L. 49.

Alan O. Sykes, 'Regulatory Protectionism and the Law of International Trade' (1999) 66:1 U. Chicago L. Rev. 31.

Alan O. Sykes, 'The (Limited) Role of Regulatory Harmonization in International Goods and Services Markets' (1999) 2 J. Int'l Econ. L. 49.

Alan O. Sykes, 'Regulatory Competition or Regulatory Harmonization? A Silly Question?' (2000) 3:2 J. Int'l Econ. L. 257.

Alan O. Sykes, 'Domestic Regulation, Sovereignty, and Scientific Evidence Requirements: A Pessimistic View' (2002) 3:2 Chicago J. Int'l L. 353.

Alan O. Sykes, 'The Least Restrictive Means' (2003) 70 U. Chicago L. Rev. 403.

Ryan David Thomas, 'Where's the Beef? Mad Cows and the Blight of the SPS Agreement' (1999) 32 Vand. J. Transnat'l L. 487.

Paul B. Thompson, 'Risk Objectivism and Risk Subjectivism' (1990) 1:3 Risk.

Faith Thomson Campbell, 'The Science of Risk Assessment for Phytosanitary Regulation and the Impact of Changing Trade Regulations' (2001) 51:2 BioScience 148.

Michael J. Trebilcock, 'An Introduction to Law and Economics' (1997) 23:1 Monash U. L. Rev. 123.

Michael Trebilcock and Robert Howse, 'Trade Liberalization and Regulatory Diversity: Reconciling Competitive Markets with Competitive Politics' (1998) 6 Eur. J. L. & Econ. 5.

M.J. Trebilcock, *et al.*, *The Choice of Governing Instrument* (A study prepared for the Economic Council of Canada, Toronto, 1982).

Lincoln Tsang, 'An Analysis of the Legal and Technical Requirements of the European Directive on Deliberative Release of Genetically Modified Organisms' (2001/2) 5:5 Bio-science Law Review 143.

United Nations University Institute of Advanced Studies, *Trading Precaution: The Precautionary Principle and the WTO* (Yokohama: UNU IAS, 2005).

United States Department of Agriculture (USDA) Economic Research Service, *An Overview of Foreign Technical Barriers to US Agricultural Exports, Staff Paper AGES#9705* (Washington DC, 1997).
United States Department of Agriculture (USDA) Economic Research Service, *A Framework for Analyzing Technical Trade Barriers in Agricultural Markets* (Washington DC: USDA, 1999).
United States Environmental Protection Agency (EPA), *Science Policy Council Handbook: Peer Review* (Washington DC: EPA, 2000).
B. Vallat and D.W. Wilson, 'The Obligations of Member Countries of the OIE (World Organisation for Animal Health) in the Organisation of Veterinary Services' (2003) 22:2 Rev. Sci. Tech. Off. Int. Epiz. 547.
David G. Victor, 'The Sanitary and Phytosanitary Agreement of the World Trade Organization: An Assessment after Five Years' (2000) 32 N.Y.U.J. Int'l L. & Pol. 865.
David G. Victor and C. Ford Runge, 'Farming the Genetic Frontier' (2002) May/June Foreign Affairs 107.
David Vogel, *The New Politics of Risk Regulation in Europe* (London: London School of Economics, 2001).
J. Martin Wagner, 'The WTO's Interpretation of the SPS Agreement has Undermined the Right of Governments to Establish Appropriate Levels of Protection Against Risk' (2000) 31:3 Law & Pol'y Int'l Bus. 855.
Wendy Wagner, 'The Science Charade in Toxic Risk Regulation' (1995) 95:7 Colum. L. Rev. 1613.
Wendy E. Wagner, 'The "Bad Science" Fiction: Reclaiming the Debate over the Role of Science in Public Health and Environmental Regulation' (2003) 66:4 Law & Contemp. Probs 63.
Vern Walker, 'The Siren Songs of Science: Toward a Taxonomy of Scientific Uncertainty for Decisionmakers' (1991) 23 Conn. L.R. 567.
Vern Walker, 'Keeping the WTO from Becoming the "World" Trans-science Organization: Scientific Uncertainty, Science Policy, and Fact-finding in the "Growth Hormones Dispute" ' (1998) Cornell Int'l L.J. 251.
Vern Walker, 'The Myth of Science as a "Neutral Arbiter" for Triggering Precautions' (2003) 26:2 B.C. Int'l & Comp. L. Rev. 197.
Lori M. Wallach, 'Accountable Governance in the Era of Globalization: The WTO, NAFTA, and International Harmonization of Standards' (2002) 50 Kansas Law Review 823.
Lynda M. Warren, 'Viewpoint: Making Use of the Best Available Science' (2004) 2 Law, Science and Policy 131.
A. Weinberg, 'Science and Trans-science' (1972) 10:2 Minerva 209.
Silvia Weyerbrock and Tian Xia, 'Technical Trade Barriers in US/Europe Agricultural Trade' (2000) 16:2 Agribusiness 235.
Jonathon B. Wiener, 'Whose Precaution After All? A Comment on the

Comparison and Evolution of Risk Regulatory Systems' (2003) 13:207 Duke J. Comp. & Int'l L. 207.

Anne Wilcock, *et al.*, 'Consumer Attitudes, Knowledge and Behaviour: A Review of Food Safety Issues' (2004) 15 Trends in Food Science & Technology 56.

Susan B.T. Wilkinson, Gene Rowe and Nigel Lambert, 'The Risks of Eating and Drinking' (2004) 5 European Molecular Biology Organization Reports S27.

J.S. Wilson and Tsunehiro Otsuki, 'To Spray or Not to Spray: Pesticides, Banana Exports, and Food Safety' (2004) 29 Food Policy 131.

Kumanan Wilson, *et al.*, *Variant Creutzfeldt-Jakob Disease and the Canadian Blood Supply: Explaining Decision-Making in the Blood System Under Conditions of Scientific Uncertainty* (Toronto: University Health Network, 2005).

David Winickoff, *et al.*, 'Adjudicating the GM Food Wars: Science, Risk, and Democracy in World Trade Law' (2005) 30 Yale J. Int'l L. 81.

David A. Wirth, 'The Role of Science in the Uruguay Round and NAFTA Trade Disciplines' (1994) 27:4 Cornell Int'l L.J. 817.

World Health Organization (WHO) and WTO, *WTO Agreements and Public Health: A Joint Study by the WHO and the WTO Secretariat* (Geneva: WHO and WTO, 2002).

Brian Wynne, 'Uncertainty and Environmental Learning: Reconceiving Science and Policy in the Preventive Paradigm' (1992) June Global Environmental Change 111.

Gary G. Yerkey, 'US Looking to Ask EU for Talks in WTO over Ban on Import of GMO Food Products', 19 Int'l Trade Rep. (BNA) 1829, 1829 (24 October, 2002).

Gary G. Yerkey, 'President Bush's High-Profile Criticism of EU over GMOs Seen Exacerbating Trade Dispute', 20 Int'l Trade Rep. (BNA) 916 (29 May, 2003).

Simonetta Zarrilli, *WTO Sanitary and Phytosanitary Agreement: Issues for Developing Countries, part of the South Centre Trade-Related Development and Equity (TRADE) Working Papers Series, No. 3* (1999).

John Ziman, 'Is Science Losing its Objectivity?' (1996) 382 Nature 751.

SECONDARY MATERIALS: BOOKS

Susan Ariel Aaronson, *Taking Trade to the Streets: The Lost History of Public Efforts to Shape Globalization* (Ann Arbor, MI: University of Michigan Press, 2001).

Natalie Avery, *Cracking the Codex: An Analysis of Who Sets World Standards* (London: National Food Alliance, 1993).

Robert Baldwin and Martin Cave, *Understanding Regulation: Theory, Strategy, and Practice* (New York: Oxford University Press, 1999).

Robert Beaglehole and Ruth Bonita, *Public Health at the Crossroads: Achievements and Prospects* (Melbourne: Cambridge University Press, 1999).

Ulrich Beck, *Risk Society* (London: Sage, 1992).

Ulrich Beck, *World Risk Society* (Cambridge, UK: Polity Press, 1999).

George A. Bermann and Petros C. Mavroidis, eds, *Trade and Human Health and Safety* (New York: Cambridge University Press, 2006) 133.

Jagdish Bhagwati, *A Stream of Windows: Unsettling Reflections in Trade, Immigration and Democracy* (Cambridge, MA: The MIT Press, 1998).

Jagdish Bhagwati and Robert E. Hudec, eds, *Fair Trade and Harmonization: Prerequisites for Free Trade? Volume 1: Economic Analysis* (Cambridge, MA: The MIT Press, 1996).

Jagdish Bhagwati and Robert E. Hudec, eds, *Fair Trade and Harmonization: Prerequisites for Free Trade? Volume 2: Legal Analysis* (Boston, MA: The MIT Press, 1996).

John Black, *Oxford Dictionary of Economics* (New York: Oxford University Press, 2002).

Albert Breton, *Competitive Governments* (Cambridge, UK: Cambridge University Press, 1996).

Stephen Breyer, *Breaking the Vicious Circle: Toward Effective Risk Regulation* (Cambridge, MA: Harvard University Press, 1993).

L. Brown (ed.), *The New Shorter Oxford English Dictionary on Historical Principles*, Vol. 2 (Oxford: Clarendon Press, 1993).

Wilson B. Brown and Jan S. Hogendorn, *International Economics in the Age of Globalization* (Peterborough, Ontario: Broadview Press, 2000).

Ian Brownlie, *Principles of International Law* (Oxford: Oxford University Press, 2003).

Rachel Carson, *Silent Spring* (Boston, MA: Houghton Mifflin Company, 1962).

Aaron Cosbey, *A Forced Evolution? The Codex Alimentarius Commission, Scientific Uncertainty and the Precautionary Principle* (Winnipeg: International Institute for Sustainable Development, 2002).

Robert W. Crandall and Lester B. Lave, eds, *The Scientific Basis of Health and Safety Regulation* (Washington DC: The Brookings Institution, 1981).

G. de Burca and J. Scott, eds, *The EU and the WTO: Legal and Constitutional Issues* (Oxford: Hart Publishing, 2001).

Gregory N. Derry, *What Science Is and How It Works* (Princeton, NJ: Princeton University Press, 1999).

Don Dewees, David Duff and Michael J. Trebilcock, *Exploring the Domain of Accident Law* (New York: Oxford University Press, 1996).

Graham Dunkley, *The Free Trade Adventure: The Uruguay Round and Globalism – A Critique* (Melbourne: University of Melbourne Press, 1997).

Graham Dunkley, *Free Trade: Myth, Reality and Alternatives* (London: Zed Books, 2004).

David L. Ebbels, *Principles of Plant Health and Quarantine* (Wallingford, UK: CABI Publishing, 2003).

Marsha A. Echols, *Food Safety and the WTO: The Interplay of Culture, Science and Technology* (New York: Kluwer Law International, 2001).

Pearl Eliadis, Margaret M. Hill and Michael Howlett, eds, *Designing Government* (Montreal and Kingston: McGill-Queen's University Press, 2005).

Mike Feintuck, *'The Public Interest' in Regulation* (New York: Oxford University Press, 2004).

Madeleine Ferrières, *Sacred Cow Mad Cow* (New York: Columbia University Press, 2006).

Baruch Fischhoff, *et al.*, *Acceptable Risk* (Cambridge, UK: Cambridge University Press, 1981).

E. Fisher, J. Jones and R. von Schomberg, eds, *Implementing the Precautionary Principle* (Cheltenham, UK and Northampton, MA: Edward Elgar, 2006).

James Gerber, *International Economics*, 2nd ed. (Boston, MA: Addison-Wesley, 2002).

Leonard Gomes, *The Economics and Ideology of Free Trade* (Cheltenham, UK and Northampton, MA: Edward Elgar, 2003).

Walter Goode, ed., *Dictionary of Trade Policy Terms* (Cambridge, UK: World Trade Organization/Cambridge University Press, 2003).

Lawrence O. Gostin, *Public Health Law: Power, Duty, Restraint* (Berkeley, CA: University of California Press, 2000).

John D. Graham and Jonathan Baert Wiener, eds, *Risk Versus Risk* (Cambridge, MA: Harvard University Press, 1995).

Barbara Herr Harthorn and Laury Oaks, eds, *Risk, Culture, and Health Inequality* (Westport, CT: Praeger, 2003).

P. van Heijnsbergen, *International Legal Protection of Wild Fauna and Flora* (Amsterdam: IOS Press, 1997).

David J. Hess, *Science Studies* (New York and London: New York University Press, 1997).

P.W. Huber, *Galileo's Revenge* (New York: Basic Books, 1991).

Mark W. Janis, *An Introduction to International Law* (New York: Aspen Publishers, 2003).

Sheila Jasanoff, *The Fifth Branch: Science Advisers as Policymakers* (Cambridge, MA: Harvard University Press, 1990).

Sheila Jasanoff, *Science at the Bar* (Cambridge, MA: Harvard University Press, 1995).

Sheila Jasanoff, *et al.*, eds, *Handbook of Science and Technology Studies* (Thousand Oaks, CA: Sage Publications, 1995).
Christian Joerges, Inger-Johanne Sand and Gunther Teubner, eds, *Transnational Governance and Constitutionalism* (Oxford: Hart, 2004) 199.
Branden B. Johnson and Vincent Covello, *The Social and Cultural Construction of Risk* (Dordrecht: Reidel, 1987).
Laura Jones, ed., *Safe Enough? Managing Risk and Regulation* (Vancouver: The Fraser Institute, 2000).
Tim Josling, Donna Roberts and Ayesha Hassan, *The Beef-Hormones Dispute and its Implications for Trade Policy* (Washington DC: USDA, 1999).
Tim Josling, Donna Roberts and David Orden, *Food Regulation and Trade* (Washington DC: Institute for International Economics, 2004).
Timothy E. Josling, Stefan Tangermann and T.K. Warley, *Agriculture in the GATT* (London: Macmillan Press, 1996).
Robert F. Kahrs, *Global Livestock Health Policy: Challenges, Opportunities, and Strategies for Effective Action* (Ames, Iowa: Iowa State Press, 2004).
Daniel Kennedy and James Southwick, eds, *The Political Economy of International Trade* (Cambridge, UK: Cambridge University Press, 2002).
Kenneth F. Kiple and Kriemhild Conee Ornelas, eds, *The Cambridge World History of Food* (Cambridge, UK: Cambridge University Press, 2000).
Merrie G. Klapp, *Bargaining with Uncertainty* (Westport, CT: Auburn House, 1992).
Noretta Koertge, ed., *A House Built on Sand: Exposing Postmodernist Myths About Science* (New York: Oxford University Press, 1997).
Helena Chmura Kraemer, Karen Kraemer-Lowe and David J. Kupfer, *To Your Health: How to Understand What Research Tells Us About Risk* (New York: Oxford University Press, 2005).
Paul Krugman, ed., *Strategic Trade Policy and the New International Economics* (London: MIT Press, 1986).
Paul R. Krugman and Maurice Obstfeld, *International Economics – Policy and Theory*, 5th ed. (Reading, MA: Addison-Wesley, 2000).
Hugh Lacey, *Is Science Value Free? Values and Scientific Understanding* (New York: Routledge, 1999).
John Leach, *A Course in Public Economics* (Cambridge, UK: Cambridge University Press, 2004).
William W. Lowrance, *Modern Science and Human Values* (New York: Oxford University Press, 1986).
Howard Margolis, *Dealing with Risk: Why the Public and the Experts Disagree on Environmental Issues* (Chicago: University of Chicago Press, 1996).
Robyn Martin and Linda Johnson, eds, *Law and the Public Dimension of Health* (London: Cavendish Publishing Limited, 2001).

Keith E. Maskus and John S. Wilson, eds, *Quantifying the Impact of Technical Barriers to Trade: Can it be Done?* (Ann Arbor, MI: The University of Michigan Press, 2001).

Deborah G. Mayo and Rachelle D. Hollander, eds, *Acceptable Evidence: Science and Values in Risk Management* (New York: Oxford University Press, 1991).

Nicholas Mercuro and Steven G. Medema, *Economics and the Law: From Posner to Post-Modernism* (Princeton, NJ: Princeton University Press, 1997).

National Research Council, *Risk Assessment in the Federal Government: Managing the Process* (Washington DC: National Research Council, 1983).

National Research Council, *Understanding Risk – Informing Decisions in a Democratic Society* (Washington DC: National Academy Press, 1996).

National Research Council, ed., *Incorporating Science, Economics, and Sociology in Developing Sanitary and Phytosanitary Standards in International Trade: Proceedings of a Conference* (Washington DC: National Academy Press, 2000).

National Research Council, *Strengthening Science at the US Environmental Protection Agency* (Washington DC: National Academy Press, 2000).

Anthony I. Ogus, *Regulation: Legal Form and Economic Theory* (New York: Oxford University Press, 1994).

Johannes Overbeek, *Free Trade versus Protectionism: A Source Book of Essays and Readings* (Cheltenham, UK and Northampton, MA: Edward Elgar, 1999).

Dennis J. Paustenbach, ed., *Human and Ecological Risk Assessment: Theory and Practice* (New York: John Wiley & Sons, Inc., 2002).

N. Perdikis, W.A. Kerr and J.E. Hobbs, 'Can the WTO/GATT Agreements on Sanitary and Phyto-Sanitary Measures and Technical Barriers be Renegotiated to Accommodate Agricultural Biotechnology' in W.H. Lesser (ed.), *Transitions in Agbiotech Economics of Strategy and Policy* (Connecticut: University of Connecticut Food Marketing Policy Center, 2000).

Oren Perez, *Ecological Sensitivity and Global Legal Pluralism* (Oxford: Hart Publishing, 2004).

Richard Perren, *The Meat Trade in Britain 1840–1914* (London: Routledge & Kegan Paul, 1978).

Pew Initiative on Food and Biotechnology, *US vs. EU: An Examination of the Trade Issues Surrounding Genetically Modified Food* (Richmond, VA: Pew Initiative on Food and Biotechnology, 2005).

Richard A. Posner, *Catastrophe: Risk and Response* (New York: Oxford University Press, 2004).

Mark R. Powell, *Science at EPA: Information in the Regulatory Process* (Washington, DC: Resources for the Future, 1999).
W. Curtis Priest, *Risks, Concerns, and Social Legislation: Forces that Led to Laws on Health, Safety, and the Environment* (Boulder, CO: Westview Press, 1988).
Thomas A. Pugel and Peter H. Lindert, *International Economics* (Boston, MA: McGraw-Hill Higher Education, 2000).
Scott C. Ratzan, *The Mad Cow Crisis: Health and the Public Good* (London: UCL Press, 1998).
Christopher Reynolds and Genevieve Howse, *Public Health: Law and Regulation* (Sydney: The Federation Press, 2004).
David Ricardo, *The Principles of Political Economy and Taxation* (London: Dent, 1917).
David Robertson and Aynsley Kellow, eds, *Globalization and the Environment: Risk Assessment and the WTO* (Cheltenham, UK and Northampton, MA: Edward Elgar Publishing Limited, 2001).
Frieder Roessler, ed., *The Legal Structure, Functions and Limits of the World Trade Order, A Collection of Essays* (London: Cameron May, 2000).
George Rosen, *A History of Public Health* (New York: MD Publications, Inc., 1958).
Pierre Salmon and Ronald Wintrobe, eds, *Competition and Structure* (Cambridge, UK: Cambridge University Press, 2000).
E.F. Schumacher, *Small is Beautiful: Economics as if People Mattered*, 1989 ed. (New York: Harper and Row, 1973).
E.F. Schumacher, *Small is Beautiful: Economics as if People Mattered – 25 Years Later . . . With Commentaries* (Vancouver: Hartley & Marks, 1999).
Upton Sinclair, *The Jungle* (Middlesex, UK: Penguin Modern Classics, 1906).
Jimmy M. Skaggs, *Prime Cut: Livestock Raising and Meatpacking in the United States 1607–1983* (College Station: Texas A&M University Press, 1986).
Paul Slovic, *The Perception of Risk* (London: Earthscan Publications Ltd, 2000).
Adam Smith, *The Wealth of Nations* (New York: The Modern Library Classics, 1776).
Richard D. Smith, *et al.*, eds., *Global Public Goods for Health: Health Economic and Public Health Perspectives* (New York: Oxford University Press, 2003).
Albert Sonnenfeld, ed., *Food: A Culinary History from Antiquity to the Present* (New York: Columbia University Press, 1999).
P.C. Stern and H.V. Fineberg, *Understanding Risk: Informing Decisions in a Democratic Society* (Washington DC: Committee on Risk Characterization, National Academy Press, 1996).

Terence P. Stewart, *The GATT Uruguay Round: A Negotiating History (1986–1992), Volume 1: Commentary* (Boston, MA: Kluwer Law International, 1993).

Terence P. Stewart, *The GATT Uruguay Round: A Negotiating History (1986–1992), Volume III: Documents* (Boston, MA: Kluwer Law International, 1993).

Terence P. Stewart, *The GATT Uruguay Round: A Negotiating History (1986–1994), Volume IV: The End Game (Part I)* (Boston, MA: Kluwer Law International, 1999).

Cass R. Sunstein, *After the Rights Revolution: Reconceiving the Regulatory State* (Cambridge, MA: Harvard University Press, 1990).

Cass R. Sunstein, ed., *Behavioural Law and Economics* (Cambridge, MA, UK: University of Cambridge Press, 2000).

Cass R. Sunstein, *Risk and Reason: Safety, Law, and the Environment* (Cambridge, MA: Cambridge University Press, 2002).

Cass R. Sunstein, *Laws of Fear – Beyond the Precautionary Principle* (Cambridge, MA: Cambridge University Press, 2005).

Cass R. Sunstein and Martha R. Nussbaum, *Animal Rights: Current Debates and New Directions* (New York: Oxford University Press, 2004).

Alan O. Sykes, *Product Standards for Internationally Integrated Goods Markets* (Washington DC: Brookings Institution, 1995).

Dick Taverne, *The March of Unreason: Science, Democracy, and the New Fundamentalism* (Oxford: Oxford University Press, 2005).

William A. Thomas, ed., *Science and Law: An Essential Alliance* (Boulder, CO: Westview Press, 1983).

Michael J. Trebilcock, *The Limits of Freedom of Contract* (Cambridge, MA: Harvard University Press, 1993).

Michael Trebilcock, *et al.*, *The Law and Economics of Canadian Competition Policy* (Toronto: University of Toronto Press, 2003).

Michael J. Trebilcock and Robert Howse, *The Regulation of International Trade*, 2nd ed. (London: Routledge, 1999).

Michael J. Trebilcock and Robert Howse, 'A Cautious View of International Harmonization: Implications from Breton's Theory of Competitive Governments', in Gianluigi Galeotti, Pierre Salmon and Ronald Wintrobe, eds, *Competition and Structure* (Cambridge: Cambridge University Press, 2000).

Michael J. Trebilcock and Robert Howse, *The Regulation of International Trade*, 3rd ed. (London: Routledge, 2005).

Tom Tyler, *Why People Obey the Law* (New Haven: Yale University Press, 1990).

David Vogel, *Trading Up: Consumer and Environmental Regulation in a Global Economy* (Cambridge, MA: Harvard University Press, 1995).

Christian Warren, *Brush with Death: A Social History of Lead Poisoning* (Baltimore, MD: Johns Hopkins University Press, 2000).

J.H.H. Weiler, ed., *The EU, the WTO and the NAFTA – Towards a Common Law of International Trade?* (New York: Oxford University Press, 2000).

C. Whipple, ed., *De Minimus Risk* (New York: Plenum Press, 1987).

World Trade Organization, *The Legal Texts: The Results of the Uruguay Round of Multilateral Trade Negotiations* (Cambridge, UK: Cambridge University Press, 1999).

SECONDARY MATERIALS: OTHER

Tyler Cowen, *Public Goods and Externalities* (2005), online: The Library of Economics and Liberty: The Concise Encyclopaedia of Economics, available online at http://www.econlib.org/library/Enc/PublicGoodsandExternalities.html (Date of Access: 2005).

European Commission, *Communication from the Commission on Impact Assessment* (Brussels, 2002, available online at http://europa.eu.int/eur-lex/en/com/cnc/2002/com2002_0276en01.pdf).

European Commission, *Communication from the Commission: Action Plan 'Simplifying and Improving the Regulatory Environment'* (Brussels, 2002, available online at http://europa.eu.int/eur-lex/en/com/cnc/2002/com2002_0278en01.pdf).

Global Knowledge Center on Crop Biotechnology, Biotechnology in Agriculture: A Lot More than GM Crops (2004), available online at: <www.isaaa.org/kc> (Date of Access: 222 February 2008).

International Law Commission, *Conclusions of the Work of the Study Group on the Fragmentation of International Law: Difficulties Arising from the Diversification and Expansion of International Law* (New York: United Nations, 2006).

McLaughlin Centre for Population Health Risk Assessment, McLaughlin Centre for Population Health Risk Assessment available online at: http://www.mclaughlincentre.ca/welcome/index.shtml (Date of Access: 12 July 2005).

Oxford English Dictionary Online (Oxford: Oxford University Press).

Julie Soloway, 'Institutional Capacity to Constrain Suboptimal Welfare Outcomes from Trade – Restricting Environmental, Health and Safety Regulation under NAFTA' (University of Toronto, 2000, unpublished).

CONFERENCE PROCEEDINGS

Zaheer Baber, 'Science in an Illiberal Democracy: The Genesis and Implications of Singapore's Biomedical Hub' (Paper presented to the

Conference on Public Science in Liberal Democracy: The Challenge to Science and Democracy, University of Saskatchewan, Saskatoon, 14–16 October 2004).
J.C. Bureau and W. Jones, 'Issues in Demand for Quality and Trade' (Paper presented to the International Agricultural Trade Research Consortium: Global Food Trade and Consumer Demand for Quality, Montreal, 2000).
Peter Cook, 'Science and Society' (Paper presented to the Conference on Public Science in Liberal Democracy: The Challenge to Science and Democracy, University of Saskatchewan, Saskatoon, 14–16 October 2004).
Graham Dunkley, 'INGOs, LINGOs, DINGOs and TRINGOs: Trade, the WTO and the Interest of Civil Society' (Paper presented to the Conference on International Trade Education and Research, Melbourne, 26–7 October 2000).
Ulrike Grote, 'Environmental and Food Safety Standards and International Trade: Concerns and Challenges for Developing Countries' (Paper presented to the International Symposium: Sustaining Food Security and Managing Natural Resources in Southeast Asia – Challenges for the 21st Century, 2002).
Ian C. Jarvie, 'The Democratic Deficit of Science and its Possible Remedies' (Paper presented to the Conference on Public Science in Liberal Democracy: The Challenge to Science and Democracy, University of Saskatchewan, Saskatoon, 14–16 October, 2004).
Grace Skogstad and Sarah Hartley, 'Science and Policy-Making: The Legitimate Conundrum' (Paper presented to the Conference on Public Science in Liberal Democracy: The Challenge to Science and Democracy, The University of Saskatchewan, Saskatoon, 15–16 October 2004).
Suzanne D. Thornsbury, 'Political Economy Determinants of Technical Barriers to US Agricultural Exports' (Paper presented to the 1999 American Agricultural Economics Association Annual Meeting 'Farm to Table: Connecting Products, Communities and Consumers', Nashville, 1999).
Michael J. Trebilcock, 'Critiquing the Critics of Economic Globalization' (Paper presented to the Law and Economics Workshop Series, Law and Economics Programme, Faculty of Law, University of Toronto, 2004).
Laurian Unnevehr and Donna Roberts, 'Food Safety and Quality: Regulations, Trade, and the WTO' (Paper presented to the International Conference on Agricultural Policy Reform and the WTO: Where Are We Heading? Capri, Italy, 2003).

SECONDARY MATERIALS: NEWSPAPER ARTICLES AND MEDIA REPORTS

Anon, 'A Short History of Hormones' *The Economist* 310:7584 (1989) 22.
Anon, 'Brie and Hormones' *The Economist* 310:7584 (1989) 21.

Malcolm W. Browne, 'In The Human Equation, Risk Perceived is Risk Endured' *The New York Times* (30 March 1980).
Deborah Cameron, 'Japanese Hungry to Know Where their Food Has Been' *Sydney Morning Herald* (25 September 2004).
Janice Castro, 'Why the Beef over Hormones? (European Ban on Beef From Hormone-injected US Cattle)' *Time* 133:3 (1989) 44.
Gloria Galloway, 'Experts Weighing Life's Many Risks' *Globe and Mail* (12 July 2005) A6.
Andrew Higgins, 'It's a Mad, Mad, Mad-Cow World' *Wall Street Journal* (12 March 2001) A13.
Ruth Rosen, 'A Physics Prof Drops a Bomb on the Faux Left' *Los Angeles Times* (23 May 1996) A11.
Paul Schukovsky, 'Ruling on Japan Boosts State Apple Growers' *Seattle Post-Intelligencer* (23 February 1999).
Nicholas Timmins, 'Hardening Attitudes to Benefits Mesh with Government Policy Social Trends' *Financial Times* (7 December 2004) 4.
Kim Willsher, 'EU Hygiene Regulations Threaten Traditional French Cheeses' *The Telegraph* (10 July 1995).

REPORTS AND PROCEEDINGS OF INTERNATIONAL ORGANIZATIONS

Codex Alimentarius Commission, *Report of the Twentieth Session of the Codex Alimentarius Commission, 28 June–7 July 1993* (Geneva: ALINORM 93/40, 1993).
Codex Alimentarius Commission, *Report of the 32nd Session of the Codex Committee on Food Additives and Contaminants 20–24 March 2000* (Beijing: Codex Alimentarius Commission, ALINORM 01/02, 2000).
Codex Alimentarius Commission, *Report of the 23rd Session of the Codex Committee on General Principles, Paris, France, 10–14 April 2006, to the Codex Alimentarius Commission 29th Session, Geneva, 3–7 July 2006* (Codex Alimentarius Commission, ALINORM 06/29/33, 2006).
League of Nations Committee of Experts for the Progressive Codification of International Law, 'The Most-Favoured-Nations Clause, reprinted' (1928) 22:(Supp. 1928) Am. J. Int'l L. 133.
World Health Organization, *The Fifty-third World Health Assembly – Resolution WHA53.15 (Food Safety)* (2000).
WTO Secretariat, Specific Trade Concerns – Note by the Secretariat (2004), WTO Doc. G/SPS/GEN/204/Rev.4.
WTO Secretariat, Specific Trade Concerns – Note by the Secretariat (2007), WTO Doc. G/SPS/GEN/204/Rev.7.

WTO, Statement by the Delegation of Argentina on Notification (1997), WTO Doc. G/SPS/N/EEC/47, G/SPS/GEN/21.

WTO, Egypt – Import Prohibition on Canned Tuna with Soybean Oil, Request for Consultations by Thailand (2000), WTO Doc. WT/DS205/1, G/L/392, G/SPS/GEN/203.

WTO, Revised Procedure to Monitor the Process of International Harmonization, WTO Doc. G/SPS/11/Rev.1, 15 November 2004.

WTO Committee on Sanitary and Phytosanitary Measures, Summary of the Meeting held on 29–30 May 1996, WTO Doc. G/SPS/R/5.

WTO Committee on Sanitary and Phytosanitary Measures, Summary of the Meeting held on 15–16 October 1997, WTO Doc. G/SPS/R/9/Rev.1.

WTO Committee on Sanitary and Phytosanitary Measures, Procedure to Monitor the Process of International Harmonization (1997), WTO Doc. G/SPS/11.

WTO Committee on Sanitary and Phytosanitary Measures, Summary of the Meeting Held on 15–16 September 1998, WTO Doc. G/SPS/R/12.

WTO Committee on Sanitary and Phytosanitary Measures, Summary of the Meeting Held on 7–8 July 1999, WTO Doc. G/SPS/R/15.

WTO Committee on Sanitary and Phytosanitary Measures, Note by the Secretariat, Revised Draft Annual Report, Procedure to Monitor the Process of International Harmonization (1999), WTO Doc. G/SPS/W/94/Rev.2.

WTO Committee on Sanitary and Phytosanitary Measures, Summary of the Meeting held on 10–11 November 1999, WTO Doc. G/SPS/R/17.

WTO Committee on Sanitary and Phytosanitary Measures, Summary of the Meeting held on 15–16 March 2000, WTO Doc. G/SPS/R/18.

WTO Committee on Sanitary and Phytosanitary Measures, Summary of the Meeting held on 21–2 June 2000, WTO Doc. G/SPS/R/19.

WTO Committee on Sanitary and Phytosanitary Measures, Note by the Secretariat, Revised Draft Annual Report, Procedure to Monitor the Process of International Harmonization (2000), WTO Doc. G/SPS/W/107/Rev.

WTO Committee on Sanitary and Phytosanitary Measures, Summary of the Meeting held on 8–9 November 2000, WTO Doc. G/SPS/R/20.

WTO Committee on Sanitary and Phytosanitary Measures, Summary of the Meeting Held on 14–15 March 2001, WTO Doc. G/SPS/R/21.

WTO Committee on Sanitary and Phytosanitary Measures, Summary of the Meeting Held on 10–11 July 2001, WTO Doc. G/SPS/R/22.

WTO Committee on Sanitary and Phytosanitary Measures, Procedure to Monitor the Process of International Harmonization, Third Annual Report (2001), WTO Doc. G/SPS/18.

WTO Committee on Sanitary and Phytosanitary Measures, Summary of the Meeting Held on 19–21 March 2002, WTO Doc. G/SPS/R/26.

WTO Committee on Sanitary and Phytosanitary Measures, Summary of the Meeting Held on 25–6 June 2002, WTO Doc. G/SPS/R/27.
WTO Committee on Sanitary and Phytosanitary Measures, Summary of the Meeting Held on 7–8 November 2002, WTO Doc. G/SPS/R/28.
WTO Committee on Sanitary and Phytosanitary Measures, Summary of the Meeting held on 24–5 June 2003, WTO Doc. G/SPS/R/30.
WTO Committee on Sanitary and Phytosanitary Measures, Summary of the Meeting held on 29–30 October 2003, WTO Doc. G/SPS/R/31.
WTO Committee on Sanitary and Phytosanitary Measures, Summary of the Meeting held on 9–10 March 2005, WTO Doc. G/SPS/R/36.
WTO Committee on Sanitary and Phytosanitary Measures, Summary of Meeting held on 24 October 2005, resumed on 1–2 February 2006, WTO Doc. G/SPS/R/39.

Index

margarine, health/trade conflict and 21–2